LONDON MATHEMATICAL SOCIETY STUDENT TEXTS

Managing editor: Dr C.M. Series, Mathematics Institute
University of Warwick, Coventry CV4 7AL, United Kingdom

1 Introduction to combinators and λ-calculus, J.R. HINDLEY & J.P. SELDIN
2 Building models by games, WILFRID HODGES
3 Local fields, J.W.S. CASSELS
4 An introduction to twistor theory, S.A. HUGGETT & K.P. TOD
5 Introduction to general relativity, L.P. HUGHSTON & K.P. TOD
6 Lectures on stochastic analysis: diffusion theory, DANIEL W. STROOCK
7 The theory of evolution and dynamical systems, J. HOFBAUER & K. SIGMUND
8 Summing and nuclear norms in Banach space theory, G.J.O. JAMESON
9 Automorphisms of surfaces after Nielsen and Thurston, A. CASSON & S. BLEILER
10 Nonstandard analysis and its applications, N. CUTLAND (ed)
11 Spacetime and singularities, G. NABER
12 Undergraduate algebraic geometry, MILES REID
13 An introduction to Hankel operators, J.R. PARTINGTON
14 Combinatorial group theory: a topological approach, DANIEL E. COHEN
15 Presentations of groups, D.L. JOHNSON
16 An introduction to noncommutative Noetherian rings, K.R. GOODEARL & R.B. WARFIELD, JR.
17 Aspects of quantum field theory in curved spacetime, S.A. FULLING
18 Braids and coverings: selected topics, VAGN LUNDSGAARD HANSEN
19 Steps in commutative algebra, R.Y. SHARP
20 Communication theory, C.M. GOLDIE & R.G.E. PINCH
21 Representations of finite groups of Lie type, FRANÇOIS DIGNE & JEAN MICHEL
22 Designs, graphs, codes, and their links, P.J. CAMERON & J.H. VAN LINT
23 Complex algebraic curves, FRANCES KIRWAN
24 Lectures on elliptic curves, J.W.S. CASSELS
25 Hyperbolic geometry, BIRGER IVERSEN
26 An Introduction to the theory of L-functions and Eisenstein series, H. HIDA
27 Hilbert Space: Compact Operators and the Trace Theorem, J.R. RETHERFORD

London Mathematical Society Student Texts 25

Hyperbolic Geometry

Birger Iversen
Aarhus University

CAMBRIDGE
UNIVERSITY PRESS

CAMBRIDGE UNIVERSITY PRESS
Cambridge, New York, Melbourne, Madrid, Cape Town, Singapore, São Paulo, Delhi

Cambridge University Press
The Edinburgh Building, Cambridge CB2 8RU, UK

Published in the United States of America by Cambridge University Press, New York

www.cambridge.org
Information on this title: www.cambridge.org/9780521435086

First published 1992
Re-issued in this digitally printed version 2008

A catalogue record for this publication is available from the British Library

ISBN 978-0-521-43508-6 hardback
ISBN 978-0-521-43528-4 paperback

CONTENTS

IV FUCHSIAN GROUPS

V FUNDAMENTAL DOMAINS

VI COVERINGS

CONTENTS

IV FUCHSIAN GROUPS

V FUNDAMENTAL DOMAINS

VI COVERINGS

VII POINCARÉ'S THEOREM

VIII HYPERBOLIC 3-SPACE

APPENDIX: AXIOMS FOR PLANE GEOMETRY 284

VII POINCARÉ'S THEOREM

VIII HYPERBOLIC 3-SPACE

INTRODUCTION

What is hyperbolic geometry? Let me try to give an answer by telling the story of the parallel axiom. I shall use modern language which will ruin part of the story but highlight the basic points.

Axioms for plane geometry

A simple set of axioms for plane geometry can be presented in the framework of metric spaces. By a line in a metric space X we understand the image of a distance preserving map $\gamma: \mathbb{R} \rightarrow X$. The three axioms of plane geometry are (the axioms are analysed in an appendix)

INCIDENCE AXIOM Through two distinct points of X there passes a unique line. The space X has at least one point.

REFLECTION AXIOM The complement of a given line in X has two connected components. There exists an isometry σ of X which fixes the points of the line, but interchanges the two connected components of its complement.

PARALLEL AXIOM Through a given point outside a given line there passes a unique line which does not intersect the given line.

Investigations of the parallel axiom by among others J.Bolyai (1802 – 1860), C.F.Gauss (1777 – 1855), N.I.Lobachevsky (1793 – 1856) show that this axiom is independent of the other axioms in the sense that there exists a plane, the so called hyperbolic plane H^2, which satisfies the first two axioms of plane geometry but not the parallel axiom. H^2 is unique in the following sense.

CLASSIFICATION THEOREM A metric space which satisfies the three axioms of plane geometry is isometric to the Euclidean plane. A space which satisfies the first two axioms but not the parallel axiom is isometric to H^2 (after rescaling[1]).

The discovery of the hyperbolic plane began with a series of attempts to prove the parallel axiom from the other axioms. Such attempts should inevitably lead to the discovery of the hyperbolic plane since a system of axioms for the hyperbolic plane consists in the incidence axiom, the reflection axiom and the negation of the parallel axiom. G.Sacheri (1667 – 1733) was the first to penetrate deeper into this universe. He was followed by J.H.Lambert (1728 – 77) who made some important speculations on the nature of a non-Euclidean geometry. By 1850 most of the properties of a new geometry were known to Bolyai, Gauss and Lobachevsky. At that time the biggest problem was the existence of such a geometry! Twenty years later this was firmly settled through the works of E.Beltrami (1835 – 1900), A.Cayley (1821 – 95) and F.Klein (1849 – 1925). For more information on the history of hyperbolic geometry see [Milnor] or [Greenberg] and the references given there.

The Poincaré half-plane In 1882 H.Poincaré (1854 – 1912) discovered a new model of H^2, which I shall now describe. The model is the upper half-plane of the complex plane equipped with the metric given by

$$\cosh \mathrm{d}(z,w) = 1 + \tfrac{1}{2}|z-w|^2\, Im[z]^{-1}\, Im[w]^{-1}$$

The lines in the Poincaré half-plane are traced by Euclidean circles with centres on the x–axis or Euclidean lines perpendicular to the x–axis. It is quite obvious that this model satisfies the incidence and reflection axioms but not the parallel axiom: Euclidean inversions furnish the isometries required by the reflection axiom.

The usefulness of the Poincaré half-plane model lies in the fact that the group of orientation preserving isometries consists of analytic transformations of the form

$$z \mapsto \frac{az+b}{cz+d} \qquad ; \begin{bmatrix} a & b \\ c & d \end{bmatrix} \in Sl_2(\mathbb{R})$$

In fact the full group of isometries of the Poincaré half-plane is the group $PGl_2(\mathbb{R})$. In particular the group $Gl_2(\mathbb{Z})$ acts as isometries on H^2, a fact which lies at the root of the applications of hyperbolic geometry to number theory.

The Poincaré disc The open unit disc D is the second most commonly used model for the hyperbolic plane. We shall not write down the metric here, but mention that the lines in this model are traced by Euclidean circles and lines orthogonal to the unit circle ∂D. The Poincaré disc has certain aesthetic qualities, as shown below by a one of the many beautiful illustrations from [Klein, Fricke].

Discrete groups Reflections in the of lines of the tesselation above generate a discrete group of isometries of H^2. The systematic study of such groups was initiated by Poincaré in his "Memoir sur les groupes fuchsienne" 1882. Starting from a discrete group Γ of isometries of H^2 he used a procedure due to Dirichlet to construct a polygon Δ with a side pairing, i.e. instructions for identification of the sides of the polygon, which allows us to reconstruct the space H^2/Γ. Then he took up the converse problem: given a hyperbolic polygon with a side pairing, when does the side pairing generate a discrete group with the given polygon as fundamental domain. Poincaré laid down necessary and sufficient conditions for a given polygon with side pairings to determine a discrete group. The theorem has a bonus: from the geometry of the polygon we can read off a complete system of generators and relations for the group. In this way the topic has had a big influence on what today is known as combinatorial group theory. The theorem of Poincaré is one of the main themes of this book. The proof of the theorem is based on some geometric ideas which I shall outline next.

New geometries Let us for a moment go back to the discussion of the parallel axiom and recall that the hyperbolic plane is quite unique. However, we find numerous new geometries in the class of hyperbolic surfaces: spaces locally isometric to H^2. We shall see that complete hyperbolic surfaces can be classified by discrete subgroups of $PGl_2(\mathbb{R})$. The main point in the proof of Poincaré's theorem is the monodromy theorem: A local isometry $f:X \to H^2$ from a complete hyperbolic surface X to the hyperbolic plane H^2 is an isomorphism.

Higher dimensions So far we have only talked about hyperbolic geometry in two dimensions, but the hyperbolic space H^n exists in all dimensions. It is the second main theme of this book to construct these spaces and to identify their isometry groups with the Lorentz group known from special relativity. Hyperbolic n-space is constructed as one of the two sheets of the hyperbola

$$x_0^2 - x_1^2 - x_3^2 - x_4^2 ..... - x_n^2 = 1$$

It turns out that the most natural framework for this construction is the theory of quadratic forms. This is actually where the books starts. We offer a general introduction to quadratic forms in Chapter I as an opener to the construction of geometries in Chapter II. It is natural to develop Euclidean and spherical geometries in a parallel fashion in order to be able to draw on analogies with these more familiar geometries. A special feature of hyperbolic geometry is the presence of a natural boundary ∂H^n. In terms of the hyperbola above, a boundary point can be thought of as an asymptote to the hyperbola, i.e. line in the cone

$$x_0^2 - x_1^2 - x_3^2 - x_4^2 ..... - x_n^2 = 0$$

It follows that the boundary ∂H^n is a sphere. In the end we find that the Lorentz group acts on the sphere as the group of Möbius transformations.

Three-dimensional geometry A specific feature of three-dimensional hyperbolic geometry is that the group $Sl_2(\mathbb{C})$ operates on H^3 (see below) in such a way as to give an isomorphism between $PGl_2(\mathbb{C})$ and the group of orientation preserving isometries of H^3, the special Lorentz group. This has the consequence that a Kleinian group, i.e. is a discrete subgroup Γ of $Sl_2(\mathbb{C})$, defines a 3-manifold H^3/Γ. The work of W.P.Thurston on 3-manifolds [Thurston] has generated new interest in hyperbolic geometry.

Exterior algebra Let us think of H^3 as a sheet of the unit hyperbola in a Minkowski space M, i.e. a four-dimensional vector space with a form of type $(-3,1)$. Once we pick an orientation of M, the space $\wedge^2$M comes equipped with the structure of a three-dimensional complex vector space with a complex quadratic form Q. The assignment of normal vector $\mathbf{n} \in \wedge^2$M to an oriented geodesic line h in H^3 creates a bijection between the complex quadric $Q = -1$ and the set of oriented geodesics in H^3.

The Hermitian model As mentioned above, each Minkowski space M comes equipped with its own copy of H^3, a sheet of the unit hyperbola. It turns out that there exist Minkowski spaces with more algebraic structures, in particular a structure of a linear representation of $Sl_2(\mathbb{C})$. We shall be interested in the space M of 2×2 Hermitian matrices, i.e. complex matrices of the form

$$\begin{bmatrix} a & b \\ \bar{b} & d \end{bmatrix} \qquad ; \mathrm{a, d} \in \mathbb{R}\ , \mathrm{b} \in \mathbb{C}$$

The restriction of the determinant to M is a quadratic form of type $(-3,1)$. The sheet of the unit hyperbola ($\mathrm{ad} - \mathrm{b}\bar{\mathrm{b}} = 1$) consisting of positive definite forms ($\mathrm{a} > 0$) is called the Hermitian model of H^3. The action of $Sl_2(\mathbb{C})$ on M is given by the formula

$$\sigma\, \mathrm{X} = \sigma \mathrm{X} \sigma^* \qquad ; \sigma \in Sl_2(\mathbb{C}),\ \mathrm{X} \in M$$

The key fact about the Hermitian model is the following trace formula: The action of $\mathrm{S} \in Gl_2(\mathbb{C})$ on H^3 can be decomposed into $\alpha\beta$ where α and β are half-turns with respect to geodesics a and b. For any such decomposition we have that

$$tr^2(S) = 4<\boldsymbol{m},\boldsymbol{n}>^2$$

where $\boldsymbol{m}$ and $\boldsymbol{n}$ denote normal vectors for the geodesics a and b.

The realisation of Minkowski space through the space M of Hermitian matrices has been used in physics for a long time. We shall introduce another tool from physics namely the Dirac algebra, which is a concrete realisation of the Clifford algebra of M. The use of Clifford algebras was suggested by the "strange formulas" from the book of [Fenchel].

Prerequisites The general prerequisites are linear algebra and a modest amount of point set topology, including compact subsets of a finite dimensional real vector space. Familiarity with the concepts of group theory and, for the understanding of Poincaré's polygon theorem, a bare minimum of combinatorial group theory ("generators and relations") is needed. Elementary topology is needed for VI.8, VII.5, VII.7, while VII.6 is a good introduction to combinatorial topology. The final chapter of the book requires familiarity with the exterior algebra.

University of Aarhus, Denmark. July 1992 Birger Iversen

ACKNOWLEDGEMENT I would like to thank Jakob Bjerregaard, Elisabeth Husum, Claus Jungersen and Pernille Pind for comments on the various notes on which the book is built. A particular thank you goes to Gunvor Juul and editor Caroline Series who helped the manuscript through its final stages.

I QUADRATIC FORMS

In the first two sections we lay down the general principles for quadratic forms and the associated orthogonal groups. The third section will classify real quadratic forms into Sylvester types, while the remaining sections will deal with real forms of specific types.

Three types of real quadratic forms are of particular interest: positive definite forms, parabolic forms and hyperbolic forms. The corresponding orthogonal groups provide us with the isometry groups of the three basis geometries: spherical geometry, Euclidean geometry and hyperbolic geometry.

For example, the orthogonal group of a hyperbolic form contains the Lorentz group as a subgroup of index 2. The Lorentz group, on the one hand, is the isometry group of hyperbolic geometry, while, on the other hand, it can be identified with the somewhat older group of Möbius transformations. We shall see that "the Möbius group is the boundary action of the Lorentz group".

I.1 ORTHOGONALITY

Let k denote a field[1] of characteristic $\neq 2$ and E a vector space over k of finite dimension n. By a quadratic form on E we understand a function $Q\colon E \to k$ homogeneous of degree 2, i.e.

1.1 $$Q(\lambda z) = \lambda^2 Q(z) \qquad ; \lambda \in k,\ z \in E$$

with the property that the symbol

1.2 $$\langle x,y\rangle = \tfrac{1}{2}(Q(x+y) - Q(x) - Q(y)) \qquad ; x,y \in E$$

[1] In this text we have applications only for the field $\mathbb{R}$ of real numbers, the field $\mathbb{C}$ of complex numbers and occasionally the field $\mathbb{Q}$ of rational numbers.

is bilinear in x and y. The relation 1.2 between the bilinear form $\langle x,y\rangle$ and the quadratic form $Q(z)$ is called the formula for *polarization*. Observe that the bilinear form $\langle x,y\rangle$ is *symmetrical* in x and y.

Proposition 1.3 Polarization gives a one to one correspondence between quadratic forms and symmetrical bilinear forms over a field k with $char(k) \neq 2$.

Proof A quadratic form $Q(z)$ satisfies the formula $Q(2z) = 4Q(z)$ as a special case of 1.1. We can now substitute z for x and y in formula 1.2 to get that

1.4 $$Q(z) = \langle z,z\rangle \qquad ; z \in E$$

which recovers the function $Q(z)$ from the symbol $\langle x,y\rangle$. Conversely, if we start with a symmetrical bilinear form $\langle x,y\rangle$ we can use formula 1.4 to define a k-valued function Q on E which is homogeneous of degree 2. Bilinearity and symmetry give us the formula

$$\langle x+y,\, x+y\rangle = \langle x,x\rangle + \langle y,y\rangle + 2\langle x,y\rangle \qquad ; x,y \in E$$

which shows that $Q(z) = \langle z,z\rangle$ is a quadratic form. □

The evaluation, $Q(x)$, of the quadratic form at a vector $x \in E$ is often referred to as the *norm* of x. Vectors $x,y \in E$ are called *orthogonal* if $\langle x,y\rangle = 0$. Linear subspaces U and V of E are called *orthogonal* if all vectors in U are orthogonal to all vectors in V. For a given linear subspace K of E we define the *total orthogonal subspace* $K^{\perp}$ by the formula

1.5 $$K^{\perp} = \{\, e \in E \mid \forall x \in K \ \langle e,x\rangle = 0 \,\}$$

Definition 1.6 A quadratic form $Q(z)$ on the finite dimensional vector space E over k is called *non-singular* if the bilinear form $\langle x,y\rangle$ satisfies

$$\forall x \in E \ (<x,y> = 0) \ \Rightarrow \ y = 0 \qquad ; y \in E$$

Otherwise expressed, the form $Q(z)$ is non-singular if and only if $E^{\perp} = 0$.

Theorem 1.7 Let E be a finite dimensional vector space equipped with a non-singular quadratic form Q. For any linear subspace K of E we have

$$\mathit{dim}\, K + \mathit{dim}\, K^{\perp} = \mathit{dim}\, E$$

Proof Let us recall that a linear form on E is nothing but a k–linear map $\xi: E \to k$. Linear forms may be added and multiplied by a constant from k to form a new vector space, the dual space E^*. Let us prove that

1.8 $$\mathit{dim}\, E^* = \mathit{dim}\, E$$

To this end we simply pick a basis $e_1,...,e_n$ for E and introduce the coordinate forms $x_1,...,x_n \in E^*$ by the formula

$$e = \sum_i x_i(e)\, e_i \qquad ; e \in E$$

Let us verify that any linear form ξ on E satisfies the formula

$$\xi = \sum_i \xi(e_i)\, x_i \qquad ; \xi \in E^*$$

To this end we simply evaluate both sides of the formula on the basis $e_1,...,e_n$. This shows that $x_1,...,x_n$ generates the vector space E^*. To see that $x_1,...,x_n$ are linearly independent, consider a relation of the form $\sum \lambda_i x_i = 0$, $\lambda_1,...,\lambda_n \in k$. Evaluate the relation on e_j, $j = 1,...,n$, and conclude that $\lambda_j = 0$.

Let us introduce the map

1.9 $$q: E \to E^* , \ q(e) = (x \mapsto \ < e,x >) \qquad ; e \in E$$

Observe that the kernel for q is $E^{\perp}$ and use that Q is non-singular to conclude that q is injective; formula 1.8 allows us to conclude that q is surjective as well. Restriction of a form along the inclusion $i: K \to E$ defines a k–linear map

$$i^*: E^* \to K^* \qquad ; \xi \mapsto \xi \circ i, \ \xi \in E^*$$

which is surjective: In the argument above pick the basis $e_1,\dots,e_n$ for E such that $e_1,\dots,e_k$ is a basis for K and observe that the coordinate forms for this basis are $x_1 \circ i,\dots,x_k \circ i$. We shall focus on the composite map

$$E \xrightarrow{q} E^* \xrightarrow{i^*} K^*$$

Since this is surjective we conclude from the dimension formula of Grassmann[2]

$$\dim E = \dim K^* + \dim \operatorname{Ker} i^*q$$

We ask the reader to show that $\operatorname{Ker} i^*q = K^{\perp}$. The result follows from the formula above in combination with 1.8. □

Let us consider a pair (E,Q) consisting of a finite dimensional vector space E and a quadratic form Q on E. By an isomorphism from (E,Q) to a second such pair (F,P) we understand a linear isomorphism $\sigma: E \xrightarrow{\sim} F$ such that $Q = P \circ \sigma$. In particular we may talk about automorphisms of (E,Q) which may be composed to form a group, the orthogonal group of (E,Q) which is denoted $O(Q)$ or just $O(E)$ when no confusion is possible.

Let us quite generally consider a quadratic form (E,Q) and pick a basis $e_1,\dots,e_n$ for the vector space E. The Gram matrix $G \in M_n(k)$ is given by

1.10 $$G_{ij} = \langle e_i, e_j \rangle \qquad ; i,j = 1,\dots,n$$

Proposition 1.11 Let Q be a non-singular quadratic form on E. An orthogonal transformation $\sigma \in O(Q)$ has determinant 1 or −1.

Proof Let us pick a basis $e_1,\dots,e_n$ for E and let A denote the matrix for σ, thus $\sigma(e_j) = \sum_i A_{ij} e_i$ for $j = 1,\dots,n$. Direct calculation gives us

$$\langle \sigma(e_i), \sigma(e_j) \rangle = \langle \textstyle\sum_k A_{ki} e_k, \sum_h A_{hj} e_h \rangle = \sum_{h,k} {}^{T}A_{ik} \langle e_k, e_h \rangle A_{hj}$$

which shows that the Gram matrix for $Q \circ \sigma$ is ${}^{T}AGA$. In particular we find that σ is an orthogonal transformation for Q if and only if the matrix A obeys

---

[2] A linear map $f: E \to F$ between finite dimensional vector spaces satisfies

$$\dim E = \dim \operatorname{Im}(f) + \dim \operatorname{Ker}(f)$$

1.12 $$^{T}A\,G\,A = G$$

Let us observe that the Gram matrix G is the matrix for the map $q:E \rightarrow E^*$ with respect to the basis $e_1,\ldots,e_n$ for E and $x_1,\ldots,x_n$ for E^*. Since Q is non-singular we conclude that *det* G $\neq 0$. The formula 1.12 gives us

$$det\ ^{T}A\ det\ G\ det\ A = det\ G$$

from which we conclude that $det^2 A = 1$, or *det* A = 1 or −1. □

The orthogonal group for the standard quadratic form on k^n

1.13 $$Q(x) = x_1^{\,2}+x_2^{\,2}+ \ldots + x_n^{\,2} \qquad ; x = (x_1,\ldots,x_n) \in k^n$$

is denoted $O_n(k)$. The subgroup of transformations with determinant 1 is denoted $SO_n(k)$. When $k = \mathbb{C}$ or any other algebraically closed field with *char*(k) $\neq 2$ any non-singular quadratic form is isomorphic to the standard form 1.13.

Theorem 1.14 Let Q be a non-singular quadratic form on the vector space E over $\mathbb{C}$ of dimension n. There exists a basis $e_1,\ldots,e_n$ for E with

$$\langle e_i,e_j\rangle = \delta_{ij} \qquad ; i,j = 1,\ldots,n$$

Proof Let us observe that Q can't be identically zero since this implies that the bilinear form $\langle x,y\rangle$ is identically zero. Thus we can find a vector $v \in E$ with $Q(v) \neq 0$. Let us write $Q(v) = \lambda^2$, $\lambda \in \mathbb{C}$, or $Q(\lambda^{-1}v) = 1$ to conclude that we can find $e \in E$ with $Q(e) = 1$. Let K denote the line spanned by e and put $F = K^{\perp}$ and observe that $E = K \oplus F$ (meaning that any vector $x \in E$ has a unique decomposition $x = f + k$, where $f \in F$ and $k \in K$). It is easily seen that the restriction of Q to F is non-singular. It is left to the reader to complete the proof by simple induction on *dim*(E). □

1.2 WITT'S THEOREM

In this section we shall be concerned with a non-singular quadratic form Q on a vector space E of dimension n over a field k with $char(k) \neq 2$. The theme is the flexibility of the orthogonal group $O(Q)$. The tool is reflections, an important class of orthogonal transformations, which we proceed to introduce.

We say that a vector $n \in E$ is non-isotropic if $Q(n) \neq 0$. A non-isotropic vector $n \in E$ defines a linear transformation τ_n of E by the formula

2.1
$$\tau_n(x) = x - 2\frac{\langle x,n\rangle}{\langle n,n\rangle} n \qquad ; x \in E$$

Direct computation gives us

$$\langle\tau_n(x),\tau_n(x)\rangle = \langle x,x\rangle \qquad ; x \in E$$

i.e. τ_n is an orthogonal transformation called reflection along n. The hyperplane orthogonal to n is fixed by τ_n while $\tau_n(n) = -n$. It follows that τ_n is an involution , i.e. $\tau_n^{\,2} = \iota$ but $\tau_n \neq \iota$, and that

2.2
$$det\ \tau_n = -1 \qquad ; n \in E,\ \langle n,n\rangle \neq 0$$

Proposition 2.3 Let Q be a quadratic form on the vector space E. For any $\lambda \in k^*$, the orthogonal group $O(Q)$ acts transitively[3] on the set

$$\{\, e \in E \mid Q(e) = \lambda \,\}$$

Proof Consider vectors $e,f \in E$ with $Q(e) = Q(f) = \lambda \neq 0$. Observe, that

$$\langle e-f,\ e+f\rangle = \langle e,e\rangle - \langle f,f\rangle = 0$$

It follows that we can't have $Q(e+f) = 0$ and $Q(e-f) = 0$ at the same time (the non-isotropic vector e is a linear combination of $e-f$ and $e+f$). If the vector $e-f$ is non-isotropic, reflection τ along this vector fixes $e+f$. Thus

$$\tau(e-f) = f-e \ , \quad \tau(e+f) = e+f$$

[3] We say that the action of a group G on a set X is transitive, if for all $x \in X$ and $y \in X$ there exists $\sigma \in G$ with $\sigma(x) = y$.

Now, add these two formulas to see that $\tau(e) = f$. If $e + f$ is isotropic, reflection ρ along this vector interchanges e and $-f$. In conclusion the orthogonal transformation $\tau_f\rho$ maps e to f as required. □

When Q is non-singular, it is also true that $O(Q)$ acts transitively on the set of <u>isotropic lines</u>, i.e. lines in E generated by isotropic vectors. In fact,

Witt's theorem 2.4 Let (E,Q) be a non-singular quadratic form. Any isomorphism σ: $(U,Q) \xrightarrow{\sim} (V,Q)$, where U and V are linear subspaces of E, can be extended to an orthogonal transformation of (E,Q).

Proof Let us first treat the case where U is non-singular. This will be done by induction on $dim(U)$. In case $dim(U) = 1$ let us pick a non-zero $u \in U$ and put $\lambda = Q(u)$. Since $Q(\sigma(u)) = \lambda$, we can apply 2.3 to extend σ to an orthogonal transformation of E.

To accomplish the induction step let us pick a non-isotropic vector $e \in U$. Since $Q(e) = Q(\sigma(e))$ we can use 2.3 to pick $\tau \in O(E)$ such that $\sigma(e) = \tau(e)$. It will suffice to extend $\tau^{-1}\sigma: U \to E$ to an orthogonal transformation of E. To recapitulate it suffices to extend $\sigma: U \to E$ under the assumption that σ fixes a non-isotropic vector $e \in U$. Let us introduce the total orthogonal space W to $L = k.e$ in U and the total orthogonal space F to L in E. Observe that W and F are non-singular and that $W \subset F$ and $\sigma(W) \subset F$. Thus we can use the induction hypothesis to extend the restriction $\sigma: W \to F$ to an orthogonal transformation of F.

Let us now assume that U is singular. Let there be given data consisting of a non-zero vector $e \in U \cap U^{\perp}$ and a linear complement R to L in U. We are going to show that there exists an isotropic vector f in E which is orthogonal to R and such that $\langle e,f \rangle = 1$. Observe that we can find $\beta \in E^*$ with $\beta(e) = 1$ and $\beta(R) = 0$; the vector $b \in E$ with $q(b) = \beta$ is orthogonal to R and satisfies $\langle b,e \rangle = 1$. Let us show that the vector $f = b - \frac{1}{2}\langle b,b \rangle e$ is isotropic

$$\langle f,f \rangle = \langle b,b \rangle - \langle b,b \rangle \langle e,b \rangle = 0$$

The isotropic vector f is orthogonal to R and satisfies $<e,f> = 1$. Let us observe that $f \notin U$ since e is orthogonal to U but $<e,f> = 1$. The space $W = U + \mathbb{R}f$ is the direct sum of the two orthogonal subspaces R and $ke + kf$. Let us perform the same construction on the data $\sigma(U)$, $\sigma(e)$, $\sigma(R)$ to find an isotropic vector $g \in E$ which is orthogonal to $\sigma(R)$ and which satisfies $<g,\sigma(e)> = 1$. Let us extend $\sigma: U \to E$ to a linear map $W \to E$ by the convention $\sigma(f) = g$. It is easy to check that the extended map preserves the inner product: use that $\sigma(W)$ is the direct sum of the orthogonal subspaces $\sigma(R)$ and $k\sigma(e) + k\sigma(f)$.

Let us finally consider the general subspace U of E. If U is non-singular we can use the first part of the proof to extend σ. If U is singular we can use the middle part of the proof a number of times until U becomes non-singular, and then apply the first part of the proof. □

The orthogonal group $O(E)$ of a non-singular form is generated by reflections along non-isotropic vectors as it follows from exercise 2.3. The number of reflections needed can be bounded by the following theorem of Elie Cartan: An orthogonal transformation of a non-singular form of dimension n can be written as a product of no more that n reflections along non-isotropic vectors.

The proof of the theorem of E. Cartan is rather complicated, see [Berger] and [Deheuvels]. Moreover, the theorem is not precise enough for our purposes. For these reasons we shall not make use of the theorem in its generality, but prove a number of more specific results along these lines: namely 4.5, 5.3, 6.11

I.3 SYLVESTER TYPES

In this section we shall deal exclusively with quadratic forms over the field $\mathbb{R}$ of real numbers. By a Euclidean vector space we understand a finite dimensional vector space E with a quadratic form Q which is positive definite, i.e. $Q(e) > 0$ for all non-zero $e \in E$. The form Q is called negative definite if $Q(e) < 0$ for all non-zero $e \in E$. Our first objective is to generalise the concept of an orthonormal basis.

Proposition 3.1 Let (E,Q) be a quadratic form on the vector space E over $\mathbb{R}$ of dimension n. There exists an orthonormal basis for E, i.e. a basis $e_1,\ldots,e_n$ with

$$<e_i,e_j> = \begin{cases} 0 & ; \ i \neq j \\ -1,\ 0,\ +1 & ; \ i = j \end{cases}$$

Proof Let us use induction with respect to $dim(E) = n$. If the form Q is identically 0, any basis for E is orthonormal. Thus we may assume the existence of an $e \in E$ with $Q(e) \neq 0$. Let us write $Q(e) = \epsilon\lambda^2$ with $\epsilon = \pm 1$ and $\lambda \in \mathbb{R}^*$. Put $e_1 = \lambda^{-1}e$ to get $Q(e_1) = \pm 1$. Let us use the induction hypothesis to find an orthonormal basis $e_2,\ldots,e_n$ for $(\mathbb{R}e_1)^{\perp}$. The basis $e_1,\ldots,e_n$ meets our requirements. □

Sylvester's theorem 3.2 Let $e_1,\ldots,e_n$ be an orthonormal basis for the quadratic form (E,Q) over $\mathbb{R}$. The numbers

$$p = \#\{\, i \mid <e_i,e_i> = -1 \,\}, \quad q = \#\{\, i \mid <e_i,e_i> = 1 \,\}$$

are independent of the orthonormal basis considered.

Proof Let us fix the basis as in the statement of the proposition, and let E_- denote the subspace of E, generated by those elements e_i from our basis for which $<e_i,e_i> = 0, -1$. Observe that $Q(e) \leq 0$ for all $e \in E_-$ and conclude that for any Euclidean subspace F of E we have $F \cap E_- = 0$. According to 3.4 we have that

$$dim(F) + dim(E_-) = dim(F + E_-) + dim(F \cap E_-)$$

This gives us $dim(F) + n - q \leq n$, i.e. $dim(F) \leq q$. It follows that

3.3 $$sup \{ dim(F) \mid \text{Euclidean } F \subseteq E \} = q$$

This shows that q is independent of the basis chosen. Let us apply this to the space (E,−Q) and conclude that p too is independent of the basis. □

With the notation of the theorem we say that the quadratic form (E,Q) has <u>Sylvester type</u> (−p,q). Observe that two quadratic forms (E,Q) and (F,R) with $dim(E) = dim(F)$ of equal Sylvester type are isomorphic.

Dimension formula 3.4 For subspaces U and V of the finite dimensional vector space E we have that

$$dim(U \cap V) + dim(U+V) = dim(U) + dim(V)$$

Proof Let us apply the Grassmann dimension formula to the linear map

$$f : U \oplus V \to E \,, \quad f(u,v) = u - v \qquad ; u \in U \,,\, v \in V$$

observing that $Ker\, f \xrightarrow{\sim} U \cap V$ and $Im\, f = U + V$. □

Discriminant inequality 3.5 Let (E,Q) be a non-singular quadratic form of Sylvester type (−s,r). For any basis $e_1,\ldots,e_n$ for E we have

$$\boxed{sign\ det_{ij} <e_i,e_j> \;=\; (-1)^s}$$

Proof Let us consider a second basis $f_1,\ldots,f_n$ related to the first basis through the transition matrix B. This gives us

$$<f_i,f_j> = <\sum_r B_{ir}e_r, \sum_s B_{js}e_s> = \sum_{r,s} B_{ir}<e_r,e_s>{}^TB_{sj}$$

From this we conclude that

$$det_{ij} <f_i,f_j> = det\ B\ det_{rs}<e_r,e_s>\ det\ {}^TB = det^2B\ det_{ij}<e_i,e_j>$$

This shows that *sign* $det_{ij}<e_i,e_j>$ is independent of the basis considered. We leave it to the reader to verify the formula in the case of an orthonormal basis. □

Symmetric matrices From a symmetric matrix $A \in M_n(\mathbb{R})$ we can construct a quadratic form on the free vector space E with basis $e_1,\dots,e_n$

$$Q(\sum x_i e_i) = \sum_{rs} a_{rs} x_r x_s$$

By the <u>Sylvester type</u> of A we understand the Sylvester type of the form Q. The Sylvester type of A can be evaluated by means of the <u>principal minors</u>

$$det\, A_1,\dots, det\, A_n$$

here A_i is the $i \times i$ matrix obtained by deleting the n–i last rows and columns of A.

Lemma 3.6 The symmetrical matrix $A \in M_n(\mathbb{R})$ is positive definite if and only if all principal minors $det\, A_1,\dots, det\, A_n$ are strictly positive.

Proof To begin let us just assume that the principal minors are different from zero and let us show how to determine the signature of A. With the notation above the subspace E_i of E generated by $e_1,\dots,e_i$ is non-singular since $det\ A_i \neq 0$. Let us pick a non-zero vector $f_i \in E_i$, orthogonal to E_{i-1}. The Gram matrices of the two bases $e_1,\dots,e_i$ and $e_1,\dots,e_{i-1},f_i$ have the same sign, thus

$$sign\ det\ A_i = sign\ <f_i,f_i>\ sign\ det\ A_{i-1}$$

In the case at hand we deduce that $<f_i,f_i> > 0$ for $i = 1,\dots,n$. □

I.4 EUCLIDEAN VECTOR SPACES

Let us consider a Euclidean vector space E , i.e. a finite dimensional real vector space equipped with a positive definite quadratic form. A vector $e \in E$ has length $|e| = \sqrt{<e,e>}$. Let us start with the inequality of

Cauchy-Schwarz 4.1 $$|<e,f>| \leq |e|\,|f| \qquad ; e,f \in E$$

Addendum: The inequality is sharp when e and f are linearly independent

Proof The inequality is trivial if e and f are linearly dependent. If e and f are linearly independent, they span a Euclidean plane and we get from 3.5 that

$$det\begin{bmatrix} <e,e> & <e,f> \\ <f,e> & <f,f> \end{bmatrix} > 0$$

from which the sharp inequality follows. □

Triangle inequality 4.2 $$|e+f| \leq |e|+|f| \qquad ; e,f \in E$$

Addendum: The triangle inequality is sharp when e and f are linearly independent.

Proof A simple direct calculation gives us

$$|e+f|^2 = |e|^2 + |f|^2 + 2|e||f| + 2(<e,f> - |e||f|)$$

and the result follows from 4.1. □

On the basis of the Cauchy–Schwartz inequality we can introduce the angle $\angle(e,f)$ between two non-zero vectors e,f by the formula

4.3 $$\cos \angle(e,f) = \frac{<e,f>}{|e|\,|f|} \qquad ; \angle(e,f) \in [0,\pi]$$

In the rest of this chapter we shall be concerned with the orthogonal group $O(E)$. The basic example of an orthogonal transformation is reflection τ in a linear hyperplane H. If $n \in E$ is a unit normal vector for H we have that

4.4 $$\tau(x) = x - 2\langle x,n\rangle n \qquad ; x \in E$$

Theorem 4.5 An orthogonal transformation σ of the Euclidean vector space E of dimension n is the product of at most n reflections through hyperplanes.

Proof We shall use induction on $n = dim(E)$. To accomplish the induction step we pick a vector $f \in E$ of unit length. If $\sigma(f) = f$ observe that σ stabilises $F = f^{\perp}$ and use the induction hypothesis to decompose σ into a product of at most n–1 reflections. In case $\sigma(f) \neq f$ observe that reflection τ along the vector $\sigma(f) - f$ interchanges f and $\sigma(f)$. It follows that $\tau\sigma$ fixes the vector f, so we can use induction to write $\tau\sigma$ as a product of at most n–1 reflections. □

Let us recall from 1.11 that an orthogonal transformation has determinant 1 or –1. It follows that the special orthogonal group

4.6 $$SO(E) = \{ \sigma \in O(E) \mid det\, \sigma = 1 \}$$

has index 2 in $O(E)$. We have seen that $O(E)$ operates transitively on the sphere

4.7 $$S(E) = \{ e \in E \mid |e| = 1 \}$$

We ask the reader to verify that $SO(E)$ acts transitively on $S(E)$, $dim\, E \geq 2$.

Euclidean plane Let us assume that $dim(\mathrm{E}) = 2$. An orthogonal transformation of E with determinant 1 is called a rotation. In order to analyse an orthogonal transformation σ of E in more detail, let us pick an orthonormal basis **i** and **j** for E; we can express that $\sigma(\mathbf{i})$ lies on the circle $S(\mathrm{E})$ by writing

$$\sigma(\mathbf{i}) = \mathbf{i}\ cos\ \theta + \boldsymbol{j}\ sin\ \theta \qquad ; \theta \in \mathbb{R}$$

It follows that $\sigma(\mathbf{j})$ is one of the unit vectors orthogonal to $\sigma(\mathbf{i})$, i.e. the vector $-\mathbf{i}\ sin\theta + \boldsymbol{j}\ cos\ \theta$ or its negative. Thus the matrix for σ takes one of the forms

4.8
$$\begin{bmatrix} cos\ \theta & -sin\ \theta \\ sin\ \theta & cos\ \theta \end{bmatrix}, \begin{bmatrix} cos\ \theta & sin\ \theta \\ sin\ \theta & -cos\ \theta \end{bmatrix} \qquad ; \theta \in \mathbb{R}$$

Evaluation of the determinants show that the first matrix defines a rotation, while the second defines a reflection. It follows from the addition formulas for the trigonometric functions that the first matrix in 4.8 defines an isomorphism between the circle group $SO(\mathrm{E})$ and the group $\mathbb{R}/2\pi\mathbb{Z}$.

Euler's proposition 4.9 An orthogonal transformation $\sigma \in SO(\mathrm{E})$ of Euclidean 3-space E is a rotation: there is a line L fixed by σ and the transformation induced by σ in the plane $\mathrm{L}^{\perp}$ is a rotation. An orthogonal transformation $\sigma \in O(\mathrm{E})$ of determinant -1 has the form $\rho\tau$ where ρ is a rotation as above with axis L and τ is reflection in the plane $\mathrm{L}^{\perp}$.

Proof Consider the characteristic polynomial for the transformation σ

$$\chi(\mathrm{t}) = det(\iota \mathrm{t} - \sigma)$$

Let us analyse a real root λ: pick a vector $\mathrm{e} \neq 0$ with $\sigma(\mathrm{e}) = \lambda \mathrm{e}$ and take norms to see that $\lambda^2 = 1$, i.e. $\lambda = 1$ or $\lambda = -1$.

If $det(\sigma) = 1$ we have $\chi(0) = -1$. Observe that $\chi(\mathrm{t})$ tends to $+\infty$ as t tends to $+\infty$ and conclude that 1 is a root for $\chi(\mathrm{t})$. Consider a line L with eigenvalue 1 and observe that σ acts on the plane $\mathrm{L}^{\perp}$ with determinant 1.

If $det(\sigma) = -1$ we have $\chi(0) = 1$. Observe that $\chi(\mathrm{t})$ tends to $-\infty$ as t tends to $-\infty$ and conclude that -1 is a root for $\chi(\mathrm{t})$. Consider a line L with eigenvalue -1 and observe that σ acts on the plane $\mathrm{L}^{\perp}$ with determinant 1. □

Theorem 4.10 Let σ be an orthogonal transformation of the Euclidean vector space E of dimension n. There exists a decomposition of E into an orthogonal sum of lines and planes stable under σ.

Proof According to lemma 4.11 below we can find a linear subspace $R \subset E$ of dimension 1 or 2 stable under σ. It follows that $R^{\perp}$ is stable under σ as well. A simple induction on *dim* E concludes the proof. □

Lemma 4.11 Let $\sigma: E \to E$ be an endomorphism of a finite dimensional real vector space E. There exists a subspace $R \subset E$ of dimension 1 or 2 stable under σ.

Proof Let us present a construction known as complexification of σ. We ask the reader to turn $E_{\mathbb{C}} = E \oplus E$ into a vector space over $\mathbb{C}$ through the convention

$$(x + iy) \cdot (e,f) = (xe - yf,\ xf + ye) \qquad ; x,y \in \mathbb{R},\ e,f \in E$$

Let us identify $e \in E$ with $(e,0) \in E_{\mathbb{C}}$ to get the formula

$$(e,f) = e + if \qquad ; \ e,f \in E$$

With this notation, we can introduce the $\mathbb{C}$-linear endomorphism $\sigma_{\mathbb{C}}: E_{\mathbb{C}} \to E_{\mathbb{C}}$ by

$$\sigma_{\mathbb{C}}(e + if) = \sigma(e) + i\sigma(f) \qquad ; \ e,f \in E$$

the details are straightforward and left to the reader. From the fundamental theorem of algebra we conclude that $\sigma_{\mathbb{C}}$ has an eigenvalue $\lambda = a + ib$. A non-trivial eigenvector $e + if$ with eigenvalue λ satisfies the equation

$$\sigma(e) + i\sigma(f) = (a + ib)(e + if)$$

From this it follows that e and f generate a subspace of E stable under σ. □

I.5 PARABOLIC FORMS

In this section we shall investigate a finite dimensional vector space F over $\mathbb{R}$ equipped with a positive quadratic form Q, which of course means that $Q(f) \geq 0$ for all $f \in F$. The point of departure is the

Cauchy-Schwarz inequality 5.1

$$\langle e,f\rangle^2 \leq \langle e,e\rangle\langle f,f\rangle \qquad ; e,f \in F$$

Proof The Cauchy–Schwarz inequality may be rewritten in determinant form

$$det \begin{bmatrix} \langle e,e\rangle & \langle e,f\rangle \\ \langle f,e\rangle & \langle f,f\rangle \end{bmatrix} \geq 0$$

This is trivial when e and f are linearly dependent. When e and f are linearly independent we get from 1.12 that the sign of the determinant is unchanged if we replace e and f by an orthonormal basis for the plane they span. □

Corollary 5.2 Let (F,Q) be a positive quadratic form. A vector $f \in F$ is isotropic if and only if $f \in F^{\perp}$, i.e. $\langle e,f\rangle = 0$ for all $e \in F$.

We shall investigate the orthogonal group $O(F)$, in particular the subgroup generated by reflections. Let us observe that a reflection τ fixes all isotropic vectors as follows from 5.2 and formula 2.1.

Theorem 5.3 Let Q be a positive quadratic form on the m–dimensional real vector space F. An isometry $\sigma \in O(F)$ which is the identity on $F^{\perp}$ can be written as a product of at most m reflections.

Proof We proceed by induction on m. Let us distinguish between two cases.

1° $Im(\sigma - \iota) \not\subset F^{\perp}$. Choose $f \in F$ such that $p = \sigma(f) - f$ is non-isotropic. Reflection π along p interchanges f and $\sigma(f)$, i.e. $\pi\sigma(f) = f$. Observe that f is non-isotropic since $\sigma(f) \neq f$ and apply the induction hypothesis to the restriction of $\pi\sigma$ to the hyperplane $f^{\perp}$.

2° $Im(\sigma - \iota) \subset F^{\perp}$. Let us decompose F into a direct sum $F = E \oplus N$ where E is Euclidean and $N = F^{\perp}$. Let us describe σ in terms of a linear map $\nu: E \to N$

$$\sigma(e+n) = e + \nu(e) + n \qquad ; \ e \in E,\ n \in N$$

We can arrange for $\nu: E \to N$ to be injective: a non-zero vector $e \in E$ of $Ker\ \nu$ is non-isotropic and fixed by σ, which allows us to use the induction hypothesis on $e^{\perp}$. We can now arrange for $\nu: E \to N$ to be bijective by replacing N by $\nu(E)$. From now on we may assume that

$$dim\ F = 2\ dim\ F^{\perp}$$

Observe that σ acts as the identity on $F^{\perp}$ and on $F/F^{\perp}$ and conclude that $det(\sigma) = 1$. Let us pick any reflection ρ in F and observe that we are back to case 1° with $\rho\sigma$ since $det(\rho\sigma) = -1$. Thus we can write

$$\rho\sigma = \rho_1 \dots \rho_s \qquad ; \ s \leq m$$

where $\rho_1, \dots, \rho_s$ are reflections. From the facts that m is even and $det(\rho\sigma) = -1$, it follows that $s < m$ as required. □

Let us say that a space (F,Q) is <u>parabolic</u> if Q is positive and the space $F^{\perp}$ is of dimension 1. The <u>isotropic line</u> $F^{\perp}$ is invariant under orthogonal transformations. It follows that we have a morphism of groups

5.4
$$\mu : O(F) \to \mathbb{R}^*$$

where the <u>multiplier</u> $\mu(\sigma) \in \mathbb{R}^*$ denotes the eigenvalue of $\sigma \in O(F)$ on the isotropic line. This allows us to introduce a subgroup of $O(F)$ of index 2

5.5
$$O_{\infty}(F) = \{ \sigma \in O(F) \mid \mu(\sigma) > 0 \}$$

Let us observe that the antipodal map $\bar{\iota}$, $x \mapsto -x$, belongs to the centre of $O(F)$ and that $\mu(\bar{\iota}) = -1$. This defines a decomposition of the orthogonal group $O(F)$ of a parabolic space F

5.6 $$O(F) = O_\infty(F) \times \mathbb{Z}/(2)$$

The line at infinity We shall pursue the study of parabolic spaces in a setting which is particular useful for Euclidean geometry. Let E be a Euclidean space of dimension n and consider the following bilinear form on $F = E \oplus \mathbb{R}$

5.7 $$\langle (e,a),(f,b)\rangle = \langle e,f\rangle \qquad e,f \in E,\ a,b \in \mathbb{R}$$

This form is easily seen to be parabolic. The isotropic line is generated by the vector $(0,1)$ and is called the line at infinity. Let us use the inner product on F to introduce the evaluation map $ev\colon F \times E \to \mathbb{R}$

$$ev((x,\xi),y) = \langle x,y\rangle + \xi \qquad ;\ x,y \in E,\ \xi \in \mathbb{R}$$

The evaluation map identifies F with the space of affine functions on E.

Similarities Let us recall that a similarity of the Euclidean space E is an affine transformation Φ of E of the form

$$\Phi(x) = \lambda\phi(x) + f \qquad ;\ \phi \in O(E),\ f \in E,\ \lambda \in \mathbb{R},\ \lambda > 0$$

Our main objective is to identify the group $Siml(E)$ of similarities with the orthogonal group $O_\infty(E)$.

Proposition 5.8 For any orthogonal transformation σ of parabolic space $F = E \oplus \mathbb{R}$, there exists a unique similarity $\tilde{\sigma} \in Siml(E)$ such that

$$ev(\sigma(f),\tilde{\sigma}(e)) = \mu(\sigma)\ ev(f,e) \qquad ;\ f \in F,\ e \in E$$

The assignment $\sigma \mapsto \tilde{\sigma}$ induces an isomorphism of groups

$$O_\infty(F) \xrightarrow{\sim} Siml(E)$$

The similarity $\lambda\phi + v$, $\phi \in O(E)$, $\lambda > 0$, $v \in E$, corresponds to $\Phi \in O_\infty(F)$ given by

$$\Phi(z,\zeta) = (\phi(z),\ -<\phi(z),v> + \lambda\zeta) \qquad ;\ z \in E,\ \zeta \in \mathbb{R}$$

Proof Let us first observe that a point $e \in E$ defines a linear form $f \mapsto ev(f,e)$ on F with evaluation 1 against the isotropic vector $(0,1)$. It is easy to see that any linear form ϕ on F with $\phi(0,1) = 1$ can be represented in this way and that such a representation is unique.

Let us return to $\sigma \in O(F)$ and observe that $\mu(\sigma)^{-1}\sigma$ is a linear automorphism of F which fixes the isotropic vector $(0,1)$. It follows that for a given $e \in E$ we can find a uniquely determined vector $\check{\sigma}(e)$ in E such that

$$ev(\sigma(f),e) = \mu(\sigma)\ ev(f,\check{\sigma}(e)) \qquad ;\ f \in F$$

Variation of $e \in E$ defines a transformation $\check{\sigma}:E \to E$. It follows at once that

$$(\sigma\tau)\check{} = \check{\tau}\check{\sigma} \quad , \quad \check{\iota} = \iota \qquad ;\ \sigma,\tau \in O(F)$$

We conclude that $\check{\sigma}$ is invertible and that the transformation $\tilde{\sigma} = \check{\sigma}^{-1}$ satisfies the required formula. In order to show that $\tilde{\sigma}$ is a similarity, we shall work our way through three special cases.

1° A vector $v \in E$ gives rise to a transformation $\theta_v \in O(F)$ given by

$$\theta_v(z,\zeta) = (z,\zeta - <v,z>) \qquad ;\ z \in E,\ \xi \in \mathbb{R}$$

Direct verification shows that $\tilde{\theta}_v(e) = e + v$, $e \in E$.

2° A real scalar $\lambda \in \mathbb{R}^*$ gives rise to a transformation $s_\lambda \in O(F)$ given by

$$s_\lambda(z,\zeta) = (z,\lambda\zeta) \qquad ;\ z \in E,\ \zeta \in \mathbb{R}$$

Direct verification gives us $\tilde{s}_\lambda(e) = \lambda e$, $e \in E$.

3° An orthogonal transformation $\sigma \in O(E)$ defines $\bar{\sigma} \in O(F)$ given by

$$\bar{\sigma}(z,\zeta) = (\sigma(z),\zeta) \qquad ;\ z \in E,\ \zeta \in \mathbb{R}$$

Another direct verification gives us $\tilde{\bar{\sigma}} = \sigma$. We leave it to the reader to show that any orthogonal transformation of F can be decomposed into a product of transformations of the three types we have just considered. □

Normal vectors Let there be given an oriented affine hyperplane (H,P) in Eucliden n-space E. By this we understand that H is an affine hyperplane in E and P is one of the open half-spaces of E bounded by H. The normal vector to (H,P) is the unit vector $\mathbf{n} \in F$ which is zero along H and positive on P. Let us observe that reflection $\tau_{\mathbf{n}}$ is given by

$$\tau_{\mathbf{n}}(z,\zeta) = (z,\zeta) - 2\langle n,z\rangle(n,\nu) \qquad ; \mathbf{n} = (n,\nu)$$

while the corresponding transformation $\tilde{\tau}_{\mathbf{n}}$ is given by the formula

$$\tilde{\tau}_{\mathbf{n}}(x) = x - 2(\langle x,n\rangle + \nu)n \qquad ; x \in E$$

which is a reflection in the affine hyperplane H, compare II.2.

Let us observe that $\sigma \in O_\infty(F)$ transforms the normal vector $\mathbf{n}$ for (H,P) into the normal vector for $(\tilde{\sigma}(H),\tilde{\sigma}(P))$ as follows from the formula

$$ev(\sigma(\mathbf{n}),\tilde{\sigma}(x)) = \mu(\sigma)ev(\mathbf{n},x) \qquad ; \ x \in E$$

Coxeter matrix Let us consider an affine simplex D, i.e. the convex hull of n+1 points $e_0,e_1,\dots,e_n$, not contained in an affine hyperplane. Let

$$\mathbf{n}_0, \mathbf{n}_1,\dots, \mathbf{n}_n$$

be the corresponding normal vectors: the evaluation of $\mathbf{n}_s$ against e_i is zero, $i \neq s$, while the evaluation against e_s is positive. From this description it follows that the normal vectors form a basis for $F = E \oplus \mathbb{R}$. Let us write

5.9
$$\lambda_0\mathbf{n}_0 + \lambda_1\mathbf{n}_1 + \dots + \lambda_n\mathbf{n}_n = (0,1)$$

and observe that the λ_is are strictly positive which follows from evaluation of formula 5.9 on the vertices of the simplex. The Coxeter matrix for D is

$$\langle \mathbf{n}_r,\mathbf{n}_s\rangle_{rs}$$

Proposition 5.10 A basis $\mathbf{n}_0,\mathbf{n}_1,\dots,\mathbf{n}_n$ of unit vectors for $E \oplus \mathbb{R}$ such that the isotropic vector (0,1) can be written in the form

$$\lambda_0\mathbf{n}_0 + \lambda_1\mathbf{n}_1 + \dots + \lambda_n\mathbf{n}_n = (0,1) \qquad ; \ \lambda_i > 0 \ , \ i = 0,\dots,n$$

defines a Euclidean simplex in E.

Proof For $s = 0,...,n$ let ξ_s denote the linear form on $E \oplus \mathbb{R}$ with $\xi_s(\mathbf{n}_s) = \lambda_s^{-1}$ and $\xi_s(\mathbf{n}_r) = 0$ for $s \neq r$. The relation above ensures that ξ_s has evaluation 1 on the isotropic vector (0,1). It follows from the remark made in the beginning of the proof of 5.7 that we can find a point $e_s \in E$ with $ev(f,e_s) = \xi_s(f)$ for all $f \in F$. The points $e_0,...,e_n$ define the simplex we are looking for. □

Proposition 5.11 Two simplices D and P in Euclidean space E are similar if and only if they have the same Coxeter matrix.

Proof We have already seen that two similar simplices have the same Coxeter matrices. With the notation above let $\mathbf{m}_0,\mathbf{m}_1,...,\mathbf{m}_n$ be the normal vectors for the simplex P enumerated such that

$$<\mathbf{n}_r,\mathbf{n}_s> = <\mathbf{m}_r,\mathbf{m}_s> \qquad ; \; r, s = 0,1,...,n$$

Let $\sigma \in O(F)$ be such that $\sigma(\mathbf{n}_r) = \sigma(\mathbf{m}_r)$, for $r = 0,...,n$. It remains to prove that $\mu(\sigma) > 0$. To this end let us apply σ to the relation 5.9 to get that

$$\lambda_0\mathbf{m}_0 + \lambda_1\mathbf{m}_1 + ... + \lambda_n\mathbf{m}_n = (0,\mu(\sigma))$$

Let us evaluate this against an interior point $p \in P$ to get that

$$\lambda_0 ev(\mathbf{m}_0,p) + ... + \lambda_n ev(\mathbf{m}_n,p) = \mu(\sigma)$$

which shows that $\mu(\sigma) > 0$ as required. □

In the rest of this section we shall be concerned with the existence of simplices with given Coxeter matrices (a_{rs}). For application to groups generated by reflections, see [Coxeter[2]] and [Bourbaki[2]], we shall be concerned with <u>acute angled</u> simplices, i.e. simplices whose matrices satisfies $\boxed{a_{rs} \leq 0 \text{ for } r \neq s}$.

Indecomposable matrices 5.12 Let $A \in M_n(\mathbb{R})$ be a positive symmetric matrix with *det* A=0, which has 1 s along the diagonal and satisfies

$$a_{rs} \leq 0 \ , \ r \neq s, \ r,s = 0,\dots,n \qquad ; A = (a_{rs})$$

If the matrix is indecomposable, in the sense that no non-trivial subset J of 0,...,n exists such that

$$a_{rs} = 0 \ \text{ for } r \in J \text{ and } s \notin J$$

then there exists a Euclidean n−simplex with A as Coxeter matrix.

Proof Let us consider a real vector space F equipped with a basis $e_0,\dots,e_n$ and a quadratic form such that

$$<e_r,e_s> = a_{rs} \qquad ; r,s = 0,\dots,n$$

By assumption the form is positive but singular. We shall eventually prove that F is parabolic. To this end we shall show that for a non-zero isotropic vector $n = \sum x_s a_s$ we have $a_s \neq 0$ for all $s = 0,1,\dots,n$. From the inequality

$$Q(\sum x_s e_s) = \sum_{r,s} a_{rs} x_r x_s \geq \sum_{r,s} a_{rs}|x_r||x_s| = Q(\sum |x_s| e_s)$$

it follows that we can assume that all coefficients of n are positive. Let us for a moment assume that the set $J = \{s \mid x_s > 0\}$ is different from 0,...,n. Let us recall from 5.2 that the isotropic vector n is orthogonal to any vector:

$$0 = <n,e_s> = \sum_{r \in J} x_r <e_r,e_s> \qquad ; s = 0,\dots,n$$

In particular we conclude that $a_{rs} = <e_r,e_s> = 0$ for $s \notin J$, contradicting the hypothesis that the matrix A is indecomposable.

Let us prove that $F^{\perp}$ has dimension 1. To this end let us introduce the space $N \subset \mathbb{R}^{n+1}$ representing $F^{\perp}$

$$N = \{(x_0,\dots,x_n) \in \mathbb{R}^{n+1} \mid Q(\sum x_s e_s) = 0\}$$

By the observations made above we conclude that the intersection between N and the hyperplane $x_0 = 0$ is zero; this implies $dim(N) = 1$. In conclusion if $\sum x_s e_s \neq 0$ is isotropic then all coefficients are non-zero of the same sign. Finally we can use an isometry of F with $E \oplus \mathbb{R}$ to construct a basis for $E \oplus \mathbb{R}$ and apply 5.10. □

I.6 LORENTZ GROUP

Let us consider a vector space F of dimension n+1 over $\mathbb{R}$ equipped with a quadratic form of Sylvester type (−n,1). We shall be concerned with the action of $\mathcal{O}(F)$ on the hyperboloid or pseudosphere

6.1
$$\mathcal{S}(F) = \{ f \in F \mid \langle f,f \rangle = 1 \}$$

Let us start our investigations by a Cauchy–Schwarz type inequality.

Lemma 6.2 Two points x and u of the hyperboloid satisfy the inequality

$$1 \le |\langle x,u \rangle| \qquad ; x,u \in \mathcal{S}(F)$$

Moreover, the inequality is sharp when x and u are linearly independent in F.

Proof When x and u are linearly dependent we have x = u or x = −u and the inequality is trivial. When x and u are linearly independent, the plane R they span is of Sylvester type (−1,1), (−1,0) or (−2,0), compare 6.3. From the fact that R contains vectors of strictly positive norm it follows that the type is (−1,1). The discriminant lemma 6.3 gives us $\langle u,u \rangle \langle x,x \rangle < \langle u,x \rangle^2$ as required. □

Discriminant lemma 6.3 Let R be a plane in F generated by two vectors e and f. The Sylvester type of R is determined by the discriminant

$$\Delta = \langle e,e \rangle \langle f,f \rangle - \langle e,f \rangle^2$$

according to the following table

$\Delta < 0$	$\Delta = 0$	$\Delta > 0$
(−1,1)	(−1,0)	(−2,0)

Proof Let us first prove that the plane R contains a vector of norm −1. Use any orthonormal basis for F to exhibit a hyperplane E of Sylvester type (−n,0). It follows from the dimension formula 3.4 that

$$dim(\mathrm{R}) + dim(\mathrm{E}) = dim(\mathrm{R} \cap \mathrm{E}) + dim(\mathrm{R} + \mathrm{E})$$

Using $dim(\mathrm{R}+\mathrm{E}) \leq n+1$ we get that $dim(\mathrm{R} \cap \mathrm{E}) \geq 1$. Thus R contains vectors of strictly negative norm. From this we conclude that the only possible Sylvester types for R are $(-1,1)$, $(-1,0)$ and $(-2,0)$. It follows from the discriminant inequality 3.5 that these three cases correspond to $\Delta < 0$, $\Delta = 0$ and $\Delta > 0$ respectively. □

A linear hyperplane in F must be of type $(-n+1,1)$, $(-n,0)$ or $(-n+1,0)$ as it follows from the argument used in the previous proof.

Proposition 6.4 The hyperboloid $\mathcal{S}(\mathrm{F})$ has two connected components. Points $x,u \in \mathcal{S}(\mathrm{F})$ are in the same connected component if and only if $\langle x,u\rangle > 0$.

Proof Let us fix the point $u \in \mathcal{S}(\mathrm{F})$ and let E denote the hyperplane orthogonal to u. Since E has type $(-n,0)$ we find that E separates $\mathcal{S}(\mathrm{F})$ into two parts

$$\mathrm{H} = \{\, x \in \mathcal{S}(\mathrm{F}) \mid \langle x,u\rangle > 0 \,\}, \quad \mathrm{H}^{-} = \{\, x \in \mathcal{S}(\mathrm{F}) \mid \langle x,u\rangle < 0 \}$$

These two parts are interchanged by reflection along u. We are going to prove that stereographic projection $g{:}\mathrm{H} \to \mathrm{E}$ with centre $-u$ maps H homeomorphically onto the open unit disc

$$\mathrm{D} = \{\, y \in \mathrm{E} \mid -\langle y,y\rangle < 1 \,\}$$

The affine line through $x \in \mathrm{H}$ and $-u$ is parametrised $(x+u)t - u$, $t \in \mathbb{R}$. Its intersection $g(x)$ with E is given by $\langle (x+u)t - u, u\rangle = 0$, which gives us

6.5 $$g(x) = \frac{x - \langle x,u\rangle u}{1 + \langle x,u\rangle}$$

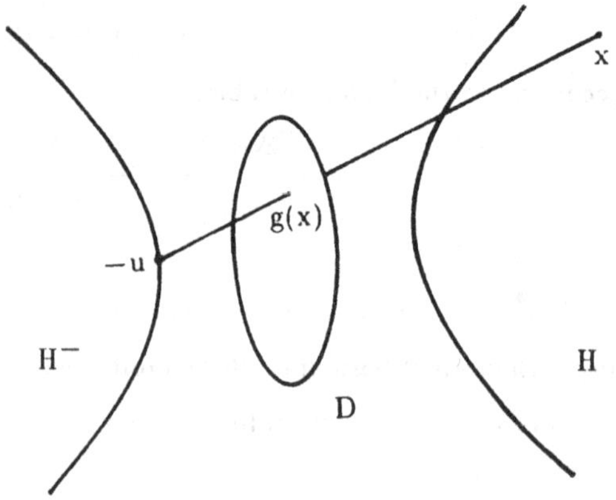

Direct computation gives us

$$\langle g(x),g(x)\rangle = \frac{1 - \langle x,u\rangle}{1 + \langle x,u\rangle}$$

From 6.2 we conclude that $<x,u> \geq 1$, so the formula above enables us to conclude that $g(x) \in D$ as required. We ask the reader to work out that the intersection between H and the affine line through the points $z \in D$ and $-u$ is given by

6.6
$$f(z) = 2\,\frac{z+u}{1+<z,z>} - u \qquad ; \; z \in D$$

These investigations show that $f{:}D \to H$ is a homeomorphism. □

Definition 6.7 By a Lorentz transformation[4] of F we understand an orthogonal transformation $\sigma \in \mathcal{O}(F)$ which preserves the connected components of the pseudosphere. The group of Lorentz transformations is called the Lorentz group and is denoted $Lor(F)$.

It follows from 6.4 that a Lorentz transformation can be recognised by the inequality

6.8
$$<x,\sigma(x)> > 0 \qquad ; \; \sigma \in Lor(F), \;\; x \in S(F)$$

The antipodal map $x \mapsto -x$ is in the centre of $\mathcal{O}(F)$ but is not a Lorentz transformation. It follows that $Lor(F)$ has index 2 in $\mathcal{O}(F)$, more precisely

6.9
$$\mathcal{O}(F) = Lor(F) \times \mathbb{Z}/(2)$$

The basic example of a Lorentz transformation is provided by reflection along a vector $c \in F$ with $<c,c> = -1$

$$\tau_c(x) = x + 2<x,c>c \qquad ; \; x \in F$$

Let us use 6.4 to check that this is a Lorentz transformation:

$$<\tau_c(x),x> = 1 + 2<x,c>^2 \qquad ; \; x \in S(F)$$

Lorentz transformations with determinant 1 are called even or special Lorentz

[4] In literature with reference to physics such a transformation is called an orthochroneous Lorentz transformation.

transformations. These transformations make up the special Lorentz group $Lor^+(F)$, which is a subgroup of $Lor(F)$ of index 2.

Let us make a detailed examination of a Lorentz transformation σ when the space F has dimension 2. Pick an orthonormal basis e,f for F with $\langle e,e\rangle = 1$ and $\langle f,f\rangle = -1$. The point $xe + tf$ lies on the hyperbola $S(F)$ if and only if $x^2 - t^2 = 1$. The branch of the hyperbola containing (1,0) can be parametrised $(\cosh s, \sinh s)$, $s \in \mathbb{R}$. Thus we can write $\sigma(e) = e\cosh s + f\sinh s$

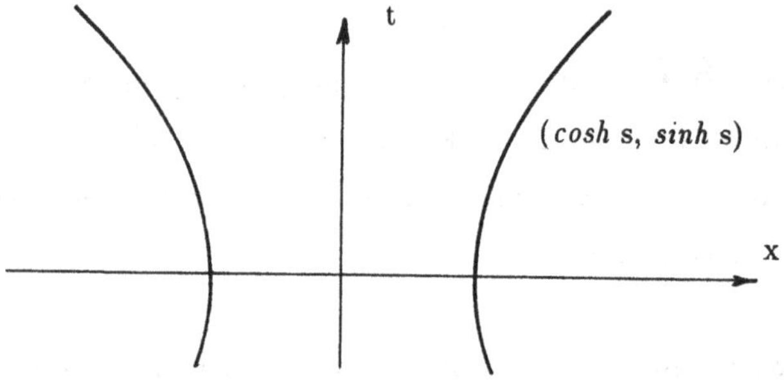

The vector $\sigma(f)$ has norm -1 and is orthogonal to $\sigma(e)$, so we conclude that $\sigma(f) = \mp(e\sinh s + f\cosh s)$. In the first case the matrix for σ is

$$\begin{bmatrix} \cosh s & -\sinh s \\ \sinh s & -\cosh s \end{bmatrix}$$

It follows that $det(\sigma) = -1$ and $tr(\sigma) = 0$. This implies that σ has eigenvalues 1 and -1. We ask the reader to verify that the basis e,f can be picked such that the matrix for σ corresponds to the case $s = 0$. This shows that σ is a reflection along a vector of norm -1. In the second case the matrix for σ is

$$L(s) = \begin{bmatrix} \cosh s & \sinh s \\ \sinh s & \cosh s \end{bmatrix} \qquad ; s \in \mathbb{R}$$

From the addition formulas for the hyperbolic trigonometric functions it follows that $L(s)$ is a morphism of groups

6.10 $$L(s+t) = L(s)\,L(t) \qquad ; s,t \in \mathbb{R}$$

This shows that the special Lorentz group is isomorphic to $\mathbb{R}$ when $dim(F) = 2$.

Theorem 6.11 Let F be a vector space of dimension n+1 equipped with a quadratic form of Sylvester type (−n,1). Any Lorentz transformation σ of F is the product of at most n+1 reflections of the type τ_c where $c \in F$ with $<c,c> = -1$.

Proof We shall proceed by induction on n. If n = 1 the result follows from our investigation above. If $n \geq 2$ let us pick a hyperplane E of F of type (−n,0) and focus on the map $E \to F$ given by

$$e \mapsto \sigma(e) - e \qquad ; \; e \in E$$

If this map is not injective we can find a vector e with norm −1 fixed by σ. The space $D = e^{\perp}$ has Sylvester type (−n+1,1) and the restriction of σ to D is a Lorentz transformation as it follows from 6.8

If the map $e \mapsto \sigma(e) - e$, $e \in E$, is injective, then the image space has dimension $n-1 \geq 2$ and must have non-zero intersection with the hyperplane E. It follows that the image space contains a vector c say of norm −1. Let us write

$$c = \sigma(e) - e \qquad ; \; e \in F, \; <e,e> < 0$$

Since $\sigma(e)$ and e have the same norm we conclude from the proof of 2.3 that reflection τ_c along c interchanges e and $\sigma(e)$. It follows that $\tau_c\sigma$ fixes the vector e of strictly negative norm. Apply the induction hypothesis to the space $e^{\perp}$ to conclude the proof. □

Let us turn to the action of the Lorentz group on the <u>isotropic cone</u> $C(F)$, i.e. the set of vectors of norm 0. The multiplicative group $\mathbb{R}^*$ acts on $C(F)-0$; the orbit space is called the <u>projectivised cone</u> of F and is denoted $PC(F)$. We may of course think of $PC(F)$ as the set of isotropic lines in F.

Proposition 6.12 Let F be a vector space of dimension n+1 equipped with a quadratic form of Sylvester type (−n,1). The projectivised cone *PC*(F) is homeomorphic to the sphere S^{n-1}.

Proof Let us return to the proof of 6.3 and introduce the unit sphere $S^{n-1} = \partial D$ in the hyperplane E orthogonal to the base vector u. We shall make use of the map ϕ:E→F given by

6.13 $$\phi(z) = 2z + u - \langle z,z\rangle u \qquad ; z \in E$$

Direct calculation using that $\langle u,z\rangle = 0$ gives us

$$\langle\phi(z),\phi(z)\rangle = (1 + \langle z,z\rangle)^2 \qquad ; z \in E$$

which shows that ϕ induces a map $\Phi: \partial U \to PC(F)$. Observe that u and $z \in \partial U$ span a plane of Sylvester type (−1,1) and that $\phi(z)$ and $\phi(-z)$ are two linearly independent isotropic vectors in that plane. Observe also that this plane is generated by $\phi(z)$ and u. From these remarks it follows that Φ is injective. In order to see that Φ is surjective note that an isotropic line L in F together with the vector u span a plane R of type (−1,1): pick a non-zero vector $v \in L$ and let us examine the sign of the discriminant

$$\langle u,u\rangle\langle v,v\rangle - \langle u,v\rangle^2 = -\langle u,v\rangle^2 < 0$$

The intersection $E \cap R$ is a line generated by a vector z say of norm −1. It follows that L is generated by one of the two vectors $\phi(z)$ and $\phi(-z)$. □

Corollary 6.14 When $dim(F) \geq 3$, the Lorentz group *Lor*(F) acts faithfully on *PC*(F).

Proof Let σ be a Lorentz transformation of F which acts trivially on *PC*(F). We shall analyse the problem in terms of the map $\Phi: S^{n-1} \to PC(F)$ introduced in the proof of 6.13. For $z \in S^{n-1}$ let $\lambda(z)$ be the eigenvalue belonging to the isotropic vector $\phi(z)$. The function $\lambda: S^{n-1} \to \mathbb{R}$ is continuous which follows from the formula

$$\langle\sigma\phi(z),u\rangle = \lambda(z)\langle\phi(z),u\rangle = 2\lambda(z) \qquad ; z \in S^{n-1}$$

The set of values of the function λ is finite simply because it is a subset of the set of eigenvalues of σ. Since $dim(\mathrm{F}) = \mathrm{n}+1 \geq 3$, we can use the connectedness of $\mathrm{S}^{\mathrm{n}-1}$ to conclude that λ is constant.

Let us prove that F is generated by $C(\mathrm{F})$. To this end we shall prove that any vector $\mathrm{f} \in \mathrm{F}$ with non-zero norm is contained in a hyperbolic plane. If $<\mathrm{f},\mathrm{f}> < 0$, then $\mathrm{f}^{\perp}$ is of type $(-\mathrm{n}+1,1)$, so we can select e orthogonal to f with $<\mathrm{e},\mathrm{e}> > 0$. When $<\mathrm{f},\mathrm{f}> > 0$ the space $\mathrm{f}^{\perp}$ has type $(-\mathrm{n},0)$ so we will have no trouble finding e orthogonal to f with $<\mathrm{e},\mathrm{e}> < 0$. Now, observe that a hyperbolic plane is generated by two isotropic vectors.

Let us observe that multiplication by a scalar λ is not an orthogonal transformation of F unless $\lambda = 1$ or $\lambda = -1$. We have already observed that multiplication by -1 interchanges the two sheets of the hyperboloid $S(\mathrm{F})$. □

Proposition 6.15 The truncated isotropic cone $C^*(\mathrm{F}) = C(\mathrm{F}) - \{0\}$ has two connected components provided that $dim(\mathrm{F}) \geq 3$. The connected components are preserved by the Lorentz group $Lor(\mathrm{F})$.

Proof Let us fix a point $\mathrm{u} \in S(\mathrm{F})$ and observe that $C^*(\mathrm{F})$ is the union of

$$C^+(\mathrm{F}) = \{\mathrm{c} \in \mathrm{F} \mid <\mathrm{c},\mathrm{u}> > 0\}, \quad C^-(\mathrm{F}) = \{\mathrm{c} \in \mathrm{F} \mid <\mathrm{c},\mathrm{u}> < 0\}$$

It follows from the proof of the previous proposition that these two sets are the connected components of $C^*(\mathrm{F})$.

Let us prove that the two connected components are unchanged if we replace the point of reference u with an other point v of the same sheet of the hyperboloid $S(\mathrm{F})$. To this end let us focus on a point $\mathrm{c} \in C^*(\mathrm{F})$ and observe that the function $\mathrm{x} \mapsto <\mathrm{x},\mathrm{c}>$, $S(\mathrm{F}) \to \mathbb{R}$, omits the value zero since the hyperplane orthogonal to c has type $(-\mathrm{n}+1,0)$, compare the remark made after 6.4. It follows that $<\mathrm{u},\mathrm{c}>$ and $<\mathrm{v},\mathrm{c}>$ have the same sign.

From the fact that $\sigma(\mathrm{u})$ and u belong to the same sheet of $S(\mathrm{F})$ it follows that $<\sigma(\mathrm{x}),\mathrm{u}>$ and $<\sigma(\mathrm{x}),\sigma(\mathrm{u})>$ have the same sign for $\mathrm{x} \in C^*(\mathrm{F})$. In particular for $\mathrm{x} \in C^+(\mathrm{F})$ we have $<\sigma(\mathrm{x}),\sigma(\mathrm{u})> = <\mathrm{x},\mathrm{u}> > 0$, and we conclude that $<\sigma(\mathrm{x}),\mathrm{u}>$ is positive; thus $\sigma(\mathrm{x}) \in C^+(\mathrm{F})$. □

Proposition 6.16 Let L denote an isotropic line in F, and let $Lor_L(F)$ denote the group of Lorentz transformations stabilising L. Restriction from F to $L^\perp$ defines an isomorphism

$$Lor_L(F) \xrightarrow{\sim} O_\infty(L^\perp)$$

where $O_\infty(L^\perp)$ denotes the subgroup of $O(L^\perp)$ made up of orthogonal transformations with positive eigenvalue on the isotropic line L, compare 5.5.

Proof Let $O_L(F)$ denote the subgroup of $O(F)$ made up of transformations which stabilises the line L. Let us prove that restriction defines an isomorphism

$$O_L(F) \xrightarrow{\sim} O(L^\perp)$$

First, restriction is surjective as it follows from Witt's theorem 2.4. – Let us analyse a $\sigma \in O_L(F)$ which restricts to the identity on $L^\perp$. To this end we introduce an n-1 dimensional subspace E of $L^\perp$ of type $(-n+1,0)$. From the fact that σ has trivial restriction to E follows that we can write $\sigma = \iota \oplus \alpha$ where α is an orthogonal transformation of $E^\perp$. The plane $E^\perp$ has type $(-1,1)$ and can be decomposed $E^\perp = L \oplus K$ where K is the second isotropic line in $E^\perp$. The transformation α preserves K and L and the eigenvalues must have the form λ, λ^{-1}, $\lambda \in \mathbb{R}^*$, since the inner product is preserved. Remembering that σ has trivial restriction to $L^\perp$ we conclude first that $\lambda = 1$ and second that $\rho = \iota$ as required. The final statement of the proposition is a consequence of 6.15 in case $dim(F) \geq 3$. In case $dim(F) = 2$ the result follows from the investigations a few lines above. □

Proposition 6.17 The special Lorentz group $Lor^+(F)$ acts transitively on the space $N(F)$ of vectors of norm -1, provided $dim(F) \geq 3$.

Proof The full orthogonal group $O(F)$ acts transitively on $N(F)$ as a consequence of Witt's theorem 2.4. To see that $Lor(F)$ acts transitively it suffices to prove that the stabiliser in $O(F)$ of a given $N \in N(F)$ is not contained in the Lorentz group. As an example we take the transformation which fixes N but transform $x \mapsto -x$ on the orthogonal complement of N in F. We ask the reader to check that the stabiliser of N in $Lor(F)$ is not contained in $Lor^+(F)$. □

The projectivised cone $PC(\mathrm{F})$ is a subset of projective space $P(\mathrm{F})$, that i.e. the set of all lines in F. Let $PS(\mathrm{F})$ denote the subset of $P(\mathrm{F})$ consisting of lines in F generated by vectors of strictly positive norm. In the natural topology of $P(\mathrm{F})$ we have $\partial PS(\mathrm{F}) = PC(\mathrm{F})$.

Let H^n denote any one of the two sheets of the hyperbola $S(\mathrm{F})$. We have a natural homeomorphism $\mathrm{H}^n \to PS(\mathrm{F})$ which assigns to a point $\mathrm{X} \in \mathrm{H}^n$ the line L in F generated by the vector X. Using this identification we can write

6.18 $$\partial \mathrm{H}^n = PC(\mathrm{F})$$

I.7 MÖBIUS TRANSFORMATIONS

Let E denote a Euclidean vector space of dimension n. A sphere $\mathcal{C}$ with centre c and radius $r > 0$ gives rise to an involution σ of $E-\{c\}$ given by

7.1
$$\sigma(x) = c + r^2 \frac{x-c}{|x-c|^2} \qquad ; \; x \in E - \{c\}$$

The transformation σ is called <u>inversion</u> in the sphere $\mathcal{C}$. It is apparent from the formula that $\sigma(x)$ lies on the ray through x emanating from c. The sphere $\mathcal{C}$ can be recovered as the fixed point set for σ.

In the following we shall consider the 1-point compactification $\hat{E}$ of E, obtained from E by adding an extra point ∞. Reflection σ in the sphere $\mathcal{C}$ can be extended to an involution $\hat{\sigma}$ of $\hat{E}$ which interchanges ∞ and the centre c of $\mathcal{C}$. It is easy to check directly that $\hat{\sigma}$ is continuous.

Let us consider an affine hyperplane H in E through the point $u \in E$. Using a unit normal vector $n \in E$ for H we can describe write Euclidean reflection in H by the formula

7.2
$$\lambda(x) = x - 2\langle x-u\, ,\, n\rangle n \qquad ; \; x \in E$$

We leave it to the reader to extend this to a continuous transformation $\hat{\lambda}$ of $\hat{E}$, which preserves the point at infinity.

In order to unify the treatment let us introduce the word <u>sphere</u> $\mathcal{C}$ in $\hat{E}$ for a Euclidean sphere in E or a subset of $\hat{E}$ of the form $\mathcal{C} = \{\infty\} \cup H$ where H is an affine hyperplane in E. The two types of transformation of $\hat{E}$ we have introduced are called <u>inversions in spheres</u> in $\hat{E}$. The subgroup of the group of homeomorphisms of $\hat{E}$ generated by inversions in spheres in $\hat{E}$ is called the <u>Möbius group</u> of E and is denoted $Möb(E)$.

The main purpose of this section is to identify the Möbius group with a Lorentz group. To this end we introduce the auxiliary space $E \oplus \mathbb{R}^2$ equipped with the bilinear form

7.3 $<(e,a,b),(f,c,d)> = -<e,f> + \frac{1}{2}(ad+bc) \qquad ; e,f \in E,\ a,b,c,d \in \mathbb{R}$

It is easily seen that this space has Sylvester type $(-n-1,1)$. Let us consider the transformation $\iota: E \to C(E \oplus \mathbb{R}^2)$ given by

7.4 $\iota e = (e,<e,e>,1) \qquad ; e \in E$

This induces a map from E into $PC(E \oplus \mathbb{R}^2)$ which identifies E with the complement of the point of $PC(E \oplus \mathbb{R}^2)$ representing the line through the point (0,1,0). Let us extend this to a bijection $\iota: \hat{E} \to PC(E \oplus \mathbb{R}^2)$, where $\iota(\infty)$ is the line through the point (0,1,0).

Theorem 7.5 The action of the Lorentz group $Lor(E \oplus \mathbb{R}^2)$ on the projectivised cone $PC(E \oplus \mathbb{R}^2)$ and the homeomorphism $\iota: \hat{E} \xrightarrow{\sim} PC(E \oplus \mathbb{R}^2)$ identifies the Möbius group $M\ddot{o}b(E)$ with the Lorentz group $Lor(E \oplus \mathbb{R}^2)$.

Proof Let us return to the sphere $\mathcal{C}$ from 7.1 and form the vector

$$C = (c,<c,c>-r^2,1)$$

We ask the reader to verify the formulas

$$<C,C> = -r^2 , \quad 2<C,\iota x> = |x-c|^2 - r^2 \qquad ; x \in E$$

From this we can deduce a commutative square

$$\begin{array}{ccc} \hat{E} & \xrightarrow{\iota} & PC(E \oplus \mathbb{R}^2) \\ \downarrow \hat{\sigma} & & \downarrow \tau_C \\ \hat{E} & \xrightarrow{\iota} & PC(E \oplus \mathbb{R}^2) \end{array}$$

More precisely we ask the reader to verify the formula

$$\sigma\,\iota\,(x) = r^2|x-c|^{-2}\,\tau_C\,\iota\,(x) \qquad ; \ x \in E - \{c\}$$

Next, we return to the reflection λ from 7.2 and introduce the vector

$$N = (n,2<n,u>,0)$$

From this we can deduce a commutative square

$$\begin{array}{ccc} \hat{E} & \xrightarrow{\iota} & PC(E \oplus \mathbb{R}^2) \\ \downarrow \hat{\lambda} & & \downarrow \tau_N \\ \hat{E} & \xrightarrow{\iota} & PC(E \oplus \mathbb{R}^2) \end{array}$$

More precisely, we ask the reader to verify that $\iota\, \tau_N = \lambda\, \iota$.

According to 6.11 we can write any Lorentz transformation as a product of transformations of the form τ_N where $N \in E \oplus \mathbb{R}^2$ with $<N,N> < 0$. Let us observe that the locus

$$\{\, x \in E \mid <\iota x,N> = 0 \,\}$$

is a sphere or an affine hyperplane: Let us write $N = (e,a,b)$; upon multiplying N by a non-zero scalar we may assume that $b = 1$ or $b = 0$.

Case $N = (e,a,1)$. Let us introduce the number $r = \sqrt{-<N,N>}$ and observe that $a = <e,e> - r^2$ to get the formula

$$2<\iota x,N> = -2<x,e> + (a + <x,x>) = |x - e|^2 - r^2$$

which shows that the locus is the circle with centre e and radius r.

Case $N=(e,a,0)$. We have $<N,N> = -<e,e>$, in particular $e \neq 0$, so we can multiply N by a constant to obtain that e is a unit vector. Choose $u \in E$ such that $<u,e> = \frac{1}{2}a$, and observe that

$$<\iota x,N> = -<x,e> + \tfrac{1}{2}a = <u - x,e>$$

which shows that the locus is the affine hyperplane through u with normal vector e. At this point it remains only to quote 6.14. □

Corollary 7.6 A Möbius transformation σ of the Euclidean space of dimension n can be written as a product of at most n+2 inversions.

Proof Any Lorentz transformation of the space $E \oplus \mathbb{R}^2$ can be written as a product of at most n+2 reflections, compare 6.11. □

Let us say that a Möbius transformation σ of E is odd or even depending on the sign of the determinant of the corresponding Lorentz transformation. An even Möbius transformation can be expressed as a product of an even number of inversions, while an odd Möbius transformations is a product of an odd number of inversions.

Corollary 7.7 The restriction to E of the group of Möbius transformations of $\hat{E}$ which fixes ∞ is the full group *Siml*(E) of Euclidean similarities of E.

Proof Let us first check through three special cases. First observe that a $\lambda>0$ gives rise to a Lorentz transformation

$$(x,a,b) \mapsto (x,\lambda a,\lambda^{-1}b)$$

which induces the dilatation $x\mapsto\lambda x$ of E. Next observe that a $\sigma \in O(E)$ is induced on E by the Lorentz transformation

$$(x,a,b) \mapsto (\sigma(x),a,b)$$

Let us finally observe that a vector $v \in E$ gives rise to the Lorentz transformation

$$(x,a,b) \mapsto (x+bv,\ a+b<v,v>+2<x,v>,\ b)$$

which induces translation $x\mapsto x+v$ on E. This shows that any similarity of E is a Möbius transformation. In order to prove the converse, we must show that these three types of transformation generate the group $Lor_L(E\oplus\mathbb{R}^2)$ consisting of Lorentz transformations which stabilise the line L generated by (0,1,0).

To this end we shall use the map $(z,\zeta)\mapsto(-z,2\zeta,0)$ to identify $E\oplus\mathbb{R}$ with the subspace $L^\perp$ of $E\oplus\mathbb{R}^2$. This allows us to express the evaluation map from I.5 in terms of the inner product on $E\oplus\mathbb{R}^2$.

7.8 $$ev((z,\zeta),x) = <(-z,2\zeta,0),\iota x> \qquad ; (z,\zeta)\in E\oplus\mathbb{R},\ x\in E$$

We ask the reader to complete the proof by means of 6.16 and 5.8. □

Discs Recall also that $N = N(E\oplus\mathbb{R}^2)$ denotes the set of vectors in $E\oplus\mathbb{R}^2$ of norm -1. A vector $U \in N$ defines a sphere $\mathcal{S} = \{x\in\hat{E} \mid <\iota x,U> = 0\}$ and decomposes the complement of $\mathcal{S}$ into the union of two connected open subsets

$$\mathfrak{D} = \{x \in \hat{E} \mid <\iota x, U> > 0\}\ , \quad \mathfrak{D}^- = \{x \in \hat{E} \mid <\iota x, U> < 0\}$$

which are called the discs bounding the sphere $\mathfrak{S}$. The set $\mathfrak{D}$ is called the disc with normal vector U.

Let us make a slight refinement of our earlier concepts by focusing on the connected component of the truncated isotropic cone $C^*(E \oplus \mathbb{R}^2)$ made up of points whose inner product with (0,1,1) is positive:

7.9

$$C^+(E \oplus \mathbb{R}^2) = \{\ (A,a,b) \mid -<A,A> + ab = 0,\ a + b > 0\ \}$$

Observe that the map $\iota: M \to C$ factors through C^+. It is useful to reinterpret the projectivised cone as

$$PC(E \oplus \mathbb{R}^2) = C^+(E \oplus \mathbb{R}^2)/\mathbb{R}_+ \qquad ; \mathbb{R}_+ =]0,+\infty[$$

Corollary 7.10 A Möbius transformation $\nu \in M\ddot{o}b(E)$ maps a sphere $\mathfrak{S}$ into a sphere $\nu(\mathfrak{S})$. Inversion in the sphere $\nu(\mathfrak{S})$ is given by $\nu\sigma\nu^{-1}$ where σ is inversion in the sphere $\mathfrak{S}$.

Proof Let us agree to denote by $\bar{\nu}$ the Lorentz transformation of $E \oplus \mathbb{R}^2$ corresponding to ν by theorem 7.5. Let U be the normal vector for one of the discs $\mathfrak{D}$ bounding the sphere $\mathfrak{S}$, we ask the reader to identify the set $\nu(\mathfrak{D})$ with the disc with normal vector $\bar{\nu}(U)$. Observe that $\bar{\nu} = \tau_U$ and use the general formula

$$\tau_{\bar{\nu}(U)} = \bar{\nu}\tau_U\bar{\nu}^{-1}$$

to conclude that $\nu\sigma\nu^{-1}$ is reflection in the sphere $\nu(\mathfrak{S})$. □

Corollary 7.11 The Möbius group $M\ddot{o}b(E)$ acts transitively on the set of discs in $\hat{E}$.

Proof We can identify the action of $M\ddot{o}b(E)$ on the set of discs in $\hat{E}$ with the action of the Lorentz group $Lor(E \oplus \mathbb{R}^2)$ on the space $N(E \oplus \mathbb{R}^2)$ of vectors of norm -1. Once this is done, the result follows from 6.16. □

Corollary 7.12 Let $\mathfrak{K}$ be a sphere in $\hat{\mathrm{E}}$. A Möbius transformation σ which fixes $\mathfrak{K}$ pointwise is either the identity or inversion in $\mathfrak{K}$ $(dim(\mathrm{E}) \geq 2)$.

Proof Let K be a normal vector for one of the discs $\mathfrak{D}$ bounding $\mathfrak{K}$ and let κ denote inversion in the sphere $\mathfrak{K}$. From $\sigma(\mathfrak{K}) = \mathfrak{K}$ we conclude that $\sigma(\mathfrak{D}) = \mathfrak{D}$ or $\sigma(\mathfrak{D}) = \kappa(\mathfrak{D})$. Upon replacing σ by $\kappa\sigma$ we may assume that $\sigma(\mathfrak{D}) = \mathfrak{D}$ or otherwise expressed $\bar{\sigma}(\mathrm{K}) = \mathrm{K}$. Let us observe that restriction of $\bar{\sigma}$ to $\mathrm{F} = \mathrm{K}^{\perp}$ acts trivially on the $PC(\mathrm{F})$; it follows from 6.14 that the restriction of $\bar{\sigma}$ to F is ι, the identity. This gives $\bar{\sigma} = \iota$, which implies that $\sigma = \iota$. □

Definition 7.13 We say that two spheres $\mathfrak{S}$ and $\mathfrak{T}$ are orthogonal if the normal vectors for the bounding discs are orthogonal[5].

Proposition 7.14 Let σ denote inversion in the sphere $\mathfrak{S}$ and τ inversion in the sphere $\mathfrak{T}$. If $\mathfrak{S} \neq \mathfrak{T}$ then the following conditions are equivalent

1° $\mathfrak{S}$ and $\mathfrak{T}$ are orthogonal

2° $\sigma\tau = \tau\sigma$

3° $\tau(\mathfrak{S}) = \mathfrak{S}$

4° $\sigma(\mathfrak{T}) = \mathfrak{T}$

Proof It is a consequence of 7.10 that the last three conditions are equivalent. To proceed, introduce normal vectors H and K for two bounding spheres. The correspondance 7.7 identifies the Möbius inversion τ with the Lorentz reflection τ_{K} along the vector K. From the formula

$$\tau_{\mathrm{K}}(\mathrm{H}) = \mathrm{H} + 2\langle\mathrm{H},\mathrm{K}\rangle\mathrm{K}$$

we conclude that $\tau_{\mathrm{K}}(\mathrm{H}) = \pm\mathrm{H}$ if and only if $\langle\mathrm{H},\mathrm{K}\rangle = 0$. This shows that condition 1° is equivalent to condition 3°. □

[5] The geometric meaning of this concept will be given in next section.

Proposition 7.15 Let A and B be points of Ê conjugated with respect to the sphere $\mathcal{S}$ (by this we mean that A and B are points outside $\mathcal{S}$ interchanged by inversion in $\mathcal{S}$). Any sphere through A and B is orthogonal to $\mathcal{S}$.

Proof Let **a** generate the isotropic line in $E \oplus \mathbb{R}^2$ represented by A and let **n** denote a normal vector for $\mathcal{S}$. The vector $\mathbf{b} = \mathbf{a} + 2\langle\mathbf{n},\mathbf{a}\rangle\mathbf{n}$ generates the isotropic line represented by B. Since $A \notin \mathcal{S}$ we have $\langle\mathbf{n},\mathbf{a}\rangle \neq 0$ and the formula for **b** shows that **n** belongs to the plane spanned by **a** and **b**. It follows that the normal vector for any sphere through A and B is orthogonal to **n**. □

Proposition 7.16 Let there be given an open disc $\mathfrak{D}$ in Ê. The group $M\ddot{o}b(\mathfrak{D})$ of Möbius transformations of E which leaves $\mathfrak{D}$ invariant is generated by inversions in spheres orthogonal to $\mathcal{S} = \partial\mathfrak{D}$.

Proof Let N denote the normal vector for $\mathfrak{D}$. A Möbius transformation σ leaves $\mathfrak{D}$ invariant if and only if the corresponding Lorentz transformation $\bar{\sigma}$ fixes N. It follows that we may identify $M\ddot{o}b(\mathfrak{D})$ with the Lorentz group of $N^{\perp}$. This group is generated by reflections τ_K where $\langle K,K\rangle = -1$ and $\langle K,N\rangle = 0$. □

Corollary 7.17 Let L denote a linear hyperplane in E and $\mathfrak{D}$ one of the half-spaces in E bounded by H. Restriction from Ê to $\hat{L}$ defines an isomorphism

$$M\ddot{o}b(\mathfrak{D}) \xrightarrow{\sim} M\ddot{o}b(L)$$

Proof Specialise the proof of 7.16 to this case. □

The composite of the inverse of 7.17 and the inclusion $M\ddot{o}b(\mathfrak{D}) \to M\ddot{o}b(E)$ is known as Poincaré extension : $M\ddot{o}b(L) \to M\ddot{o}b(E)$.

I.8 INVERSIVE PRODUCT OF SPHERES

Let E denote a Euclidean vector space of dimension $n \geq 2$. We shall make a closer study of the intersection of two spheres $\mathcal{S}$ and $\mathcal{T}$ in $\hat{E}$. This will be done through a numerical character $\mathcal{S}*\mathcal{T}$. Let us represent the spheres by their standard equations

8.1

$$\mathcal{S}: \quad -2\langle x,f\rangle + b\langle x,x\rangle + a = 0 \qquad ; \; -\langle f,f\rangle + ab < 0$$
$$\mathcal{T}: \quad -2\langle x,g\rangle + d\langle x,x\rangle + c = 0 \qquad ; \; -\langle g,g\rangle + cd < 0$$

<u>The inversive product</u> of $\mathcal{S}$ and $\mathcal{T}$ is given by

8.2

$$\mathcal{S}*\mathcal{T} = \frac{|\langle f,g\rangle - \frac{1}{2}ad - \frac{1}{2}bc|}{\sqrt{\langle f,f\rangle - ab}\sqrt{\langle g,g\rangle - cd}}$$

From the investigations made in the previous section it follows that the inversive product of $\mathcal{S}$ and $\mathcal{T}$ is preserved under Möbius transformations:

8.3

$$\sigma(\mathcal{S})*\sigma(\mathcal{T}) = \mathcal{F}*\mathcal{G} \qquad ; \; \sigma \in \mathcal{M\ddot{o}b}(E)$$

The relative positions of two distinct spheres $\mathcal{S}$ and $\mathcal{T}$ can be read off $\mathcal{S}*\mathcal{T}$:

8.4

$\mathcal{S}*\mathcal{T} < 1$	$\mathcal{S}*\mathcal{T} = 1$	$\mathcal{S}*\mathcal{T} > 1$
$\#(\mathcal{S} \cap \mathcal{T}) > 1$	$\#(\mathcal{S} \cap \mathcal{T}) = 1$	$\#(\mathcal{S} \cap \mathcal{T}) = 0$

Proof Let H and K be normal vectors for discs bounding $\mathcal{S}$ and $\mathcal{T}$. These two vectors span a plane R, and we can make the identification

$$\mathcal{S} \cap \mathcal{T} = \mathcal{PC}(R^{\perp})$$

The geometry is ruled by the discriminant

$$\Delta = \langle H,H\rangle\langle K,K\rangle - \langle H,K\rangle^2 = 1 - (\mathcal{S}*\mathcal{T})^2$$

From the discriminant lemma 6.3 we deduce that

$\mathcal{S}*\mathcal{T} < 1$: $\quad \Delta < 0$, the plane R has Sylvester type $(-2,0)$ which implies that

$R^\perp$ has Sylvester type $(-n+1,1)$. Thus $PC(R^\perp)$ consists of at least two points.

$\mathcal{S}*\mathcal{T} = 1$: $\Delta = 1$, the plane R has Sylvester type $(-1,0)$ which implies that $R^\perp$ has Sylvester type $(-n,0)$. Thus $\#PC(R^\perp) = 1$.

$\mathcal{S}*\mathcal{T} > 1$: $\Delta < 0$, the plane R has Sylvester type $(-1,0)$ which implies that $R^\perp$ has Sylvester type $(-n,0)$. Thus $PC(R^\perp) = \emptyset$. □

Proposition 8.5 The Möbius group $M\ddot{o}b(E)$ acts transitively on pairs of spheres $(\mathcal{S},\mathcal{T})$ with given inversive products.

Proof This follows easily from Witt's theorem 2.4. □

Let us investigate an ordinary point $x \in E$ of intersection of the spheres $\mathcal{S}$ and $\mathcal{T}$, $x \in \mathcal{S} \cap \mathcal{T}$. To this end we pick a disc $\mathcal{D}$ bounding $\mathcal{S}$ and let $n_x(\mathcal{D}) \in E$ denote the outward pointing unit normal to $\mathcal{S}$ at x. Similarly, let $\mathcal{E}$ be a disc bounding $\mathcal{T}$ and let $n_x(\mathcal{E}) \in E$ denote the outward pointing unit normal to $\mathcal{T}$ at x. We have that

8.6
$$\mathcal{S}*\mathcal{T} = |\langle n_x(\mathcal{D}),n_x(\mathcal{E})\rangle|$$

We shall prove more precisely that the inner product of the normal vectors $N(\mathcal{D}),N(\mathcal{E}) \in \mathcal{N}(E \oplus \mathbb{R}^2)$ is given by

8.7
$$\boxed{\langle N(\mathcal{D}),N(\mathcal{E})\rangle = -\langle n_x(\mathcal{D}),n_x(\mathcal{E})\rangle}$$

Proof Let us suppose that $\mathcal{S}$ is an ordinary sphere in E and that $\mathcal{D}$ is its interior. In the equation 8.1 for $\mathcal{S}$ we can take $b = 1$ which gives us that f is the centre for $\mathcal{S}$, while the radius r is given by $r^2 = \langle f,f\rangle - a$. Thus
$$n_x(\mathcal{D}) = r^{-1}(x-f) \quad , \quad N(\mathcal{D}) = -r^{-1}(f,a,1)$$
When $\mathcal{T}$ is an ordinary sphere with radius s and centre g we can write
$$n_x(\mathcal{E}) = s^{-1}(x-g) \ , \ N(\mathcal{E}) = -s^{-1}(g,c,1)$$

where $c = \langle g,g\rangle - s^2$. Direct calculation gives us

$$\langle N(\mathcal{D}),N(\mathcal{S})\rangle = r^{-1}s^{-1}(-\langle f,g\rangle + \tfrac{1}{2}a + \tfrac{1}{2}c)$$

Then using 8.1 with $b = 1$ and $d = 1$ to eliminate a and c we get that

$$\langle N(\mathcal{D}),N(\mathcal{S})\rangle = r^{-1}s^{-1}(\langle x,f\rangle - \langle f,g\rangle + \langle x,g\rangle - \langle x,x\rangle) = -\langle n_x(\mathcal{D}),n_x(\mathcal{S})\rangle$$

When $\mathcal{T}$ is an affine hyperplane through x with $n_x(\mathcal{S}) = n$ we have that

$$n_x(\mathcal{S}) = n \quad , \quad N(\mathcal{S}) = (n,2\langle x,n\rangle,0)$$

Maintaining that $\mathcal{S}$ is an ordinary sphere as above, we get that

$$\langle N(\mathcal{D}),N(\mathcal{S})\rangle = r^{-1}(-\langle f,n\rangle + \langle x,n\rangle) = -\langle n_x(\mathcal{D}),n_x(\mathcal{S})\rangle$$

The case where both $\mathcal{S}$ and $\mathcal{T}$ are affine hyperplanes is now obvious. □

In the rest of this section we shall limit ourselves to the two-dimensional case. First a rather amusing example due to Jacob Steiner.

Steiner's alternatives 8.8 Let $\mathcal{A}$ and $\mathcal{B}$ be circles in the Euclidean plane of which $\mathcal{B}$ contains $\mathcal{A}$ in its interior. It is possible to find a kissing chain of n circles each touching $\mathcal{A}$ on the outside and $\mathcal{B}$ on the inside, if and only if

$$\mathcal{A}*\mathcal{B} = 1 + 2\tan^2\frac{\pi}{n}$$

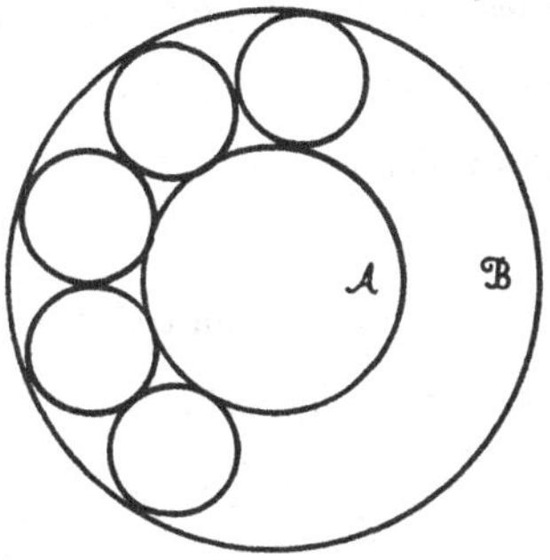

Proof According to 8.5 we may assume that the circles $\mathcal{A}$ and $\mathcal{B}$ are concentric with radii a and b, $a < b$. The condition for the existence of a chain of kissing circles can easily be worked out by Euclidean trigonometry. The result is

$$\sin\frac{\pi}{n} = \frac{1-\theta}{1+\theta} \qquad ; \quad \theta = \frac{a}{b}$$

From this formula we deduce without difficulty that

$$\frac{a}{b} = \frac{1 - \sin\frac{\pi}{n}}{1 + \sin\frac{\pi}{n}}$$

On the other hand we find by direct computation

$$\mathcal{A} * \mathcal{B} = \frac{1}{2}\left(\frac{a}{b} + \frac{b}{a}\right)$$

A simple straightforward calculation concludes the proof. □

Definition 8.9 By a <u>pencil</u> of circles in the Euclidean plane E we understand a two-dimensional subspace R of $E \oplus \mathbb{R}^2$. The actual circles in the pencil are the circles in $\hat{E}$ with normal vector in R.

Proposition 8.10 Let R be a pencil of circles in $\hat{E}$. The set of circles in the pencil can be described as follows, depending on the Sylvester type

(−2,0) : circles through two given distinct points P and Q

(−1,0) : circles through a given point P orthogonal to a given circle $\mathcal{N}$ through P.

(−1,1) : circles which conjugate two given distinct points P and Q.

Proof Let us check through the three possible Sylvester types.

Case (−2,0). The orthogonal space $R^{\perp}$ has type (−1,1) and contains precisely two isotropic lines P and Q say. We can recover R as the set of vectors orthogonal to P and Q. In geometric terms, a circle $\mathcal{S}$ belongs to the pencil if and only if it passes through P and Q, compare the proof of 7.15.

Case (−1,1). The space R contains exactly two isotropic lines. Let us observe that a circle $\mathcal{S}$ with normal vector N conjugates P and Q if and only if $N \in R$. Alternatively, let us treat the case $P = \infty$ and $Q = 0$ directly. The circles in $\hat{E}$ which conjugate these two points are the ordinary circles in E with centre 0. Notice that through each point of $\hat{\mathbb{C}} - \{0,\infty\}$ passes a unique circle of this pencil.

Case $(-1,0)$. Let us observe that the dual pencil $R^{\perp}$ has the same type and that $R \cap R^{\perp} = P$ is an isotropic line. Pick a circle $\mathcal{N}$ in the pencil $R^{\perp}$ and observe that the circles in R are precisely the circles through P, orthogonal to $\mathcal{N}$. □

Example 8.11 Consider the following families of circles in $\mathbb{R}^2$

$$x^2 + y^2 - b + 2\lambda x = 0 \qquad ; \lambda \in \mathbb{R}$$

1° $b > 0$. All circles pass through $(0,\sqrt{b})$ and $(0,-\sqrt{b})$

2° $b = 0$. The equation for the circle can be written

$$(x - \lambda)^2 + y^2 = \lambda^2$$

which shows that all circles pass through (0,0) and are tangent to the y-axis at this point.

3° $b < 0$. Put $b = -k^2$ and rewrite the equation as follows

$$(x - \lambda)^2 + y^2 = \lambda^2 - k^2$$

For $\lambda = k$, the point $(k,0)$ is the only solution, and similarly $\lambda = -k$ gives $(-k,0)$. The system is, in fact, orthogonal to the pencil of circles through $(k,0)$ and $(-k,0)$.

I.9 RIEMANN SPHERE

Let us introduce the Riemann sphere $\hat{\mathbb{C}}$ as the 1-point compactification of of the complex plane $\mathbb{C}$. This compactification can be realised in a number of ways, but we shall use a construction from projective geometry which displays the underlying algebra in a natural way.

Consider the scalar action of the complex multiplicative group $\mathbb{C}^*$ on the complex vector space $\mathbb{C}\times\mathbb{C}$, and let $\hat{\mathbb{C}}$ denote the orbit space for the induced action on the complement of the origin. The equivalence class of a pair of complex numbers (z,w) $\neq$ (0,0) is denoted $\begin{bmatrix} z \\ w \end{bmatrix}$. Thus

$$\begin{bmatrix} z\alpha \\ w\alpha \end{bmatrix} = \begin{bmatrix} z \\ w \end{bmatrix} \qquad ; \ \alpha \in \mathbb{C}^*$$

The group $Gl_2(\mathbb{C})$ acts on the Riemann sphere in virtue of the formula

9.1
$$\begin{bmatrix} a & b \\ c & d \end{bmatrix}\begin{bmatrix} z \\ w \end{bmatrix} = \begin{bmatrix} az+bw \\ cz+dw \end{bmatrix} \qquad ; \ ad - bc \neq 0$$

Let us remark that scalar matrices act trivially on $\hat{\mathbb{C}}$, thus we are dealing with an action of $PGl_2(\mathbb{C}) = Gl_2(\mathbb{C})/\mathbb{C}^*$ on $\hat{\mathbb{C}}$. In order to calculate the action of the inverse matrix S^{-1} we can use the <u>cofactor</u> matrix $S^{\check{}}$ given by

9.2
$$\begin{bmatrix} a & b \\ c & d \end{bmatrix}^{\check{}} = \begin{bmatrix} d & -b \\ -c & a \end{bmatrix}$$

This is justified by the well known formula $S\,S^{\check{}} = S^{\check{}}\,S = \iota\, det\, S$.

Proposition 9.3 The action of the group $PGl_2(\mathbb{C})$ on $\hat{\mathbb{C}}$ is <u>triply transitive</u>: for any two triples A,B,C and P,Q,R, each consisting of three distinct points, there is a unique $\sigma \in PGl_2(\mathbb{C})$ which transforms the first triple into the second.

Proof Let the points A,B,C be represented by the vectors **a**, **b**, **c** in $\mathbb{C}\times\mathbb{C}$. By assumption, the first two vectors are linearly independent so we can write

$$\mathbf{c} = \mathbf{a}\lambda + \mathbf{b}\mu \qquad ; \lambda,\mu \in \mathbb{C}$$

Since λ and μ are different from zero we can change **a** and **b** such that $\mathbf{c} = \mathbf{a}+\mathbf{b}$. Quite similarly, we can represent the points P,Q,R by vectors **p**, **q**, $\mathbf{p}+\mathbf{q}$. The linear automorphism σ of $\mathbb{C}\times\mathbb{C}$ with $\sigma(\mathbf{a}) = \mathbf{p}$ and $\sigma(\mathbf{b}) = \mathbf{q}$ induces the required Möbius transformation.

It remains to investigate a linear automorphism τ of $\mathbb{C}\times\mathbb{C}$ which induces the identity on $\hat{\mathbb{C}}$. By assumption, all non-zero vectors of $\mathbb{C}\times\mathbb{C}$ are eigenvectors for τ. It follows easily that τ is multiplication by a scalar $\mu \in \mathbb{C}^*$. □

Let us introduce the point at <u>infinity</u> $\infty = \begin{bmatrix} 1 \\ 0 \end{bmatrix}$ and observe that we can identify the complement $\hat{\mathbb{C}} - \{\infty\}$ with $\mathbb{C}$ once we identify the complex number z with the point $\begin{bmatrix} z \\ 1 \end{bmatrix}$ of $\hat{\mathbb{C}}$. Let us investigate the transformation σ from 9.1 in this terminology. When $\sigma(\infty) = \infty$ we have $c = 0$. If $c \neq 0$ we find that $\sigma(\infty) = \frac{a}{c}$ and $\sigma^{-1}(\infty) = -\frac{d}{c}$. To recapitulate

9.4 $$\sigma(-\tfrac{d}{c}) = \infty \ , \ \sigma(\infty) = \tfrac{a}{c} \ , \quad \sigma(z) = \frac{az+b}{cz+d} \qquad ; z \in \mathbb{C},\ z \neq -\tfrac{d}{c}$$

Theorem 9.5 The group $Gl_2(\mathbb{C})$ acts on $\hat{\mathbb{C}}$ as even Möbius transformations. More precisely, this action induces an isomorphism of groups

$$PGl_2(\mathbb{C}) \xrightarrow{\sim} M\ddot{o}b^+(\mathbb{C})$$

Proof Let us first observe that the group of upper triangular matrices acts on $\hat{\mathbb{C}}$ fixing ∞ and induces the group of even similarities of $\mathbb{C}$. In general, the matrix identity $(c \neq 0)$

$$\begin{bmatrix} a & b \\ c & d \end{bmatrix} = \begin{bmatrix} 1 & \frac{a}{c} \\ 0 & 1 \end{bmatrix} \begin{bmatrix} 0 & 1 \\ 1 & 0 \end{bmatrix} \begin{bmatrix} c & d \\ 0 & b-\frac{ad}{c} \end{bmatrix}$$

leads us to consider the transformation $z \mapsto z^{-1}$ of $\hat{\mathbb{C}}$. This transformation is the composite of inversion in the unit circle and reflection $z \mapsto \bar{z}$ in the x-axis as follows from the basic formula

$$z^{-1} = \bar{z}\,|z|^{-2} \qquad ; z \in \mathbb{C}^*$$

This concludes the proof of the fact that $Gl_2(\mathbb{C})$ acts through even Möbius transformations. In order to conclude the proof, observe that $PGl_2(\mathbb{C})$ acts transitively on $\hat{\mathbb{C}}$ and that the group of even Möbius transformations fixing ∞ induces the group of even similarities on $\mathbb{C}$, compare 7.7. □

Cross ratio We shall introduce a fundamental invariant of four distinct points (P,Q,R,S) of the Riemann sphere $\hat{\mathbb{C}}$. To this end let us represent each of the four points by 2×1 matrices **p**, **q**, **r**, **s** and define the <u>cross ratio</u> as the point of $\hat{\mathbb{C}}$ given by

9.6
$$[P,Q,R,S] = \left[\begin{array}{c} det[\mathbf{p},\mathbf{r}] \cdot det[\mathbf{q},\mathbf{s}] \\ det[\mathbf{p},\mathbf{s}] \cdot det[\mathbf{q},\mathbf{r}] \end{array}\right]$$

where the symbol $[\mathbf{p},\mathbf{q}]$ denotes the 2×2 matrix with first column **p** and second column **q**. For a Möbius transformation σ we have

$$det\,[\sigma(\mathbf{p}),\sigma(\mathbf{q})] = det(\sigma)\; det\,[\mathbf{p},\mathbf{q}]$$

From this it follows that the cross ratio is invariant under Möbius transformations

9.7
$$[\sigma(P),\sigma(Q),\sigma(R),\sigma(S)] = [P,Q,R,S] \qquad ; \sigma \in Gl_2(\mathbb{C})$$

At this point let us remark that the cross ratio [P,Q,R,S] is well defined as long as P,Q,R,S represent at least three distinct points of $\hat{\mathbb{C}}$. This allows us to fix three distinct points P,Q,R of $\hat{\mathbb{C}}$ and make a free variation of the fourth point S. The result is a Möbius transformation:

Proposition 9.8 Let P,Q,R denote three distinct points of $\hat{\mathbb{C}}$. The map

$$S \mapsto [P,Q,R,S] \qquad ; S \in \hat{\mathbb{C}}$$

is an even Möbius transformation, which transforms P,Q,R into ∞,0,1.

Proof Let us rewrite formula 9.6 as follows

$$[P,Q,R,S] = \begin{bmatrix} -q_2 \,det[\mathbf{p},\mathbf{r}] & q_1 \,det[\mathbf{p},\mathbf{r}] \\ -p_2 \,det[\mathbf{q},\mathbf{r}] & p_1 \,det[\mathbf{q},\mathbf{r}] \end{bmatrix} \begin{bmatrix} s_1 \\ s_2 \end{bmatrix}$$

Observe that the 2×2 matrix has determinant $det[\mathbf{p},\mathbf{r}]\,det[\mathbf{q},\mathbf{r}]\,det[\mathbf{p},\mathbf{q}] \neq 0$. Direct calculation reveals that this transformation maps P,Q,R into $\infty,0,1$. □

Proposition 9.8 in connection with 9.3 offers an implicit definition of the cross product as the even Möbius transformation which maps P,Q,R into $\infty,0,1$. In particular we find that

9.9 $$[\infty,\ 0,\ 1,\ S] = S \qquad ;\ S \in \hat{\mathbb{C}}$$

Proposition 9.10 Two 4-touples of distinct points of the Riemann sphere P,Q,R,S and X,Y,Z,W can be transformed into another by an even Möbius transformation if and only if

$$[P,Q,R,S] = [X,Y,Z,W]$$

Proof It follows from 9.7 that the condition is necessary. Conversely, if the two cross ratios are identical, let σ and τ be the Möbius transformations given by

$$\sigma(P,Q,R) = (\infty,0,1)\ , \qquad \tau(X,Y,Z) = (\infty,0,1)$$

According to formulas 9.7 and 9.9 we get that

$$[P,Q,R,S] = [\infty,0,1,\sigma(S)] = \sigma(S)$$

and similarly, $[X,Y,Z,W] = \tau(W)$. Thus the Mobius transformation $\tau^{-1}\sigma$ maps S to W as required. □

Proposition 9.11 An odd Möbius transformation $\beta \in M\ddot{o}b(\mathbb{C})$ satisfies

$$[\sigma(P),\sigma(Q),\sigma(R),\sigma(S)] = [P,Q,R,S]^{-}$$

whenever P,Q,R,S represent four distinct points of $\hat{\mathbb{C}}$.

Proof In order to generate the full Möbius group we need $PGl_2(\mathbb{C})$ and one single odd Möbius transformation, for example complex conjugation

$$\begin{bmatrix} z \\ w \end{bmatrix} \mapsto \begin{bmatrix} \bar{z} \\ \bar{w} \end{bmatrix}$$

The formula is easily verified for complex conjugation. □

Let us record two symmetry properties of the cross ratio. The verification is immediate and is left to the reader.

9.12 $$[P,Q,R,S] = [R,S,P,Q] \quad , \quad [P,Q,R,S] = [Q,P,S,R]$$

Proposition 9.13 Any $\sigma \in PGl_2(\mathbb{C})$ is conjugated to a transformation given by a matrix of the following form

$$\begin{bmatrix} a & 0 \\ 0 & 1 \end{bmatrix} , \begin{bmatrix} 1 & 1 \\ 0 & 1 \end{bmatrix} \qquad ; \ a \in \mathbb{C}^*$$

Proof Let σ be given by the matrix $S \in Gl_2(\mathbb{C})$. Note that a non-zero vector $v \in \mathbb{C} \times \mathbb{C}$ is an eigenvector for S if and only if $[v] \in \hat{\mathbb{C}}$ is a fixed point for σ. From this we conclude that $\sigma \neq \iota$ has one or two fixed points on $\hat{\mathbb{C}}$.

When σ has two distinct fixed points A and B, let τ be an even Möbius transformation with $\tau(\infty) = A$ and $\tau(0) = B$. The transformation $\tau^{-1}\sigma\tau$ will fix 0 and ∞. It follows that a representing matrix must be diagonal. After multiplication by a complex number we get a matrix of the first type. When σ has only one fixed point A, choose an auxiliary point $C \neq A$ and let τ be the even Möbius transformation given by

$$\tau(\infty) = A \ , \quad \tau(0) = C \quad , \quad \tau(1) = \sigma(C)$$

Observe that the transformation $\tau^{-1}\sigma\tau$ fixes ∞ and maps 0 to 1. It follows that this transformation can be represented by a matrix of the form

$$\begin{bmatrix} a & 1 \\ 0 & 1 \end{bmatrix} \qquad ; \ a \in \mathbb{C}^*$$

But the matrix can only have one eigenvalue so we must have a = 1. □

Let us introduce an important invariant of a matrix $S \in Gl_2(\mathbb{C})$

9.14
$$tr^2\, S = \frac{(tr\, S)^2}{det\, S} \qquad ; \ S \in Gl_2(\mathbb{C})$$

Observe that tr^2 S depends only on the conjugacy class of S in $PGl_2(\mathbb{C})$. In particular, we may introduce the invariant $tr^2\sigma$ of an even Möbius transformation σ.

Proposition 9.15 Let σ and τ be even Möbius transformations, both different from ι. Then σ is conjugated to τ in $PGl_2(\mathbb{C})$ if and only if $tr^2\sigma = tr^2\tau$.

Proof With the notation of 9.14 we have the following explicit formulas

$$tr^2\begin{bmatrix} 1 & 1 \\ 0 & 1 \end{bmatrix} = 4 \ , \quad tr^2\begin{bmatrix} a & 0 \\ 0 & 1 \end{bmatrix} = 2 + a + a^{-1} \qquad ; \ a \in \mathbb{C} - \{0,1\}$$

Notice, that the value of $a + a^{-1}$ is unchanged if a is replaced by a^{-1}. On the other hand we have the matrix formulas

$$\begin{bmatrix} 0 & -1 \\ 1 & 0 \end{bmatrix}\begin{bmatrix} a & 0 \\ 0 & 1 \end{bmatrix}\begin{bmatrix} 0 & 1 \\ -1 & 0 \end{bmatrix} = \begin{bmatrix} 1 & 0 \\ 0 & a \end{bmatrix} \sim \begin{bmatrix} a^{-1} & 0 \\ 0 & 1 \end{bmatrix}$$

The remaining details are left to the reader. □

I EXERCISES

EXERCISE 1.1 Let E be a finite dimensional vector space equipped with a non-singular quadratic form.

1° Let F be a non-singular subspace of E, show that $E = F \oplus F^{\perp}$ and that

$$(x,y) \mapsto (x,-y) \qquad ; x \in F, y \in F^{\perp}$$

defines an isometry $\sigma \in O(E)$ with $\sigma^2 = \iota$.

2° Conversely, let $\sigma \in O(E)$ be given with $\sigma^2 = \iota$. Show that $E = E_+ \oplus E_-$ where E_+ is the eigenspace for 1 and E_- is the eigenspace for −1. Hint: Use that

$$x = \tfrac{1}{2}(x - \sigma(x)) + \tfrac{1}{2}(x + \sigma(x)) \qquad ; x \in E$$

3° Given an isometry $\sigma \in O(E)$ with $\sigma^2 = \iota$. Show that the fixed point space E_σ for σ is a non-singular subspace of E and that $\sigma(x) = -x$ for $x \in E_\sigma{}^{\perp}$.

EXERCISE 1.2 1° Let (E,R) and (F,S) be quadratic spaces. Show that the function

$$Q(e,f) = R(e) + S(f) \qquad ; (e,f) \in E \oplus F$$

defines a quadratic form on the direct sum $E \oplus F$ of E and F. The form Q on $E \oplus F$ is denoted $(E,R) \perp (F,S)$

2° Let (D,Q) be a quadratic form and E and F orthogonal subspaces of D such that $D = E + F$ and $E \cap F = 0$. Show that (D,Q) is isomorphic to the form $(E,R) \perp (F,S)$ where R and S denote the restrictions of Q to E and F respectively.

EXERCISE 2.1 By a <u>hyperbolic</u> <u>plane</u> over k, we understand a quadratic form (F,P) which has a basis consisting of two isotropic vectors e and f with $\langle e,f \rangle \neq 0$.

1° Show that all hyperbolic planes over k are isomorphic.

2° Let (E,Q) be a non-singular quadratic form which contains an isotropic vector $e \neq 0$. Show that e is contained in a hyperbolic plane $V \subseteq E$.

Hint: Choose $g \in E$ with $\langle e,g \rangle = 1$ and observe that $f = 2g - \langle g,g \rangle e$ is isotropic with $\langle e,f \rangle = 2$.

3° Under the assumptions of 2°, show that the equation $Q(x) = a$ has a solution x in E for all $a \in E$.

EXERCISE 2.2 Let us consider a non-singular quadratic form (E,Q). We say that a subspace F of E is <u>isotropic</u> if the restriction of Q to F is identically zero.

1° Given maximal isotropic subspaces C and D of E. Show there exists a $\sigma \in O(E)$ such that $\sigma(C) = D$. Hint: Use Witt's theorem.

2° Show that an isotropic subspace F of E is contained in a maximal isotropic subspace. Hint: Use Witt's theorem.

3° Let q denote the dimension of a maximal isotropic subspace of E. Show that (E,Q) is isomorphic to a quadratic form of the type

$$H_1 \perp \dots \perp H_q \perp F$$

with $H_1,\dots,H_q$ hyperbolic planes and F non-singular without isotropic vectors.

EXERCISE 2.3 Let Q be a quadratic form on the vector space E, and let

$$O_p(E) = \{\sigma \in O(E) \mid \sigma(e) = e \; ; \; e \in E^{\perp}\}$$

1° Show that $O_p(E)$ is a normal subgroup of $O(E)$ and that it contains all reflections along non-isotropic vectors.

2° Given $\lambda \in k^*$, show that $O_p(E)$ acts transitively on the set

$$\{\, e \in E \mid Q(e) = \lambda \,\}$$

Hint: Use the proof of proposition 2.3. 3° Show that any $\sigma \in O_p(E)$ can be written as a product of at most $2\,dim\,E$ orthogonal reflections.

4° When $dim\,E = 2$ show that any $\sigma \in O_p(E)$ can be written as a product of one or two orthogonal reflections.

EXERCISE 2.4 Consider the quadratic form on $\mathbb{R}^3$ with Gram matrix G (below)

$$G = \begin{bmatrix} 0 & 0 & 1 \\ 0 & 0 & 0 \\ 1 & 0 & 0 \end{bmatrix} \quad A = \begin{bmatrix} 1 & 0 & 0 \\ 1 & 1 & 0 \\ 0 & 0 & 1 \end{bmatrix} \quad R = \begin{bmatrix} 0 & 0 & -1/s \\ -r & 1 & -r/s \\ -s & 0 & 0 \end{bmatrix}, \quad \begin{matrix} r,s \in \mathbb{R} \\ s \neq 0 \end{matrix}$$

1° Show that any orthogonal reflection of $\mathbb{R}^3$ has a matrix of the form R above.

2° Show that the matrix A represents an orthogonal transformation of $\mathbb{R}^3$ which can't be written as a product of two orthogonal reflections.

EXERCISE 3.1 1° Show that the determinant *det* is a quadratic form on $M_2(\mathbb{R})$ and that its polarisation is given by

$$\langle A,B\rangle = \tfrac{1}{2}\, tr\, AB^{\vee} \qquad ; A,B \in M_2(\mathbb{R})$$

where the symbol $B^{\vee}$ denotes the cofactor matrix of B given by

$$\begin{bmatrix} a & b \\ c & d \end{bmatrix}^{\vee} = \begin{bmatrix} d & -b \\ -c & a \end{bmatrix}$$

2° Show that the Sylvester type of *det* on $M_2(\mathbb{R})$ is $(-2,2)$.

3° Show that

$$\langle A,B\rangle = \tfrac{1}{2}(\ tr\, A\ tr\, B - tr\, AB\ )$$

EXERCISE 4.1 Let E denote a Euclidean space of dimension 3.

1° Given two planes P and Q, show that there exists a plane R whose normal vector is orthogonal to the normal vectors of P and Q.

2° Show that an even isometry $\sigma \in SO(E)$ can be written $\sigma = \kappa\lambda$, where κ and λ are half-turns in lines K and L (reflections in K and L in the notation from Exercise 1.1).

3° Show that any two half-turns are conjugated in $SO(E)$.

EXERCISE 4.2 Show that $SO_3(\mathbb{R})$ is a simple group, in the sense that any normal subgroup $N \neq \{\iota\}$ equals $SO_3(\mathbb{R})$. Hint: Use the previous exercise to see that it suffices to show that N contains a half-turn. Use the next exercise actually to find a half-turn in N.

EXERCISE 4.3 Let E be a Euclidean vector space of dimension 3 and let $\sigma \in SO(E)$ be a rotation with angle $\geq \pi/2$.

1° Show that there exists a line L through 0 such that L and $\sigma(L)$ are orthogonal. Hint: Observe that the angle between L and $\sigma(L)$ is a continuous function of L.

2° Let λ be half-turn with respect to a line L with the property that L and $\sigma(L)$ are orthogonal. Show that $\sigma\lambda\sigma^{-1}\lambda$ is a half-turn. Hint: Show that $\sigma(L)$ is stable under $\sigma\lambda\sigma^{-1}\lambda$ with eigenvalue -1.

EXERCISE 4.4 1° Show that the Killing form

$$<X,Y> = tr(X\ ^TY) \qquad ; X,Y \in M_n(\mathbb{R})$$

is a symmetrical bilinear form on $M_n(\mathbb{R})$ and that the corresponding quadratic form satisfies the Cauchy–Schwarz inequality

$$|<X,Y>| \le |X|\ |Y| \qquad ; X,Y \in M_n(\mathbb{R})$$

2° Show that the Killing form is invariant under $O_n(\mathbb{R})$ in the sense that

$$<gXg^{-1}, gXg^{-1}> = <X,X> \qquad ; X \in M_n(\mathbb{R}),\ g \in O_n(\mathbb{R})$$

EXERCISE 6.1 1° Show that $Lor^+(F)$ acts transitively on each of the sheets H^n of the hyperboloid $S(F)$.

2° Show that the special Lorentz group $Lor^+(F)$ is connected.

EXERCISE 6.2 Show that the orthogonal group $O(F)$ has four connected components.

EXERCISE 7.1 Justify the classical construction by ruler and compass of inversion B of the point A in the sphere with centre O and radius r

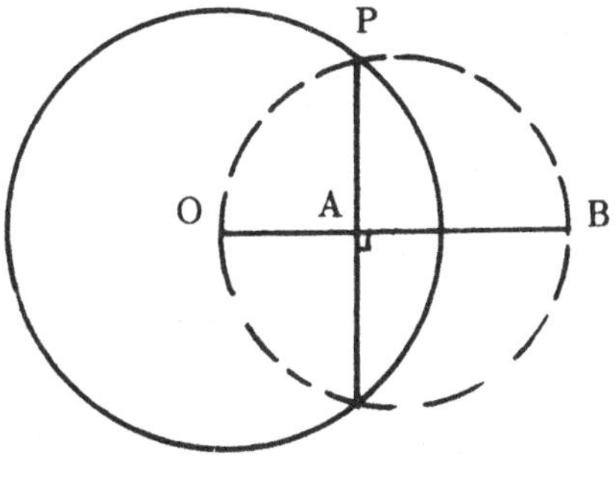

Hint: Note that ΔOAP is similar to ΔOPB and deduce that $d(O,A)d(O,B) = r^2$. Alternatively, observe that the circle with diameter BP is orthogonal to the initial circle, and passes through the point A; compare I.7.14 and I.8.7.

EXERCISE 7.2 1° Show that dilatation with centre 0 and ratio $r>0$ can be realised as the composite of reflections in the Euclidean spheres with centre 0 and radius 1 and $\sqrt{r}$.

2° Show that parallel translation in E can be written as a composite of Euclidean reflections in two parallel affine planes.

EXERCISE 7.3 (Ptolemy's theorem) Let E be an Euclidean vector space of dimension n. Show that n+2 points $x_1,\ldots,x_{n+2}$ of E lie on a sphere if and only if

$$det_{i,j}\ d(x_i,x_j)^2 = 0$$

where d denotes the Euclidean distance, compare II.2.1. Hint: With the notation of I.7.4 show that $\langle \iota x, \iota y\rangle = \frac{1}{2}\, d(x,y)^2$ for $x,y \in E$.

EXERCISE 8.1 Let us consider Euclidean discs $\mathcal{U}$ and $\mathcal{V}$ with centres u and v and radii r and s. Show that

$$\mathcal{U}*\mathcal{V} = \frac{|u-v|^2 - r^2 - s^2}{2rs}$$

EXERCISE 8.2 Let us consider a square matrix with n+2 rows and n+2 columns

diagonal entries: -1

entries elsewhere: 1

$$\begin{bmatrix} -1 & 1 & 1 & 1 \\ 1 & -1 & 1 & 1 \\ 1 & 1 & -1 & 1 \\ 1 & 1 & 1 & -1 \end{bmatrix}$$

1° Show that this matrix defines a quadratic form of type $(-n-1,1)$

2° Let E be a Euclidean n-space. Show that there exists n+2 spheres in $\hat{E}$ which are two and two tangent to each other. Hint: Pick a basis for $E \oplus \mathbb{R}^2$ with the above matrix as Gram matrix.

EXERCISE 8.3 Let R denote a pencil of circles in the Euclidean plane E, and let G denote the group generated by reflections in the circles of the pencil.

1° Show that the action of G on R identifies G with a subgroup of $O(R)$, and identify this subgroup depending on the type of the pencil.

2° Show that a product $\alpha\beta\gamma$ of three reflections in circles of the pencil is itself a reflection in a circle from the pencil.

EXERCISE 9.1 1° Show that $A \in Gl_2(\mathbb{C})$ with $tr\,A = 0$ defines a Möbius transformation on $\hat{\mathbb{C}}$, which is an involution.

2° Show that all involutions in $M\ddot{o}b(\mathbb{C})$ can be represented in this way.

EXERCISE 9.2 Let P,Q,R denote three distinct points of $\hat{\mathbb{C}}$. Show that the point $S \in \hat{\mathbb{C}}$ belongs to the circle through P,Q,R if and only if $[P,Q,R,S] \in \hat{\mathbb{R}}$.

EXERCISE 9.3 Let P,Q,R,S be four distinct points on a circle $\mathcal{K}$. Show that

$$[P,Q,R,S] < 0$$

if and only if the pair {P,Q} separates the pair {R,S}; meaning that P and Q belongs to different connected components of $\mathcal{K} - \{R,S\}$.

EXERCISE 9.4 By a <u>half-turn</u> we understand an even Möbius transformation $\sigma \neq \iota$ with $\sigma^2 = \iota$.

1° Show that a half-turn has precisely two fixed points on $\hat{\mathbb{C}}$.

2° Show that there exists a unique half-turn fixing two given points of $\hat{\mathbb{C}}$.

3° Let there be given four distinct points P,Q,R,S of $\hat{\mathbb{C}}$. Show that there exists a unique $\sigma \in PGl_2(\mathbb{C})$ which transposes P and Q and transposes R and S. In the following this half-turn will be denoted (P,Q)(R,S). Observe that this symbol is still meaningful in case R = S.

4° Show that an even Möbius transformation of $\mathbb{C}$ can be written as a product of two half-turns; otherwise expressed, show that a given triple A,B,C of points of $\hat{\mathbb{C}}$ can be transformed into another given triple P,Q,R by a product of two half-turns. Hint: In case $A \neq Q$ and $B \neq P$ let D denote the image of C by $(A,Q) \circ (B,P)$ and consider the transformation

$$(R,D)(P,Q) \circ (A,Q)(B,P)$$

When $A = P$ and $B = Q$, use the transformation $(A,B)(R,R) \circ (A,B)(C,R)$.

EXERCISE 9.5 1° Show that the cross ratio of four distinct points is a complex number distinct from 0 and 1 which obeys the rules

$$[P,Q,R,S] = [Q,P,R,S]^{-1} = [P,Q,S,R]^{-1}$$

2° Five distinct points P,Q,R,S,T of $\hat{\mathbb{C}}$ satisfy the cancellation rule

$$[P,Q,R,S]\ [P,Q,S,T] = [P,Q,R,T]$$

EXERCISE 9.6 Fix three distinct points P,Q,R of $\hat{\mathbb{C}}$ and consider the rational function λ on $\hat{\mathbb{C}}$ given by $\lambda(S) = [P,Q,R,S]$. Show that permutations of P,Q,R give the six rational functions $\lambda,\ \frac{1}{\lambda},\ 1-\lambda,\ \frac{1}{1-\lambda},\ \frac{\lambda-1}{\lambda},\ \frac{\lambda}{\lambda-1}$

EXERCISE 9.7 Let $\sigma,\tau \in PGl_2(\mathbb{C})$ be transformations represented by matrices $S,T \in Gl_2(\mathbb{C})$. Show that $tr[\sigma,\tau] = tr\ STS^{-1}T^{-1}$ is independent of S and T.

2° Show that $tr[\sigma,\tau] = 2$ if and only if σ and τ have a common fixed point on the Riemann sphere $\hat{\mathbb{C}}$. Hint: Use the normal forms for S from exercise 9.1.

3° Give an example of transformations σ and τ with $tr[\sigma,\tau] = -2$.

EXERCISE 9.8 For $\sigma \in PGl_2(\mathbb{C})$ let $Fix(\sigma)$ denote the set of fixed points for σ on $\hat{\mathbb{C}}$. Show that if $\tau \in PGl_2(\mathbb{C})$ commutes with σ then $\sigma(Fix(\tau)) = \tau(Fix(\sigma))$.

2° Show the converse of 1°. In particular, show that $Fix(\sigma) = Fix(\tau)$ implies that σ and τ commute.

3° Find the condition for two half-turns to commute.

EXERCISE 9.9 Show that the evaluation map

$$PGl_2(\mathbb{C}) \to \hat{\mathbb{C}} \times \hat{\mathbb{C}} \times \hat{\mathbb{C}} \qquad ; \sigma \mapsto (\sigma(\infty),\sigma(0),\sigma(1))$$

is a homeomorphism of $PGl_2(\mathbb{C})$ onto the open subset W of $\hat{\mathbb{C}} \times \hat{\mathbb{C}} \times \hat{\mathbb{C}}$ consisting of triples of distinct points of $\hat{\mathbb{C}}$. Hint. With the notation of II.9.6 show that the following Möbius transformation maps $(\infty,0,1)$ onto (P,Q,R)

$$\begin{bmatrix} p_1 det[q,r] & -q_1 det[p,r] \\ p_2 det[q,r] & -q_2 det[p,r] \end{bmatrix}$$

II GEOMETRIES

We are now ready to introduce the three basic geometries: Euclidean, spherical and hyperbolic geometry. Using the algebra from the previous chapter we shall work our way through the three geometries and identify their isometries and geodesic[1] curves. In the case of hyperbolic geometry we present three "non-linear models" : the Klein disc model, the Poincaré disc model and the Poincaré half-space model.

II.1 GEODESICS

Let us recall that a metric space is a set X equipped with a symmetrical function $d: X \times X \to \mathbb{R}$ with the property that $d(A,B) = 0$ if and only if $A = B$ and which satisfies the

Triangle inequality 1.1 $d(A,C) \leq d(A,B) + d(B,C) \quad ; A,B,C \in X$

A map $\sigma: X \to Y$ from one metric space to another is distance preserving if

1.2 $$d(\sigma(A),\sigma(B)) = d(A,B) \qquad ; A,B \in X$$

It is clear from the definition that a distance preserving map is injective. A distance preserving map $\sigma: X \to Y$ which is surjective is called an isometry. Observe that the inverse map of an isometry is itself an isometry. It follows that the set of isometries of a given metric space form a group. We shall say that metric spaces X and Y are isometric if there exists an isometry of X onto Y.

---

[1]The geometries we are going to study all carry the structure of metric space. The notion of a geodesic curve in a metric space, as introduced in section II.1, is consistent with the notion of a geodesic curve in Riemannian geometry.

The basic example of a metric space is the set $\mathbb{R}$ of real numbers equipped with the standard metric $d(x,y) = |y - x|$, $x,y \in \mathbb{R}$. It will be important for us to know all isometries of $\mathbb{R}$.

Proposition 1.3 An isometry $\sigma:\mathbb{R}\to\mathbb{R}$ is either a translation $x \mapsto x+a$ or a reflection $x \mapsto b-x$.

Proof For a given isometry σ of $\mathbb{R}$, let τ be the reflection or translation which agrees with σ at 0 and 1. We are going to prove that $\sigma = \tau$. Suppose for a moment that $c \in \mathbb{R}$ is a point where $\tau(c) \neq \sigma(c)$: but this implies that the point $a = \sigma(0) = \tau(0)$ is the midpoint of $\sigma(c)$ and $\tau(c)$ since

$$d(\sigma(c),a) = d(\sigma(c),\sigma(0)) = d(c,0) = d(\tau(c),\tau(0)) = d(\tau(c),a)$$

Similarly, $b = \sigma(1) = \tau(1)$ is the midpoint of $\sigma(c)$ and $\tau(c)$, contradicting $a \neq b$. □

Definition 1.4 Let $J \subseteq \mathbb{R}$ be an interval and X a metric space. A curve $\gamma:J\to X$ is said to be a geodesic curve if each point $c \in J$ has a neighbourhood $U \subset J$ such that the restriction of $\gamma:U\to X$ is distance preserving.

Lemma 1.5 Let u be a point of the open interval J of $\mathbb{R}$ and $\gamma:J\to\mathbb{R}$ a geodesic curve with $\gamma(u)=0$. We can find $\epsilon = \pm 1$ such that

$$\gamma(t) = \epsilon(t-u) \qquad ; t \in J$$

Proof Let us choose an open neighbourhood $U \subseteq J$ of a given point u such that the restriction of γ to U is distance preserving. We conclude from the proof of 1.3 that the restriction of γ to U is an affine function. It follows that γ is differentiable on U with constant derivative. Using that $\mathbb{R}$ is connected we conclude that γ' is globally constant on J. At this point the result follows from elementary calculus. □

II.2 EUCLIDEAN SPACE

Let E denote an n-dimensional Euclidean space. The scalar product of vectors e and f will be denoted $<e,f>$, while the length of a vector c is given by $|c| = \sqrt{<c,c>}$. The distance between two points A and B of E is defined by

2.1
$$d(A,B) = |A - B| \qquad ; A,B \in E$$

This turns E into a metric space as follows from Cauchy–Schwarz, I.4.1.

Let us observe that a point $v \in E$ and a vector e of norm 1 defines a geodesic curve $t \mapsto te + v$, $t \in \mathbb{R}$, whose image is an affine line. Conversely

Proposition 2.2 Any geodesic curve $\gamma : \mathbb{R} \to E$ has the form

$$\gamma(t) = et + v \qquad ; t \in \mathbb{R}$$

where v is a point of E and e is a vector of norm 1.

Proof Let us consider a point $a \in \mathbb{R}$ and pick an open interval J around a such that the restriction of γ to J is distance preserving. It follows from the sharp triangle inequality I.4.2 that for any three distinct parameters $r,s,t \in J$ the points $\gamma(r)$, $\gamma(s)$,$\gamma(t)$ lie on an affine line. If we fix two distinct points $c,d \in J$ we find that $\gamma(J)$ is contained in the affine line through $\gamma(c)$ and $\gamma(d)$. It follows from 1.5 that we can find a vector e of norm 1 such that

$$\gamma(t) = e(t-a) + \gamma(a) \qquad ; t \in J$$

This implies that γ is a differentiable curve whose derivative $\gamma'(t)$ is locally constant. Since $\mathbb{R}$ is connected, we conclude that $\gamma'(t)$ is independent of $t \in \mathbb{R}$. It follows from elementary calculus that γ has the required form. □

Let us find all isometries of the Euclidean space E. A basic example is reflection τ in an affine hyperplane H of E. In terms of a unit normal vector n to

H and a point $u \in H$, the action of τ is given by

2.3 $$\tau(x) = x - 2\langle x-u, n\rangle n \qquad ; x \in E$$

The isometry τ fixes H while the image $\tau(x)$ of a point $x \notin H$ can be described by observing that the affine line through x and $\tau(x)$ is perpendicular to H and that H intersects this line in the midpoint of x and $\tau(x)$.

Lemma 2.4 Let there be given two sequences $A_1,...,A_p$ and $B_1,...,B_p$ of points of E such that

$$d(A_i,A_j) = d(B_i,B_j) \qquad ; i,j = 1,...,p$$

Then there exists an isometry σ of E composed of at most p reflections such that

$$\sigma(A_i) = B_i \qquad ; i = 1,...,p$$

Proof Let us recall that the <u>bisecting hyperplane</u> for two distinct points A and B of E is given by

$$\{ X \in E \mid d(X,A) = d(X,B) \}$$

It is easily seen that reflection in the bisecting hyperplane interchanges A and B.

We shall prove the lemma by induction on p. Observe that we have just taken care of the case $p = 1$. To accomplish the induction step, suppose that an isometry ρ composed of at most $p-1$ reflections has been found to transform $A_1,...,A_{p-1}$ into $B_1,...,B_{p-1}$. When $\rho(A_p) = B_p$ we can take $\sigma = \rho$. When $\rho(A_p) \neq B_p$, observe that

$$d(\rho(A_p),B_i) = d(\rho(A_p),\rho(A_i)) = d(A_p,A_i) = d(B_p,B_i) \qquad ; i = 1,...,p-1$$

which means that the points $B_1,...,B_{p-1}$ all lie on the hyperplane bisecting $\rho(A_p)$ and B_p. If τ denotes reflection in this hyperplane then the isometry $\sigma = \tau\rho$ serves our purpose. □

By a Euclidean <u>simplex</u> in the n-dimensional space E we understand a sequence $A_0,A_1,...,A_n$ of points of E not contained in an affine hyperplane.

Lemma 2.5 Let there be given a Euclidean simplex $A_0,A_1,\dots,A_n$ of E. If two isometries α and β of E agree on $A_0,\dots,A_n$ then $\alpha = \beta$.

Proof Put $\sigma = \beta^{-1}\alpha$ and assume that P is a point of E where $\sigma(P) \neq P$. From

$$d(\sigma(P),A_i) = d(\sigma(P),\sigma(A_i)) = d(P,A_i) \qquad ; i = 0,\dots,n$$

it follows that the points $A_0,\dots,A_n$ all belong to the affine hyperplane bisecting P and $\sigma(P)$. This contradicts that the points $A_0,\dots,A_n$ form a simplex. □

Corollary 2.6 Let σ be an isometry of E which fixes an affine hyperplane K pointwise. Then, either σ is reflection in K or $\sigma = \iota$.

Proof Let us assume that $\sigma \neq \iota$ and pick $P \in E$ with $\sigma(P) \neq P$. It follows from the proof of 2.5 that K equals the perpendicular bisector for $\sigma(P)$ and P. Let us form the composite of τ, reflection in K, and σ to get an isometry $\tau\sigma$ which fixes P and the points of K. We conclude from 2.5 that $\tau\sigma = \iota$. □

Theorem 2.7 Any isometry β of a Euclidean space of dimension n can be written as the product of at most n+1 reflections.

Proof Let us pick a Euclidean simplex $A_0,\dots,A_n$ in E. We can use lemma 2.4 to pick an isometry σ which is the composite of at most n+1 reflections with

$$\sigma(A_i) = \beta(A_i) \qquad ; i = 0,\dots,n$$

It follows from lemma 2.5 that $\sigma = \beta$. □

We are now in a position to describe all isometries of Euclidean space. First of all a vector $e \in E$ defines translation τ_e given by the formula

2.8
$$\tau_e(x) = e + x \qquad ; x \in E$$

The translations form a subgroup $T(\mathrm{E})$ of the group $Isom(\mathrm{E})$ of all isometries.

Lemma 2.9 Any isometry ρ of Euclidean space E can be written

$$\rho = \tau_e\, \sigma \qquad ; e \in \mathrm{E}\ , \sigma \in O(\mathrm{E})$$

The orthogonal transformation σ is called the <u>linearisation</u> of ρ and is denoted $\vec{\rho}$.

Proof Let $Isom'(\mathrm{E})$ denote the set of isometries of E which can be written in the form $\tau_e\sigma$ as above. We ask the reader to verify the formula

$$\sigma\, \tau_e\, \sigma^{-1} = \tau_{\sigma(e)} \qquad ; e \in \mathrm{E},\ \ \sigma \in O(\mathrm{E})$$

We can now verify that $Isom'(\mathrm{E})$ is closed under composition

$$\tau_v\, \sigma\, \tau_e\, \rho = \ \tau_v\, \tau_{\sigma(e)}\, \sigma\rho = \tau_{v+\sigma(e)}\, \sigma\rho$$

Observe also that $Isom'(\mathrm{E})$ contains all reflections in hyperplanes in E as follows by inspection of the formula 2.3. Thus we get $Isom(\mathrm{E}) = Isom'(\mathrm{E})$ from 2.7. □

The linearisation $\rho \mapsto \vec{\rho}$ of an isometry plays an important role in Euclidean geometry. Let us notice some of the basic features. First of all

2.10
$$\overrightarrow{(\phi \circ \psi)} = \vec{\phi} \circ \vec{\psi} \qquad ; \ \ \phi, \psi \in Isom(\mathrm{E})$$

Observe that for $\phi \in Isom(\mathrm{E})$ we have $\vec{\phi} = \iota$ if and only if ϕ is a translation. This fact can be recaptured by saying that the following sequence of groups is <u>exact</u>

2.11
$$0 \ \to\ T(\mathrm{E}) \ \to\ Isom(\mathrm{E}) \ \to\ O(\mathrm{E}) \ \to\ 0$$

As a final general remark on the structure of $Isom(\mathrm{E})$, we ask the reader to verify

2.12
$$\sigma\, \tau_v\, \sigma^{-1} = \tau_{\vec{\sigma}(v)} \qquad ; \sigma \in Isom(\mathrm{E}),\ v \in \mathrm{E}$$

We shall classify isometries of Euclidean space of dimension 2 and 3. The discussion is based on the following innocent looking theorem.

Classification theorem 2.13 An isometry ψ of Euclidean space E can be decomposed into a product of two commuting isometries

$$\psi = \tau_v \circ \phi = \phi \circ \tau_v$$

where τ_v is a translation along a vector $v \in E$ and ϕ is an isometry with a fixed point. Such a decomposition of ψ is unique.

Proof Let us decompose $\psi = \tau_e \sigma$ as in 2.9. The transformation ϕ, we are looking for have a fixed point u, say, in E. This means that we are looking for a transformation of the form $\phi = \tau_u \sigma \tau_{-u}$, $u \in E$. From 2.12 we find that the condition for ϕ to commute with a translation τ_v, $v \in E$, is that $\vec{\phi}(v) = v$ or $\sigma(v) = v$. Thus we are looking for two vectors u and v in E such that

$$\tau_e \sigma = \tau_v \tau_u \sigma \tau_u^{-1} \ , \quad \sigma(v) = v$$

Using $\sigma \tau_u^{-1} = \tau_{-\sigma u}\, \sigma$ this may be rewritten as

$$e = v + (u - \sigma u) \qquad ; v \in Ker(\sigma - \iota), \ u \in E$$

Such a decomposition of $e \in E$ is always possible by the following lemma. Uniqueness of v comes from the same source. □

Lemma 2.14 A transformation $\sigma \in O(E)$ gives rise to an orthogonal decomposition of E

$$E = Ker(\sigma - \iota) \oplus Im(\sigma - \iota)$$

Proof Let us first show that the two subspaces $Ker(\sigma - \iota)$ and $Im(\sigma - \iota)$ of E are orthogonal to each other

$$\langle x, \sigma(y) - y \rangle = \langle x, \sigma(y) \rangle - \langle x, y \rangle = \langle \sigma(x), \sigma(y) \rangle - \langle x, y \rangle = 0$$

This gives us $Ker(\sigma - \iota) \bigcap Im(\sigma - \iota) = 0$. From Grassmann's dimension formula applied to the linear map $\sigma - \iota$ we get that

$$dim\ Ker(\sigma - \iota) + dim\ Im(\sigma - \iota) = n$$

The result follows from the dimension formula I.3.4. □

Let us return to the classification theorem 2.13. The fact that τ_v and ϕ commute can be written in the form

$$\phi(x+v) = \phi(x) + v \qquad ; \; x \in E$$

This shows that the fixed point set for ϕ is stabilised by the translation τ_v.

Let us apply the classification theorem to dimensions 2 and 3. As a consequence of our investigations of $O(E)$ made in I.4.8 and I.4.9 we get

Euclidean plane 2.15 An even isometry of the plane is a rotation or a translation. An odd isometry is a glide reflection, i.e. an isometry $\tau\kappa$ composed of a reflection κ in a line k and a translation τ along k. The special case where $\tau = \iota$ is a reflection.

Euclidean space 2.16 An even isometry is a screw composed of a rotation and a translation along the axis of rotation. An odd isometry is either a rotary reflection, composed of a reflection in a plane and a rotation with axis perpendicular to the same plane, or a glide reflection , composed of a reflection in a plane and a translation along that plane.

II.3 SPHERES

Let us consider a Euclidean space F of dimension n+1. We shall introduce a metric on its unit sphere $S^n = \mathcal{S}(F)$. To begin, observe that the Cauchy–Schwarz inequality I.4.1 gives us

$$<P,Q> \in [-1,1] \qquad ; P,Q \in S^n$$

This allows us to define the spherical distance d(P,Q) between two points P and Q of S^n through the formula

3.1
$$\cos d(P,Q) = <P,Q> \qquad ; d(P,Q) \in [0,\pi]$$

It is clear that the spherical distance d(P,Q) is symmetrical in P and Q. Let us show that two distinct points P and Q have non-zero distance. If P and Q are linearly independent, the Cauchy–Schwarz inequality I.4.1 is sharp and we find that $<P,Q> \in]-1,1[$. If P and Q are linearly dependent, we have $P = -Q$ which gives $<P,Q> = -1$ and $d(P,Q) = \pi$. The triangle inequality will be proved in 3.6 after an introduction to spherical trigonometry.

Definition 3.2 By a tangent vector to a point $A \in S^n$ we understand a vector $T \in F$ with $<T,A> = 0$. A tangent vector of norm 1 is called a unit tangent vector. The space of tangent vectors to S^n at A forms a linear hyperplane $T_A(S^n)$ of F called the tangent space to the sphere S^n at A.

Lemma 3.3 Let A and B be points of the sphere S^n. A unit tangent vector U to S^n at A can be chosen such that

$$B = A \cos d(A,B) + U \sin d(A,B)$$

Proof Let us at first suppose that A and B are linearly independent. Let U be a unit tangent vector at A in the plane spanned by A and B, and let us write

$$B = xA + yU \qquad ; x,y \in \mathbb{R}$$

Using $\langle A,U\rangle = 0$ we get $x^2 + y^2 = 1$. Thus we can determine s such that

$$B = A \cos s + U \sin s \qquad ; s \in [-\pi,\pi]$$

Notice that the formula is unchanged under the substitution $(s,U) \mapsto (-s,-U)$, and conclude that we can write B as above with $s \in [0,\pi]$. Observe that

$$\cos d(A,B) = \langle A,B\rangle = \cos s$$

and conclude that $s = d(A,B)$. If A and B are linearly dependent we have $B = A$ or $A = -B$. It follows that $\sin d(A,B) = 0$ and that any unit tangent vector U to S^n at A will serve our purpose. □

Let us consider a <u>spherical</u> <u>triangle</u> ΔABC. By this we understand three points A,B,C of S^n linearly independent in the ambient space F. The following notation is standard

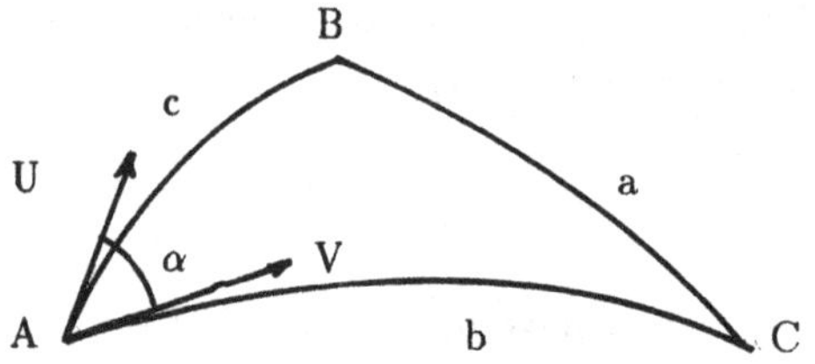

$a = d(B,C)$

$b = d(A,C)$

$c = d(A,B)$

Let us use lemma 3.3 to pick tangent vectors U and V at A such that

$$B = A \cos c + U \sin c \quad , \quad C = A \cos b + V \sin b$$

We can use this to introduce $\angle A = \alpha$ as the angle between U and V

3.4
$$\cos \alpha = \langle U,V\rangle \qquad ; \alpha \in]0,\pi[$$

We are now in a position to prove the basic formula of spherical trigonometry.

Cosine relation 3.5

$$\boxed{\cos a = \cos b \cos c + \sin b \sin c \cos \alpha}$$

Proof We ask the reader to verify the determinant formula

$$\begin{vmatrix} \langle A,A\rangle & \langle B,A\rangle \\ \langle A,C\rangle & \langle B,C\rangle \end{vmatrix} = \sin c \sin b \begin{vmatrix} \langle A,A\rangle & \langle U,A\rangle \\ \langle A,V\rangle & \langle U,V\rangle \end{vmatrix}$$

Hint: Observe that the left hand side is zero if B is replaced by A or if C is replaced by A. Evaluation of the determinants using 3.4 gives us 3.5. □

Triangle inequality 3.6

Three points A,B,C in S^n satisfy

$$d(A,B) \le d(A,C) + d(C,B)$$

This inequality is sharp when A,B,C are linearly independent in F.

Proof When A,B,C are linearly independent we can use the cosine formula with $\cos\alpha < 1$ to get the inequality

$$\cos a < \cos(c-b)$$

When c−b is positive we deduce that $a > c-b$ which gives $c < a+b$. Observe that this inequality is trivial when c−b is negative. This concludes the proof of the sharp triangle inequality. When A,B,C lie in a plane we can modify the trigonometry to maintain the cosine formula 3.5 allowing for $\alpha = 0,\pi$. We can deduce the triangle inequality from 3.5 using the inequality $\cos\alpha \le 1$. □

An alternative proof of the triangle inequality including the sharp triangle inequality can be found in exercise III.5

Let us observe that the metric we have introduced on the sphere S^n defines the topology on S^n induced from the ambient Euclidean space F as follows from

3.7 $$|P-Q| = \sqrt{2-2\langle P,Q\rangle} = \sqrt{2-2\cos d(P,Q)}$$

Proposition 3.8 Any geodesic curve $\gamma:\mathbb{R}\to S^n$ can be written in the form

$$\gamma(t) = A\,cos\,t + T\,sin\,t \qquad ; \ t\in\mathbb{R}$$

where $A\in S^n$ and T is a unit tangent vector to S^n at A.

Proof Let us investigate γ in a neighbourhood of a point $u\in\mathbb{R}$. To this end we pick an open interval J around u of length less than π and such that the restriction of γ to J is distance preserving. It follows from 3.6 that for any three distinct parameters $r,s,t\in J$, the points $\gamma(r),\gamma(s),\gamma(t)$ are linearly dependent. In particular if we fix two distinct points $c,d\in J$ such that $\gamma(c)$ and $\gamma(d)$ are linearly independent, we find that $\phi(J)$ is contained in the plane R spanned by $\gamma(c)$ and $\gamma(d)$. Put $A=\gamma(u)$ and pick a unit tangent vector $T\in T_A(S^n)$ contained in R and consider the curve

$$\theta(s) = A\,cos\,s + T\,sin\,s \qquad ; \ s\in\,]-\pi,\pi[$$

It follows by a simple direct calculation using 3.3 that θ induces an isometry between $]-\pi,\pi[$ and the set of points $X\in R\cap S^n$ with $d(A,X)<\pi$. According to 1.5 we can find $\epsilon=\pm1$ such that

$$\gamma(t) = A\,cos\,\epsilon(t-u) + T\,sin\,\epsilon(t-u) \qquad ; \ t\in J$$

We conclude that γ is continuously differentiable on J with $\gamma'(u)=\epsilon T$, thus

3.9
$$\gamma(t) = \gamma(u)\,cos(t-u) + \gamma'(u)\,sin(t-u) \qquad ; \ t\in J$$

Let us show that two geodesic curves $\sigma,\gamma\colon\mathbb{R}\to S^n$ which coincide in a neighbourhood of 0 are identical. To this end we introduce the set

$$\{\ t\in\mathbb{R} \mid \sigma(t)=\gamma(t)\ ,\ \sigma'(t)=\gamma'(t)\ \}$$

Observe that the set is closed since $\gamma,\sigma,\gamma',\sigma'$ are continuous. The set is open as well, as follows from formula 3.9. Since the set is non-empty, by assumption, we get from the connectedness of $\mathbb{R}$ that $\gamma=\sigma$. Apply this to the given geodesic curve γ and the geodesic curve σ given by the right hand side of 3.9 with $u=0$, and the result follows. □

Isometries of the sphere An orthogonal transformation $\sigma \in O(F)$ induces an isometry of the sphere S^n. We intend to show that any isometry of the sphere has this form; but first a lemma.

Lemma 3.10 Let there be given two sequences $A_1,...,A_p$ and $B_1,...,B_p$ of points of S^n such that

$$d(A_i,A_j) = d(B_i,B_j) \qquad ; \, i,j = 1,...,p$$

Then there exists an isometry σ composed of at most p hyperplane reflections with

$$\sigma(A_i) = B_i \qquad ; \, i = 1,...,p$$

Proof Consider two distinct points A and B of S^n, let us analyse the set

$$\{ \, P \in S^n \mid d(A,P) = d(B,P) \, \}$$

From the definition of the spherical distance it follows that this is the intersection between S^n and the linear hyperplane K orthogonal to the vector $A-B$. Orthogonal reflection in K will interchange A and B. We can now conclude the proof by the method used in the Euclidean case, see the proof of 2.4. □

Theorem 3.11 Any isometry β of the sphere S^n is induced by an orthogonal transformation of the ambient Euclidean space.

Proof Let us pick a spherical simplex in S^n, i.e. a sequence $A_0,...,A_n$ of points of S^n linearly independent in F. We can use lemma 3.10 to pick an isometry σ which is the composite of at most n+1 reflections in hyperplanes with

$$\sigma(A_i) = \beta(A_i) \qquad ; \, i = 0,...,n$$

We can conclude the proof by the method used in the Euclidean case, see 2.5. □

II.4 HYPERBOLIC SPACE

Let us consider a real vector space F of dimension n+1 equipped with a quadratic form of Sylvester type (−n,1). The hyperboloid $\mathcal{S}$(F) consisting of all vectors of norm 1 has two sheets as we have seen in I.6.3. There is no way in general to distinguish the two sheets of the hyperboloid, but let H^n denote any one of them. We are going to introduce a metric on H^n.

The point of departure is the inequality $\langle P,Q\rangle \geq 1$ obeyed by any two points P and Q of H^n, compare I.6.2 and I.6.4. We define the hyperbolic distance d(P,Q) between P and Q to be the positive real number d(P,Q) such that

4.1 $$\cosh d(P,Q) = \langle P,Q\rangle \qquad ; P,Q \in H^n$$

It is clear that the hyperbolic distance d(P,Q) is symmetrical in P and Q. Let us verify that $d(P,Q) > 0$ for $P \neq Q$: observe that the vectors P and Q are linearly independent and conclude from I.6.2 that $\langle P,Q\rangle > 1$. Let us now turn to the triangle inequality.

Triangle inequality 4.2 Any three points A,B,C in H^n satisfy

$$d(A,B) \leq d(A,C) + d(C,B)$$

The triangle inequality is sharp in case A,B,C are linearly independent in F.

Proof Consider points A,B,C on H^n and put $a = d(B,C)$, $b = d(C,A)$ and $c = d(A,B)$. We shall calculate the determinant of the Gram-matrix I.1.10 of A,B,C using the standard abbreviations $ch\,a = cosh\,a$ and $sh\,a = sinh\,a$.

4.3 $$\Delta = det \begin{bmatrix} 1 & ch\,a & ch\,b \\ ch\,a & 1 & ch\,c \\ ch\,b & ch\,c & 1 \end{bmatrix}$$

Direct expansion after the first row of Δ gives us

$$\Delta = (1 - ch^2c) - ch\,a(ch\,a - ch\,b\ ch\,c) + ch\,b(ch\,a\ ch\,c - ch\,b) =$$
$$1 - ch^2a - ch^2b - ch^2c + 2\,ch\,a\ ch\,b\ ch\,c\ =$$
$$(ch^2b - 1)(ch^2c - 1) - (ch\,b\ ch\,c - ch\,a)^2 -$$
$$sh^2b\ sh^2c - (ch\,b\ ch\,c - ch\,a)^2 =$$
$$(ch\,a - ch\,b\,ch\,c + sh\,b\ sh\,c)(ch\,b\ ch\,c + sh\,b\ sh\,c - ch\,a) =$$
$$[ch\,a - ch(b - c)]\,[ch(c + b) - ch(a)]\ =$$
$$4\,sh\tfrac{1}{2}(a + b - c)\ sh\tfrac{1}{2}(a + c - b)\ sh\tfrac{1}{2}(a + b + c)\ sh\tfrac{1}{2}(c + b - a)$$

This gives us the following factorisation of the discriminant

4.4 $$\boxed{\Delta = 4\,sh\,p\ sh(p - a)\ sh(p - b)\ sh(p - c)} \qquad ;\ p = \tfrac{1}{2}(a + b + c)$$

Let us assume that the vectors A,B,C are linearly independent in the ambient space F. It follows that they generate a subspace of F of type (−2,1) and we conclude from I.3.5 that $\Delta > 0$. If $c \geq a$ and $c \geq b$ we have that

$$p - a = \tfrac{1}{2}(c - a) + \tfrac{1}{2}b > 0\ ,\quad p - b = \tfrac{1}{2}(c - b) + \tfrac{1}{2}a > 0$$

Using $\Delta > 0$ and $p > 0$ we conclude from the factorisation 4.4 that $p - c > 0$. It follows that $a + b - c = 2(p - c) = 0$.

If the vectors A,B,C are linearly dependent in the ambient space F, then $\Delta = 0$. If A,B,C do not represent the same point then $p \neq 0$ and we conclude from the factorisation 4.4 that $p - a = 0$, $p - b = 0$ or $p - c = 0$. The remaining details are left to the reader. □

Definition 4.5 By a <u>tangent vector</u> T to a point $A \in H^n$ we understand a $T \in F$ with $<T,A> = 0$. A tangent vector of norm −1 is called a <u>unit tangent</u> vector. The space of tangent vectors to H^n at A forms a hyperplane of F of type (−n,0) which is called the <u>tangent space</u> to H^n at A and is denoted $T_A(H^n)$.

The tangent space $T_A(H^n)$ inherits a quadratic form from the ambient space F. It is easily seen that the tangent space has Sylvester type (−n,0)

Lemma 4.6 Let A and B be points of H^n. It is possible to choose a unit tangent vector U to H^n at A such that

$$B = A\ cosh\ d(A,B) + U\ sinh\ d(A,B)$$

Proof If $A = B$ we can use any unit tangent vector U at A. If $A \neq B$, the two vectors are linearly independent and the plan R they span has Sylvester type $(-1,1)$: according to the discriminant lemma I.6.3 only three types are possible $(-1,1)$, $(-1,0)$, $(-2,0)$, but only the first type contains vectors of norm 1. It follows that we can pick a unit tangent vector U in R and write

$$B = xA + yU \qquad ; x,y \in \mathbb{R}$$

Using $<A,U> = 0$ we conclude that $x^2 - y^2 = 1$. Thus we can find s such that

$$B = A\ cosh\ s + U\ sinh\ s \qquad ; s \in \mathbb{R}$$

Notice, that the formula is unchanged under the substitution $(U,s) \mapsto (-U,-s)$ and conclude that we can arrange for the formula to hold with $s \geq 0$. Observe that

$$cosh\ d(A,B) = <A,B> = cosh\ s$$

and conclude that $s = d(A,B)$ as required. □

Geodesics Starting from a point $A \in H^n$ and a unit tangent vector T to H^n at A we can construct a curve $\phi: \mathbb{R} \to H^n$ by the explicit formula

$$\phi(s) = A\ cosh\ s + T\ sinh\ s \qquad ; s \in \mathbb{R}$$

Let us calculate the hyperbolic distance between $\gamma(s)$ and $\gamma(t)$, $s,t \in \mathbb{R}$

$$cosh\ d(\gamma(t),\gamma(s)) = <\gamma(t),\gamma(s)> = cosh(t-s) = cosh|t-s|$$

From this we conclude that

$$d(\gamma(t),\gamma(s)) = |t-s| \qquad ; s,t \in \mathbb{R}$$

This shows that $\phi: \mathbb{R} \to H^n$ is a geodesic curve. The image of the curve is contained in the plane R spanned by A and T; in fact it follows from lemma 4.6 that the image of γ is $H^n \cap R$.

Proposition 4.7 Any geodesic curve $\gamma:\mathbb{R}\to H^n$ can be written in the form

$$\gamma(t) = A\,cosh\;t + T\,sinh\;t \qquad ;\; t\in\mathbb{R}$$

where $A \in H^n$ and T is a unit tangent vector to H^n at A.

Proof Let us fix a point $u \in \mathbb{R}$ and pick an open interval J around u such that the restriction of γ to J is distance preserving. It follows from the sharp triangle inequality 4.2 that for any three distinct parameters $r,s,t \in J$, the points $\gamma(r),\gamma(s),\gamma(t)$ are linearly dependent. In particular if we fix two distinct points $c,d \in J$ such that $\gamma(c)$ and $\gamma(d)$ are linearly independent we find that $\phi(J)$ is contained in the plane R spanned by $\gamma(c)$ and $\gamma(d)$. Put $A = \gamma(u)$ and pick a unit tangent vector $T \in T_A(H^n)$ contained in R. It follows from the discussion preceding 4.7 that we can use lemma 1.5 to find $\epsilon = \pm 1$ such that

$$\gamma(t) = A\,cosh\;\epsilon(t-u) + T\,sinh\;\epsilon(t-u) \qquad ;\; t\in J$$

This implies that γ is continuously differentiable on J with $\gamma'(u) = \epsilon T$. Thus

$$\gamma(t) = \gamma(u)\,cosh(t-u) + \gamma'(u)\,sinh(t-u) \qquad ;\; t\in J$$

The merits of this formula is to reconstruct γ in a neighbourhood of $u \in \mathbb{R}$ from the values of $\gamma(u)$ and $\gamma'(u)$. We ask the reader to conclude the proof by the method used in the proof of 3.8. □

In recapitulation, a geodesic curve $\gamma:\mathbb{R}\to H^n$ is distance preserving and not merely locally distance preserving. By a <u>geodesic line</u> or <u>hyperbolic line</u> in H^n we understand the image of a geodesic curve $\gamma:\mathbb{R}\to H^n$. We have seen that through two distinct points of H^n there passes a unique geodesic line.

Isometries A Lorentz transformation $\sigma \in Lor(F)$ preserves the two sheets of the unit hyperbola S(F), in fact it induces an isometry of hyperbolic space H^n. We intend to show that this procedure induces an isomorphism between the Lorentz group and the group of isometries of H^n.

Lemma 4.8 Let there be given two sequences $A_1,\dots,A_p$ and $B_1,\dots,B_p$ of points of H^n such that

$$d(A_i,A_j) = d(B_i,B_j) \qquad ; i,j = 1,\dots,p$$

Then there exists an isometry σ composed of at most p Lorentz reflections with

$$\sigma(A_i) = B_i \qquad ; i = 1,\dots,p$$

Proof Consider two distinct points A and B of H^n and let us analyse the perpendicular bisector for A and B

$$\{ P \in H^n \mid d(A,P) = d(B,P) \}$$

From the definition of the hyperbolic distance it follows that this is the intersection between H^n and the linear hyperplane orthogonal to the vector $N = A - B$. We can use I.6.2 and I.6.4 to conclude that

$$<N,N> = 2 - 2<A,B> < 0$$

Reflection along N will interchange A and B. We can now conclude the proof by the method used in the proof of 2.4. □

Theorem 4.9 An isometry β of the hyperbolic space H^n is induced by a Lorentz transformation of the ambient space F.

Proof Let us pick a hyperbolic simplex, i.e. a sequence $A_0,\dots,A_n$ of points of H^n linearly independent in F. We can use lemma 4.8 to pick an isometry σ which is the composite of at most n+1 Lorentz reflections such that

$$\sigma(A_i) = \beta(A_i) \qquad ; i = 0,\dots,n$$

It follows that $\sigma = \beta$, compare the proof of 2.5. □

Definition 4.10 An isometry σ of the hyperbolic space H^n is called even or odd if the corresponding Lorentz transformation is even or odd.

II.5 KLEIN DISC

In this section we shall construct a model of hyperbolic n-space based on the open unit disc D^n of a Euclidean space E of dimension n. To begin we observe that the auxiliary space $E \oplus \mathbb{R}$ carries a quite natural quadratic form of Sylvester type (−n,1): the corresponding bilinear form is given by

5.1 $$<(A,a),(B,b)> = -<A,B> + ab \qquad ; A,B \in E, \quad a,b \in \mathbb{R}$$

For hyperbolic space H^n we take the sheet of the unit hyperbola in $E \oplus \mathbb{R}$ consisting of points (A,a) with $-<A,A> + a^2 = 1$ and $a > 0$. We shall make use of the parametrisation $p:D^n \to H^n$ given by the formula

5.2 $$p(A) = \frac{(A,1)}{\sqrt{1-<A,A>}} \qquad ; A \in D^n$$

Geometrically, the point p(A) is the intersection of H^n and the line through (A,1) and 0. The Klein model of hyperbolic n-space is D^n equipped with the metric obtained by transporting the hyperbolic metric along $p: D^n \to H^n$. It follows from 5.2 and 4.1 that the metric in the Klein model is given by

5.3 $$\cosh d(A,B) = \frac{1-<A,B>}{\sqrt{(1-<A,A>)(1-<B,B>)}} \qquad ; A,B \in D^n$$

The geodesic lines in the Klein model are precisely traces on D of ordinary affine lines in E. This can be seen from the description of the geodesic lines H^n given in 4.7 and the geometric origin of the parametrisation $p:D^n \to H^n$.

The boundary ∂H^n (in the sense of 6.18) can be parametrised by the Euclidean boundary of ∂D^n by assigning to a point $A \in \partial D^n$ the isotropic line in $E \oplus \mathbb{R}$ through the point (A,1).

II.6 POINCARÉ DISC

Let us once again start with an n-dimensional Euclidean space E and its open unit disc D^n. We shall make use of the auxiliary space $E \oplus \mathbb{R}$ equipped with the hyperbolic form given in 5.1 and the sheet H^n of the unit hyperboloid in $E \oplus \mathbb{R}$ passing through (0,1). This time we shall consider the parametrisation $f: D^n \to H^n$ given by the formula

6.1

$$f(P) = \frac{(2P, 1+<P,P>)}{1-<P,P>} \qquad ; P \in D^n$$

From the geometric point of view, this is the stereographic projection with centre (0,−1) already considered in I.6.6. The result of transporting the hyperbolic metric along f is given by

6.2

$$\boxed{\cosh d(P,Q) = 1 + 2\frac{|P-Q|^2}{(1-|P|^2)(1-|Q|^2)}} \qquad ; P,Q \in D^n$$

The open unit disc D^n equipped with this metric is called the Poincaré disc.

Theorem 6.3 The group $Möb(D^n)$ of Möbius transformations of E which leaves D^n invariant acts as the group of isometries of the Poincaré disc D^n. In particular, any isometry of D^n extends to a Möbius transformation of E.

Proof Let us first record from I.6.5 that $f^{-1}: H^n \to D^n$ is given by the formula

$$f^{-1}(A,a) = A(1+a)^{-1} \qquad ; (A,a) \in H^n$$

Let τ denote reflection along a vector (N,n) in $E \oplus \mathbb{R}$ of norm −1. Let us transform τ to D^n. We ask the reader to verify the formula

$$f^{-1}\tau f(P) = \frac{P + (n + n<P,P> - 2<P,N>)N}{|N-nP|^2} \qquad ; P \in D^n$$

When $n = 0$, we have $<N,N> = 1$ and $f^{-1}\tau f$ is Euclidean reflection along N.

When $n \neq 0$, we have $1+n^2 = \langle N,N \rangle$ and $f^{-1}\tau f$ is given by the formula

$$f^{-1}\tau f(P) = \frac{P + (|nP-N|^2 - 1)Nn^{-1}}{|nP-N|^2} = Nn^{-1} + n^{-2}\frac{P - Nn^{-1}}{|P-Nn^{-1}|^2} \quad ; P \in D^n$$

This shows that $f^{-1}\tau f$ is induced by inversion in the Euclidean sphere $\mathcal{N}$ with centre Nn^{-1} and radius n^{-1}. Let us observe that

$$(n^{-1})^2 + 1 = \langle Nn^{-1}, Nn^{-1} \rangle$$

which shows that $\mathcal{N}$ is orthogonal to $S^{n-1} = \partial D^n$. The result follows by combining I.7.16 with II.4.9 and I.6.11. □

Lemma 6.4 An isometry ϕ of the Poincaré disc $D^n \subset E$ is induced by an orthogonal transformation of E if and only if $\phi(0) = 0$.

Proof An isometry of D can be represented by a Möbius transformation ϕ of $\hat{E}$ which commutes with reflection σ in the unit sphere, I.7.15. When $\phi(0) = 0$, let us evaluate the formula $\sigma\phi = \phi\sigma$ at ∞ to get $\sigma(\phi(\infty)) = \phi(0) = 0$ or $\phi(\infty) = \infty$. It follows from I.7.7 that ϕ is a Euclidean similarity of E. Using once again that $\phi(0) = 0$ and $\phi(D) = D$, we conclude that ϕ is orthogonal. □

Proposition 6.5 An isometry ϕ of the Poincaré disc $D^n \subset E$ can be written $\rho\mu$, where $\mu \in O(E)$ and ρ is reflection in a sphere orthogonal to $S^{n-1} = \partial D^n$.

Proof Let us focus on the point $P = \phi(0) \in D^n$. If $P = 0$, the result follows from 6.4. If $P \neq 0$, put $R = \sigma(P)$ where σ is inversion in the sphere S^{n-1}. Let $\mathfrak{R}$ be a circle with centre R orthogonal to S^{n-1}. We ask the reader to verify that

$$\mathfrak{R} * S^{n-1} = \frac{1}{2r}\,|\rho(0) - \sigma(R)|$$

Observe that $\sigma(R) = P$ and conclude that $\rho(0) = P$ or $\rho(P) = 0$. It follows that $\rho\phi$ fixes 0 and we conclude from the previous lemma that $\rho\phi \in O(E)$. □

II.7 POINCARÉ HALF-SPACE

Let us start from a Euclidean vector space L of dimension n−1 and form the Euclidean space $E = L \oplus \mathbf{R}$. We shall be concerned with the <u>upper half-space</u> E^+ consisting of points (p,h) of E with $h > 0$. Inversion σ in a sphere with radius $\sqrt{2}$ and centre in the south pole $S = (0,-1)$ will map E^+ onto the unit disc D^n

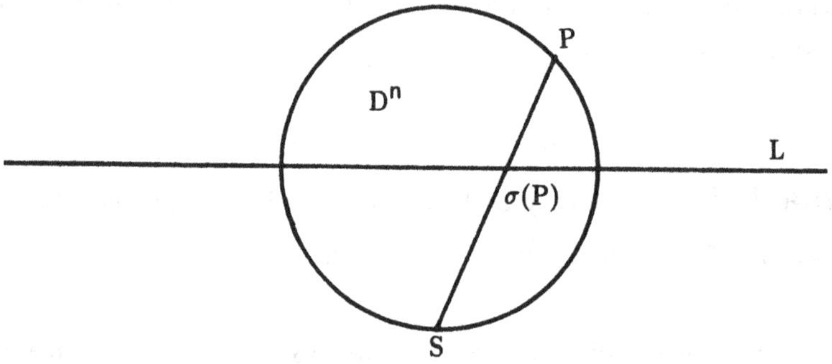

Observe that σ induces a stereographic projection of ∂D^n onto the boundary $\hat{L}$ of the upper half-space. The explicit formula I.7.1 for $\sigma: E^+ \to D^n$ may be written

7.1
$$\sigma(p,h) = (0,-1) + 2\,\frac{(\,p\,,\,1+h)}{|p|^2 + (1+h)^2} \qquad ;\ p \in L,\ h \in \mathbf{R}$$

Let us use σ to transport the metric of the Poincaré disc onto the upper half-space E^+. Explicit calculation based on 6.2 and 7.1 gives us

7.2
$$\boxed{\cosh d(P,Q) = 1 + \frac{|P-Q|^2}{2hk}} \qquad ;\ P = (p,h),\ Q = (q,k)$$

The open half-space E^+ equipped with this metric is called the <u>Poincaré half-space</u>.

Theorem 7.3 The group $M\ddot{o}b(E^+)$ of Möbius transformations of E which leaves E^+ invariant acts as the group of isometries of the Poincaré half-space. In particular, any isometry of E^+ can be extended to a Möbius transformation of E.

Proof Let σ denote the inversion considered in 7.1. Conjugation by σ in $M\ddot{o}b(\mathrm{E})$ transforms the subgroup $M\ddot{o}b(\mathrm{D}^n)$ into the subgroup $M\ddot{o}b(\mathrm{E}^+)$. This allows us to transform theorem 6.3 into 7.3. □

Corollary 7.4 The group of isometries of the Poincaré half-space E^+ is isomorphic to the full Möbius group of $\mathrm{L} = \partial \mathrm{E}^+$.

Proof This results from a simple combination of I.7.16 and II.7.3. □

The geodesics in the Poincaré half-space E^+ are traced by circles in $\hat{\mathrm{E}}$ orthogonal to the boundary $\partial \mathrm{E}^+$.

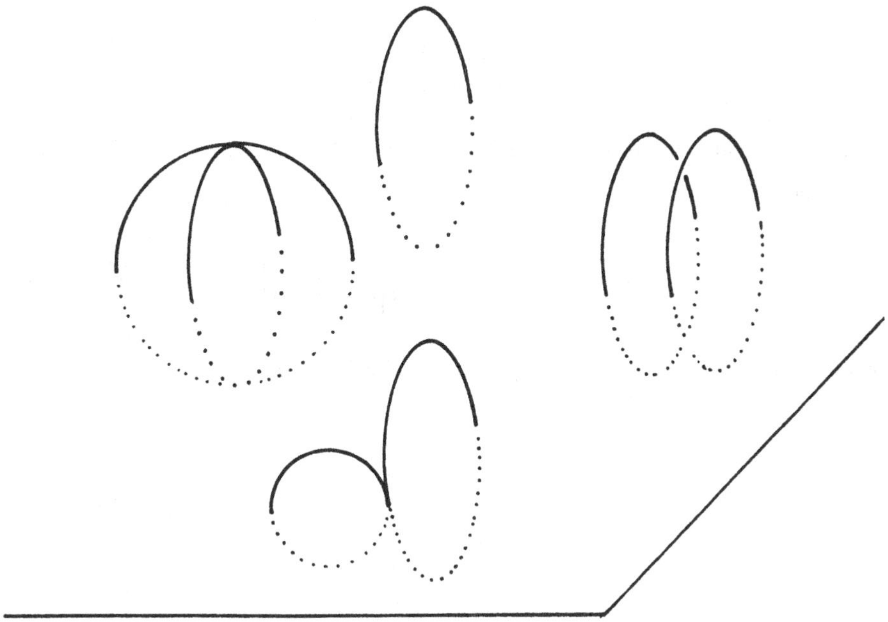

II.8 POINCARÉ HALF-PLANE

The open upper half-plane H^2 of C is a model of for the hyperbolic plane when we specify the metric according to 7.2

8.1
$$\cosh d(z,w) = 1 + \frac{|w-z|^2}{2\, Im[w]\, Im[z]} \qquad ; \ z,w \in H^2$$

For the convenience of the reader we shall give two different formulas

8.2
$$\sinh\tfrac{1}{2}d(z,w) = \frac{|w-z|}{2\sqrt{Im[w]\, Im[z]}}, \qquad \cosh\tfrac{1}{2}d(z,w) = \frac{|w-\bar{z}|}{2\sqrt{Im[w]\, Im[z]}}$$

In order to describe the isometries of H^2 recall from I.9 that $Gl_2(\mathbb{R})$ acts on C–R through the formula

$$\begin{bmatrix} a & b \\ c & d \end{bmatrix} z = \frac{az+b}{cz+d}$$

We ask the reader to work out the following useful expression for the imaginary part of this fraction.

8.3
$$Im\left[\frac{az+b}{cz+d}\right] = \frac{Im[z]}{|cz+d|^2} \det\begin{bmatrix} a & b \\ c & d \end{bmatrix} \qquad ; \ \begin{bmatrix} a & b \\ c & d \end{bmatrix} \in Gl_2(\mathbb{R})$$

This leads us to define the following action of $Gl_2(\mathbb{R})$ on H^2

8.4
$$\sigma z = \begin{cases} \dfrac{az+b}{cz+d}, & \det\sigma > 0 \\[2mm] \dfrac{a\bar{z}+b}{c\bar{z}+d}, & \det\sigma < 0 \end{cases} \qquad ; \ \sigma = \begin{bmatrix} a & b \\ c & d \end{bmatrix}$$

Proposition 8.5 The action of $Gl_2(\mathbb{R})$ on H^2 identifies $PGl_2(\mathbb{R})$ and the group $Isom(H^2)$ of isometries of H^2.

Proof The formula 8.4 in fact defines an action of $Gl_2(\mathbb{R})$ on $\hat{\mathbb{C}}$, which preserves H^2. By 7.3 it suffices to prove that any Möbius transformation of $\hat{\mathbb{C}}$ which preserves H^2 has this form. By Poincaré extension I.7.16 it suffices to identify $PGl_2(\mathbb{R})$ and the Möbius group of the subspace $\mathbb{R}$ of $\mathbb{C}$. This can be done by modifying the construction in I.9 to bring $PGl_2(\mathbb{R})$ to act on $\hat{\mathbb{R}}$ in a triply transitive manner. □

Proposition 8.6 The geodesics in H^2 are traced by circles in $\hat{\mathbb{C}}$ orthogonal to $\hat{\mathbb{R}}$. Expressed in another way, the geodesics are traced by Euclidean circles with centres on the x-axis $\mathbb{R}$ and Euclidean lines perpendicular to the x-axis.

Proof The curve $\gamma: \mathbb{R} \to H^2$ given by $\gamma(s) = ie^s$, $s \in \mathbb{R}$, is distance preserving:

$$\cosh(s-t) = \tfrac{1}{2}(e^{s-t} + e^{t-s}) = \frac{e^{2t}+e^{2s}}{2e^t e^s} = 1 + \frac{(e^t - e^s)^2}{2e^t e^s} = \cosh d(ie^s, ie^t)$$

It follows that $\mathcal{Y}$, the positive y-axis, is a geodesic. Let us observe that $PGl_2(\mathbb{R})$ acts transitively on the set of circles in $\hat{\mathbb{C}}$ orthogonal to $\hat{\mathbb{R}}$ as follows from the fact that $PGl_2(\mathbb{R})$ acts transitively on $\hat{\mathbb{R}}$. The action of $PGl_2(\mathbb{R})$ on the set of geodesics in H^2 is transitive as well as follows from 8.5 and 4.9. □

Let us amplify the fact that the isometries of H^2 are induced by Möbius transformations by asking the reader to verify that the distance can be written

8.7 $$\cosh d(z,w) = 1 + 2\,|[z,\bar{w},w,\bar{z}]| \qquad ; z,w \in H^2,\ z \neq w$$

Alternatively, let us introduce the circle $\mathcal{L}$ through z and w orthogonal to $\hat{\mathbb{R}}$ and let z^* and w^* denote the points of intersection between the circle $\mathcal{L}$ and $\hat{\mathbb{R}}$, defined such that z,w and z^*,w^* give the same orientation to the geodesic $H \cap \mathcal{L}$.

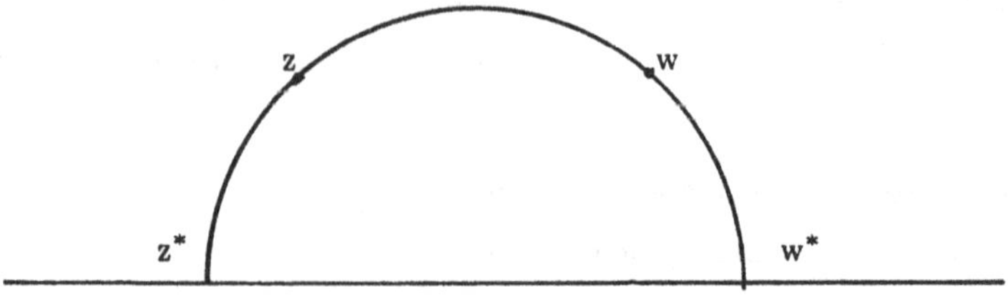

For the distance between the points z and w we have

8.8 $$d(z,w) = log\ [z,w,w^*,z^*] \qquad ; z,w \in H^2,\ z \neq w$$

Proof Upon performing a Möbius transformation we may assume that z and w are points of $\mathcal{Y}$ with $Im[z] < Im[w]$. This gives $z^* = 0$ and $w^* = \infty$ and

$$[z,w,\infty,0] = \tfrac{w}{z}$$

Let us write $z = ie^s$ and $w = ie^t$ $(s < t)$ and conclude from the proof of 8.6 that $d(z,w) = t - s$. Let us put this together

$$log\ [z,w,\infty,0] = log\ \tfrac{w}{z} = log\ e^{t-s} = t - s = d(z,w)$$

and the formula follows. □

Let us take a closer look at the action of a given $\sigma \in Sl_2(\mathbb{R})$ on H^2 using the theory of pencils of circles developed in I.8. We shall make use of the invariant $tr^2\sigma = (tr\sigma)^2$. The discussion will be resumed in III.4 and IV.2.

Following Felix Klein we shall divide these transformations into three classes depending on the invariant $tr^2\sigma$.

Elliptic (rotation): $tr^2\sigma < 4$. The transformation σ has a fixed point $A \in H^2$ and fixes the pencil of circles orthogonal to the pencil of geodesics through A. A circle from this pencil which is contained in H^2 is called a hyperbolic circle with centre A. Hyperbolic circles with centre A can be characterised as the orbits for the group of elliptic isometries with fixed point A.

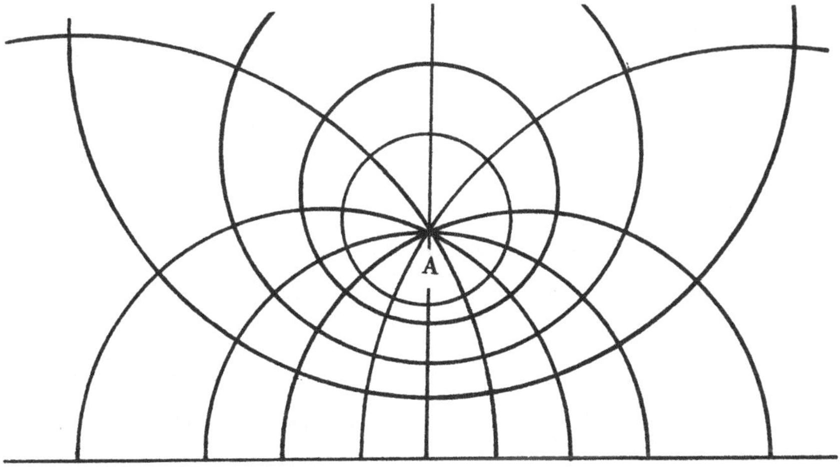

Hyperbolic (translation) $tr^2\sigma > 4$. The isometry σ has two distinct fixed points A and B on ∂H^2 and leaves the geodesic h with ends A and B invariant. The pencil of circles through A and B is left invariant by σ. The trace on H^2 of a circle from this pencil is called a <u>hypercycle</u> for the geodesic h through A and B. The hypercycles can be identified with the orbits of the group of even isometries which leave A and B invariant.

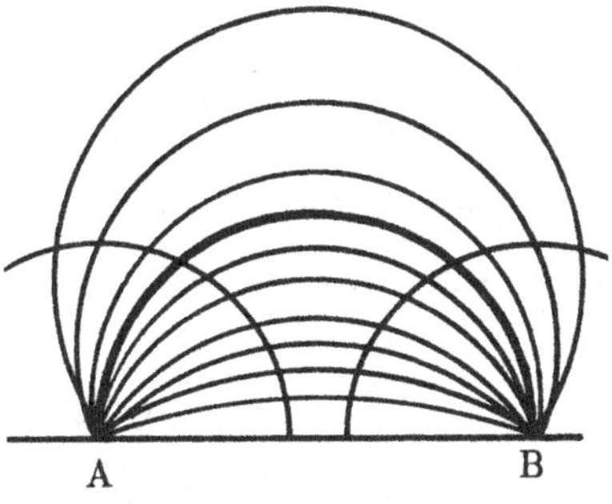

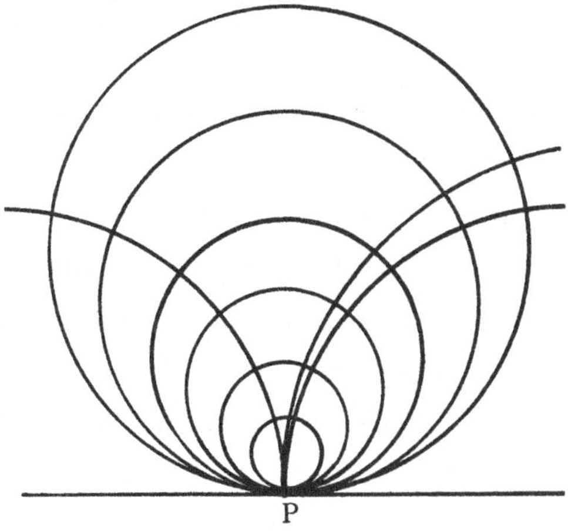

Parabolic (horolation) $tr^2\sigma = 4$. The transformation σ has a unique fixed point P on ∂H^2. The pencil of circles through P and tangent to the x-axis is invariant under σ. A circle from this pencil which is contained in H^2 is called a <u>horocycle</u> with centre Q. The horocycles can be characterised as the orbits for the group of parabolic transformations with fixed point $P \in \partial H^2$.

II EXERCISES

EXERCISE 1.1 Let X and Y be metric spaces. For points $P = (x,y)$ and $Q = (\xi,\eta)$ on $X \times Y$ let us define the distance by

$$d(P,Q) = \sqrt{d(x,\xi)^2 + d(y,\eta)^2}$$

Show that this defines a metric on $X \times Y$.

EXERCISE 2.1 Give a direct proof of the fact that an isometry f of Euclidean space E with $f(0) = 0$ is linear. Hint: Use polarisation to show that

$$\langle f(x),f(y)\rangle = \langle x,y\rangle \qquad ; x,y \in E$$

Conclude from this that vectors of the form below have norm zero

$$f(x+y) - f(x) - f(y) \ , \quad f(rx) - rf(x) \qquad ; x,y \in E,\ r \in \mathbb{R}$$

EXERCISE 2.2 Let σ and ρ be rotations of the Euclidean plane with distinct centres A and B.

1° Justify the following procedure for the calculation of $\rho\sigma$: Write $\rho = \beta\gamma$ and $\sigma = \gamma\alpha$ where γ is reflection in the line through A and B and β is reflection in a line b through A and α is reflection in a line a through B. If a and b are parallel we find that $\rho\sigma = \beta\alpha$ is a translation. If a and b intersect in C we find that $\rho\sigma = \beta\alpha$ is a rotation with centre C.

2° Suppose that ρ and σ are rotations with the same angle $-2\pi/3$ but distinct centres A and B. Show that $\rho\sigma$ is a rotation with angle $2\pi/3$ and that its centre C forms an equilateral triangle with A and B.

3° Let ΔABC be a triangle in the Euclidean plane. Let A^* denote the point in the plane such that ΔA^*BC is equilateral and such A and A^* lie on opposite sides of the line through B and C. We let θ_A denote the rotation about the circumcentre O_A of ΔA^*BC which transforms B to C. Define similar transformations θ_C and θ_B and show that $\theta_C\theta_B\theta_A = \iota$. Hint: Show that $\theta_C\theta_B\theta_A$ is rotation with centre B and angle 0.

4° With the notation above, show that the circumcentres O_A, O_B and O_C form an equilateral triangle. Hint: Use the method from 1° to calculate $\theta_B\,\theta_A$.

EXERCISE 3.1 Given two lines k and l in F through the origin. Let us define the acute angle $\angle(k,l) \in [0,\frac{\pi}{2}]$ between k and l by the formula

$$\cos \angle(k,l) = |\langle \mathbf{k},\mathbf{l} \rangle|$$

where $\mathbf{k}$ and $\mathbf{l}$ are unit vectors generating k and l.

1° Show that the acute angle defines a metric on projective space $P(\mathrm{F})$, the set of lines in F. Hint: Interpret $P(\mathrm{F})$ as the orbit space for the action of the antipodal map on the sphere $S(\mathrm{F})$.

2° Show that through two distinct points of $P(\mathrm{F})$ passes a unique geodesic.

EXERCISE 3.2 Let S^2 denote the unit sphere in $\mathbb{R}^3$ and let the standard spherical coordinates $\sigma: \mathbb{R}^2 \to S^2$ be given by

$$\sigma(\theta,\phi) = (\sin\theta \cos\phi,\ \sin\theta \sin\phi,\ \cos\theta)$$

1° Show that a continuously differentiable curve $\gamma: [a,b] \to S^2$ which does not pass through the north and the south pole can be written $\gamma = \sigma(\theta,\phi)$, where $\theta,\phi:[a,b] \to \mathbb{R}$ are continuously differentiable functions.

2° With the notation above show that the length of the curve γ is given by

$$l(\gamma) = \int_a^b \sqrt{\dot\theta^2 + \sin^2\theta\, \dot\phi^2}\ dt$$

and deduce from this that

$$l(\gamma) \geq |\theta(b) - \theta(a)|$$

3° Let us consider the colatitude θ as a continuous function $\theta:S^2 \to [0,\pi]$. Show that a continuously differentiable curve $\gamma:[a,b] \to S^2$ satisfies the inequality

$$l(\gamma) \geq |\theta(\gamma(b)) - \theta(\gamma(a))|$$

Hint: If $0 < \theta(\gamma(a)) \leq \theta(\gamma(b)) < \pi$, let $c \in [a,b]$ be the last time $\theta(\gamma(t))$ takes the value $\theta(\gamma(a))$ and $d \in [a,b]$ the first time $\theta(\gamma(t))$ takes the value $\theta(\gamma(b))$ and apply the inequality from 2° to the restriction of γ to the subinterval $[c,d]$.

4° Let P, Q be points on the sphere S^2. Show that the spherical distance is

$$d(P,Q) = \inf_\gamma l(\gamma)$$

where γ runs through the set of piecewise continuously differentiable curves γ on S^2 connecting P with Q. Hint: Arrange for P and Q to have the same longitude.

EXERCISE 3.3 In $\mathbb{R}^3$ let the standard cylinder be given by

$$Cyl = \{ (x,y,z) \mid x^2+y^2=1 \}$$

For p,q $\in$ *Cyl* define the distance d(p,q) to be

$$d(p,q) = inf_\gamma\ l(\gamma)$$

where γ runs through the set of piecewise continuously differentiable curves on the cylinder *Cyl*.

1° Show that the distance defined above is a metric on *Cyl*.

2° Recall that the standard cylinder coordinates $\kappa:\mathbb{R}^2 \to Cyl$ are given by

$$\kappa(\theta,r) = (\cos\theta, \sin\theta, r)$$

Show that a continuously differentiable curve $\gamma:[a,b] \to Cyl$ can be factored $\gamma = \kappa\mu$ where $\mu:\mathbb{R} \to \mathbb{R}^2$ is a continuously differentiable curve.

3° With the notation above show that the length of the curve γ on *Cyl* is

$$l(\gamma) = l(\mu)$$

4° Show that the distance between points p and q on *Cyl* is given by

$$d(p,q) = \inf_{P,Q;\ p=\kappa(P), q=\kappa(Q)} d(P,Q)$$

5° Show that $\kappa:\mathbb{R}^2 \to Cyl$ is a local isometry, or more precisely that there exists a constant k>0 such that for every $P \in \mathbb{R}^2$ the disc with centre P and radius k is mapped bijectively onto the metric disc with centre $p = \kappa(P)$ and radius k.

6° Show that an affine line in $\mathbb{R}^2$ is mapped by κ onto a geodesic in *Cyl* and that all geodesics on *Cyl* have this form.

7° Let there be given points p and q on *Cyl* with $d(p,q) = l$. Show that we can find a geodesic curve $\gamma:\mathbb{R} \to Cyl$ with $\gamma(0) = p$ and $\gamma(l) = q$.

EXERCISE 3.4 We shall be concerned with the <u>torus</u> $S^1 \times S^1$ considered as metric space as described in exercise 1.1. The canonical map $\chi:\mathbb{R} \to S^1$ induces a map

$$\nu: \mathbb{R}^2 \to S^1 \times S^1, \quad \nu(x,y) = (\chi(x),\chi(y)) \qquad ; x,y \in \mathbb{R}$$

1° For points p and q on the torus $S^1 \times S^1$, show that

$$d(p,q) = \inf_{P,Q;\ p=\nu(P),\ q=\nu(Q)} d(P,Q)$$

2° Show that $\nu:\mathbb{R}^2 \to S^1 \times S^1$ is a local isometry, or more precisely that there exists a constant $k > 0$ such that for all $P \in \mathbb{R}^2$ the disc with centre P and radius k is mapped bijectively onto the metric disc with centre $p = \nu(P)$ and radius k.

3° Show that an affine line in $\mathbb{R}^2$ is mapped by ν onto a geodesic in $S^1 \times S^1$, and that all geodesics on $S^1 \times S^1$ have this form.

4° Let there be given points p and q on $S^1 \times S^1$ with $d(p,q) = d$. Show that we can find a geodesic curve $\gamma: \mathbb{R} \to S^1 \times S^1$ with $\gamma(0) = p$ and $\gamma(d) = q$.

EXERCISE 3.5 For real numbers a,b,c consider the determinant g given by

$$g = det \begin{bmatrix} 1 & cos\,a & cos\,b \\ cos\,a & 1 & cos\,c \\ cos\,b & cos\,c & 1 \end{bmatrix}$$

1° Show that

$$g = 4\, sin(\pi - p)\, sin(p-a)\, sin(p-b)\, sin(p-c) \qquad ; \; p = \tfrac{1}{2}(a+b+c)$$

2° For real numbers a, b, $c \in]0,\pi[$ show that $g \geq 0$ if and only if $a \leq b+c$, $b \leq a+c$, $c \leq a+b$ and $a+b+c \leq 2\pi$.

3° Show that the inequality $g > 0$ occurs precisely when the remaining four inequalities all are sharp.

4° Use this to give an alternative proof of the triangle inequality II.3.6

EXERCISE 5.1 For $d \in D^n$, $d \neq 0$, let d^* denote the mirror image of d in the unit sphere S^{n-1} and let κ_d denote reflection in the circle with centre d^* orthogonal to S^{n-1}. Let ρ_d denote Euclidean reflection in the hyperplane orthogonal to d and put $\tau_d = \kappa_d \rho_d$.

1° Extend the definition of the symbol τ_d to all of D^n by the convention $\tau_0 = \iota$, and show that any $\mu \in M\ddot{o}b(D^n)$ can in a unique way be written

$$\mu = \tau_d \sigma \qquad ; \; d \in D^n \,,\; \sigma \in O(E)$$

2° Use the previous result to establish a homeomorphism

$$M\ddot{o}b(D^n) \xrightarrow{\sim} D^n \times O(E)$$

The topology on $M\ddot{o}b(D^n)$ is uniform convergence on S^{n-1}. (See [Beardon] §3.7.)

EXERCISE 8.1 Show that $\sigma \in Gl_2(\mathbb{R})$ acts on H^2 as reflection in a geodesic if and only if $det\,\sigma < 0$ and $tr\,\sigma = 0$.

III HYPERBOLIC PLANE

In this chapter we shall make a detailed study of the hyperbolic plane H^2: classification of isometries and trigonometry. The even isometries of H^2 can be divided into three classes: rotations, translations and horolations. This division follows the natural occurrence of geodesics in three types of pencils. We shall keep track of the pencils by assigning a normal vector to an oriented geodesic. This will be dealt with in a specific model for H^2 referred to as the $sl_2(\mathbb{R})$-model. The big advantage of the $sl_2(\mathbb{R})$-model over say the Poincaré models is a very rich vector calculus [1].

III.1 VECTOR CALCULUS

In this section we shall present a very useful three-dimensional real vector space with an inner product of Sylvester type (−2,1) with a good deal more algebraic structure. The underlying real vector space is $sl_2(\mathbb{R})$, the space of real 2×2 matrices with trace 0

$$sl_2(\mathbb{R}) \;=\; \{\ \mathrm{R} \in M_2(\mathbb{R}) \mid tr\,\mathrm{R} = 0\ \}$$

The square of a matrix $\mathrm{R} \in sl_2(\mathbb{R})$ is a scalar matrix, in fact

1.1 $$\mathrm{R}^2 = -\,\iota\ det\,\mathrm{R} \qquad ;\ \mathrm{R} \in sl_2(\mathrm{R})$$

In view of the importance we offer direct verification of 1.1

$$\begin{bmatrix} a & b \\ c & -a \end{bmatrix}\begin{bmatrix} a & b \\ c & -a \end{bmatrix} = (a^2 + bc)\begin{bmatrix} 1 & 0 \\ 0 & 1 \end{bmatrix}$$

---

[1] An interpretation of the vector calculus on $sl_2(\mathbb{R})$ can be given in terms of Clifford algebra. See the remarks at the end of section III.1.

We shall be concerned with the symmetrical bilinear form

1.2 $$\langle R,S\rangle = -\tfrac{1}{2}\, tr(RS) \qquad ; \quad R,S \in sl_2(\mathbb{R})$$

The corresponding quadratic form is the determinant as follows from 1.1

1.3 $$det\, R = \langle R,R\rangle \qquad ; \quad R \in sl_2(\mathbb{R})$$

Otherwise expressed, the inner product 1.2 is obtained by polarisation of the determinant, compare I.1.2. Let us use this to "polarise" the identity 1.1

$$(R+S)^2 - R^2 - S^2 = -2\langle R,S\rangle\, \iota$$

An elementary calculation gives us the formula

1.4 $$RS + SR = -2\,\langle R,S\rangle\, \iota \qquad ; \quad R,S \in sl_2(\mathbb{R})$$

It is time to show that the Sylvester type of our inner product is $(-2,1)$. To this end we ask the reader to verify that the following three matrices form an orthonormal basis for $sl_2(\mathbb{R})$

$$\begin{bmatrix} 1 & 0 \\ 0 & -1 \end{bmatrix}, \begin{bmatrix} 0 & 1 \\ 1 & 0 \end{bmatrix}, \begin{bmatrix} 0 & 1 \\ -1 & 0 \end{bmatrix}$$

The determinants of these matrices are -1, -1, $+1$.

It is now time to introduce an important tri-linear form *vol* on $sl_2(\mathbb{R})$. The evaluation of the form *vol* on three vectors K,L,M is given by

1.5 $$vol(K,L,M) = -\tfrac{1}{2}\, tr(KLM) \qquad ; \quad K,L,M \in sl_2(\mathbb{R})$$

This is actually an <u>alternating form</u>, in the sense that the form annihilates any triple where two of the vectors are equal: for example if $K = M$ we get from $K^2 = -\iota\, det\, K$

$$tr(KLK) = tr(LK^2) = -\, tr(L\, det\, K) = -\, det\, K\, tr\, L = 0$$

The alternating form 1.5 is a volume form for $sl_2(\mathbb{R})$ in the sense that it takes the value 1 or -1 whenever K,L,M form an orthonormal basis. To see this, we ask the reader to verify that *vol* has evaluation 1 against the basis above.

Let us introduce a third opeation on $sl_2(\mathbb{R})$, the wedge product of two vectors K and L from $sl_2(\mathbb{R})$, given by the formula

1.6 $$K \wedge L = \tfrac{1}{2}(KL - LK) \qquad ; \quad K,L \in sl_2(\mathbb{R})$$

The three elements of structure are related through the formula

1.7 $$\langle K \wedge L, M \rangle = vol(K,L,M) \qquad ; \; K,L,M \in sl_2(\mathbb{R})$$

as follows from straightforward computation:

$$\langle K \wedge L, M \rangle = -\tfrac{1}{4}\, tr(KLM - LKM) = \tfrac{1}{2}(vol(K,L,M) - vol(L,K,M)) = vol(K,L,M)$$

We shall need one more formula relating the inner product and the wedge product

1.8 $$A \wedge (B \wedge C) = \langle A,C \rangle B - \langle A,B \rangle C \qquad ; \; A,B,C \in sl_2(\mathbb{R})$$

This formula can be deduced directly from the definitions using formula 1.4:

$$\begin{aligned} &4(\langle A,C \rangle B - \langle A,B \rangle C) = \\ &-(AC+CA)B - B(AC+CA) + (AB+BA)C + C(AB+BA) = \\ &ABC - ACB + CBA - BCA = 4\, A \wedge (B \wedge C) \end{aligned}$$

The inner product of two wedge products can be evaluated by the formula

1.9 $$\langle A \wedge B, C \wedge D \rangle = \langle A,C \rangle \langle B,D \rangle - \langle A,D \rangle \langle B,C \rangle \; ; \; A,B,C,D \in sl_2(\mathbb{R})$$

Let us verify this formula. From formula 1.8 we deduce that

$$\langle A \wedge B, C \wedge D \rangle = vol(A,B,C \wedge D) = vol(B,C \wedge D,A) = \langle B \wedge (C \wedge D), A \rangle$$

We find from 1.8 that $B \wedge (C \wedge D) = \langle B,D \rangle C - \langle B,C \rangle D$ and 1.9 follows.

Let us present a very useful volume formula. For two triples of vectors A_1,A_2,A_3 and B_1,B_2,B_3 in $sl_2(\mathbb{R})$ we have

1.10

$$\boxed{vol(A_1,A_2,A_3)\ vol(B_1,B_2,B_3) = det_{ij} <A_i,B_j>}$$

Proof Let us first fix A_1,A_2,A_3. Both sides of the formula are alternating in B_1,B_2,B_3 so we may assume that B_1,B_2,B_3 is a fixed positively oriented orthonormal basis. We can make a variation of A_1,A_2,A_3 and observe, that both sides are alternating in the A_1,A_2,A_3. Thus it suffices to treat the case where the A_1,A_2,A_3 and B_1,B_2,B_3 form the same orthonormal basis. In this case the result follows from the fact that $sl_2(\mathbb{R})$ has Sylvester type $(-2,1)$. □

Clifford algebra 1.11 Let us add some remarks for readers familiar with the formalism of Clifford algebra. For reference see [Deheuvels] or the opening chapter of [Gilbert,Murray], but it is our personal opinion that the arguments of Clifford algebra are best learned through concrete examples.

The real algebra $M_2(\mathbb{C})$ is easily seen to be generated by the linear subspace $i\,sl_2(\mathbb{R})$. The square of an element X of this subspace is a real scalar. In fact X^2 is a real quadratic form of $X \in i\,sl_2(\mathbb{R})$ as follows from 1.1:

$$(iS)^2 = det\ S \qquad ; \quad S \in sl_2(\mathbb{R})$$

Observe that $dim_{\mathbb{R}} M_2(\mathbb{C}) = 2^3$ and conclude from general principles that $M_2(\mathbb{C})$ is a concrete realisation of the Clifford algebra of the quadratic space $sl_2(\mathbb{R})$ of Sylvester type $(-2,1)$. A good many of the definitions and arguments of this section conform with the general notions of Clifford algebra. This is particularly true for the definition of the wedge product 1.6 and the arguments presented in section III.4.

III.2 PENCILS OF GEODESICS

Let us introduce hyperbolic 2–space H^2 as a definite sheet of the unit hyperboloid in the space $sl_2(\mathbb{R})$. A point of the unit hyperboloid is represented by a matrix of the form

$$A = \begin{bmatrix} a & b \\ c & -a \end{bmatrix} \qquad ; a,b,c \in \mathbb{R}, \quad det\,A = 1$$

Observe that the entry $c \neq 0$ which follows from $-a^2 - bc = 1$. Thus the sign of entry c distinguishes between the sheets of the hyperboloid. We let H^2 denote the sheet consisting of matrices A as above with $\boxed{c>0}$.

Normal vectors

Let us recall from II.4.7 that a geodesic curve $\gamma:\mathbb{R}\to H^2$ has the explicit parametrisation

$$\gamma(t) = A\ cosh\ t + T\ sinh\ t \qquad ; t \in \mathbb{R}$$

where $A = \gamma(0)$ and $T = \gamma'(0)$ is a vector orthogonal to A with $<T,T> = -1$. With this notation the <u>normal vector</u> to γ is

2.1
$$N = \gamma'(t) \wedge \gamma(t) \qquad ; t \in \mathbb{R}$$

The normal vector N is independent of time t as follows from the parametrisation of γ given above. Observe that N is orthogonal to A and T and and that the norm of N is -1 :

$$<T \wedge A, T \wedge A> = <T,T><A,A> - <A,T><A,T> = -1$$

The tangent vector $\gamma'(t)$ can be recovered from N by the formula

2.2
$$\gamma'(t) = \gamma(t) \wedge N \qquad ; t \in \mathbb{R}$$

This follows from 2.1 using the general formula 1.8. Let us remark that the normal vector N depends on the orientation of the geodesic curve γ. Thus the normal vector of an unoriented geodesic is determined up to a sign only.

Proposition 2.3 The complement of a geodesic h in H^2 has two connected components. Reflection τ along a normal vector N for h will fix h pointwise but interchange the two connected components of its complement.

Proof Let N be a normal vector for h. The function $X \mapsto \langle X,N\rangle$, $X \in H^2$, is zero along h but non-zero elsewhere. This divides the complement of h in H^2 into the open sets U and V where this function is positive, resp. negative. The sets U and V are interchanged by τ since $\langle X,N\rangle = -\langle \tau(X),N\rangle$ as follows from

$$\tau(X) = X + 2\langle X,N\rangle N \qquad ; \ X \in H^2$$

It remains to prove that U is connected. Consider two distinct points A and B of U. We can use II.4.6 to pick a unit tangent T to H^2 at A such that

$$B = A\, \cosh d + T \sinh d \qquad ; \ d = d(A,B)$$

Let us form the scalar product of N and $\gamma(t) = A \cosh t + T \sinh t$, $t \in \mathbb{R}$, to get

$$\langle \gamma(t),N\rangle = \cosh t\ (\langle A,N\rangle + \langle T,N\rangle \tanh t)$$

When $\langle T,N\rangle \geq 0$ we find that $\gamma(t) \in U$ for all $t \in \mathbb{R}$. When $\langle T,N\rangle < 0$ we can use the standard properties of the function $\tanh t$ to find a number $r \in \mathbb{R}$ such that $\langle \gamma(t),N\rangle < 0$ for all $t > r$ and $\langle \gamma(t),N\rangle < 0$ for all $t < r$. It follows that $\gamma(t) \in U$ for all $t < r$ and $\gamma(t) \in V$ for all $t > r$. This shows that the geodesic arc [A,B] is entirely contained in U. □

The two connected components of the complement of h are called the <u>sides</u> of h. Once the geodesic h is oriented by a normal vector N, we can distinguish between the two sides of h: a geodesic curve through a point $A \in h$ with tangent vector N at A passes from the <u>negative side</u> of h to the <u>positive side</u> of h (warning: the function $X \mapsto \langle X,N\rangle$ takes negative values on the positive side of h).

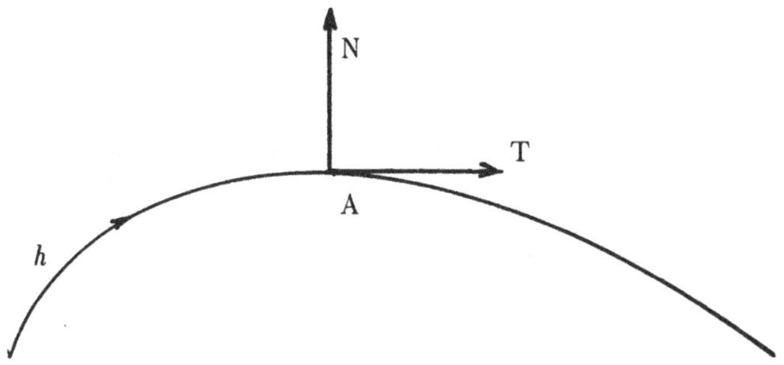

Oriented angles Let us recall that the tangent space of H^2 at a point A is the space $T_A(H^2)$ of vectors in $sl_2(\mathbb{R})$ orthogonal to the vector A. We say that a pair of tangent vectors X,Y are positively oriented if $vol(A,X,Y) > 0$. Given a unit tangent vector $S \in T_A(H^2)$, then S and $S \wedge A$ form a positively oriented orthonormal basis for $T_A(H^2)$ as follows from 1.7 and 1.9. Given two unit tangent vectors S and T to H^2 at the point $A \in H^2$, the oriented angle $\angle_{or}(S,T)$ between S and T is defined by

2.4 $$T = S \cos \angle_{or}(S,T) + S \wedge A \sin \angle_{or}(S,T) \qquad ; \angle_{or}(S,T) \in \mathbb{R}/2\pi\mathbb{Z}$$

In accordance with Euclidean geometry we shall make use of a second kind of angle, namely the interior angle $\angle_{int}A \in]0,2\pi[$ of a polygon Δ at the vertex A.

Intersecting geodesics We shall investigate the intersection between two oriented geodesics h and k through the point $A \in H^2$. Let us introduce the directional angle α from h to k as indicated below

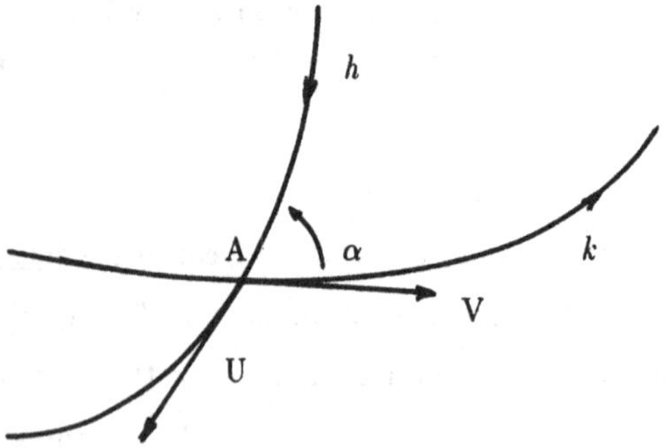

$$\alpha = \angle_{or}(V,-U)$$

$$-U = V \cos\alpha + V \wedge A \sin\alpha$$

Using 1.7 and 1.8 we deduce that $\langle U,V \rangle = \cos\alpha$ and $U \wedge V = A \sin\alpha$. For the corresponding normal vectors $H = U \wedge A$ and $K = V \wedge A$ we deduce from 1.8 that

2.5 $$\langle H,K \rangle = \cos \alpha \quad , \quad H \wedge K = A \sin \alpha$$

In particular, we find that $|\langle H,K \rangle| < 1$. Conversely

Proposition 2.6 Two distinct geodesics h and k in H^2 intersect if and only if their normal vectors H and K satisfy $|\langle H,K\rangle| < 1$.

Proof Suppose that the normal vectors H and K satisfy $|\langle H,K\rangle| < 1$. Consider the discriminant Δ for the plane E spanned by H and K

$$\Delta = \langle H,H\rangle\langle K,K\rangle - \langle H,K\rangle^2 = 1 - \langle H,K\rangle^2$$

Observe that $\Delta > 0$ and conclude from the discriminant lemma I.6.3 that E has Sylvester type $(-2,0)$. It follows that the line $E^{\perp}$ has type $(0,1)$ and that the point A of intersection between H^2 and $E^{\perp}$ lies on h and k. The converse implication follows from 2.5. □

We say that geodesics h and k are <u>perpendicular</u> if they meet at a point A of H^2 at an angle $\frac{\pi}{2}$.

Corollary 2.7 Geodesics h and k in H^2 are perpendicular if and only if their normal vectors H and K satisfy $\langle H,K\rangle = 0$.

Proof Two geodesics which meet at a right angle will have $\langle H,K\rangle = 0$ as it follows from 2.5. Conversely, if $\langle H,K\rangle = 0$ the two geodesics will meet according to 2.6. The angle at which they meet must be a right angle, 2.5. □

Geodesics with a common perpendicular Let us investigate two geodesics h and k both perpendicular to the same geodesic l. The points of intersection with l will be denoted A and B respectively. Let L be the normal vector for l corresponding to the orientation of l from A towards B. We shall make use of the normal vector $H = A \wedge L$ for h and the normal vector $K = -B \wedge L$ for k, compare 2.2.

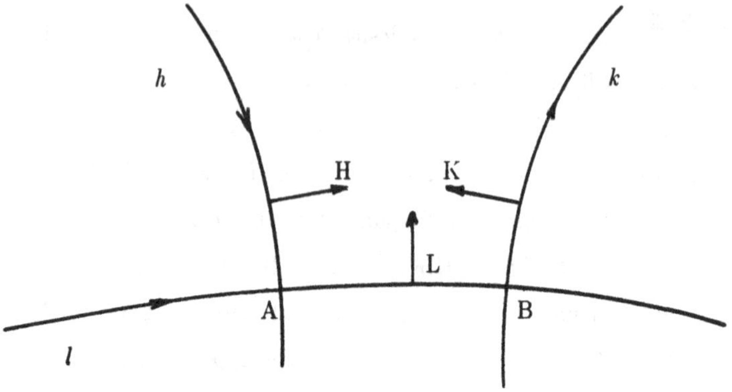

$$H = A \wedge L$$
$$K = -B \wedge L$$

Let us evaluate the inner product between the normal vectors H and K

$$<H,K> = <A,L><L,B> - <A,B><L,L> = <A,B>$$

This gives us the first of the following two formulas

2.8 $$<H,K> = \cosh d(A,B) \quad , \qquad vol(H,L,K) = -\sinh d(A,B)$$

In order to prove the second formula, observe first that $L \wedge K = B$ as follows from 1.8 using $K = L \wedge B$. Next, observe that $vol(H,L,K) = <H,B>$ to get that

$$B = A \cosh d(A,B) + H \sinh d(A,B)$$

This gives $<H,B> = <H,H> \sinh d(A,B) = -\sinh d(A,B)$ as required.

Theorem 2.9 Two distinct geodesics h and k are perpendicular to the same geodesic if and only if their normal vectors H and K satisfy $|<H,K>| > 1$. The common perpendicular geodesic is unique whenever it exists.

Proof Observe that H and K span a plane E, and form the discriminant

$$\Delta = <H,H><K,K> - <H,K>^2 = 1 - <H,K>^2$$

It follows that $\Delta<0$ if and only if $|<H,K>| > 1$. If $\Delta < 0$ we conclude from the discriminant lemma I.6.3 that E has Sylvester type $(-1,1)$. It follows that $E^{\perp}$ contains a vector L of norm -1. The geodesic l with normal vector L is perpendicular to h and k.

Conversely, consider a geodesic l perpendicular to both h and k. It follows from 3.3 that a normal vector L for l is orthogonal to both H and K, i.e. $L \in E^{\perp}$. This makes L unique up to a sign. □

Lines with a common end A geodesic h in H^2 spans a linear plane of type (−1,1). The two isotropic lines in this plane are called the ends of h. Let us observe that a normal vector H for h is orthogonal to the the two ends of h.

Proposition 2.10 Two distinct geodesics h and k in H^2 have a common end if and only if their normal vectors H and K satisfy $|<H,K>| = 1$.

Proof Observe that H and K span a plane E, and form the discriminant

$$\Delta = <H,H><K,K> - <H,K>^2 = 1 - <H,K>^2$$

If h and k have a common end S we must have $S = E^{\perp}$. In particular, we conclude that the line $E^{\perp}$ is isotropic. It follows from the discriminant lemma I.6.3 that the plane E has type (−1,0) and that $\Delta = 0$, i.e. $|<H,K>| = 1$. Conversely, if $|<H,K>| = 1$ we get $\Delta = 0$ and we get from I.6.3 that $E^{\perp}$ is an isotropic line, in fact a common end for h and k. □

Pencils of geodesics A set $\mathcal{P}$ of geodesics in H^2 is called a pencil if there exists a linear plane P in $sl_2(\mathbb{R})$ such that $\mathcal{P}$ equals the set of geodesics with normal vectors in P. Let us recall from I.6.4 that a linear plane P in $sl_2(\mathbb{R})$ is of Sylvester type (−2,0), (−1,1) or (−1,0). It follows by inspection that P is generated by its vectors of norm −1. In other words, the plane P is generated by the normal vectors of geodesics in the pencil it determines. Let us examine the three Sylvester types separately.

Type (−2,0). The line $P^{\perp}$ has type +1 and intersects H^2 at a point A, say. The pencil $\mathcal{P}$ is the set of geodesics through the point A.

Type (−1,1). The line $P^{\perp}$ has type −1 and is generated by the normal vector of a geodesic h. The pencil $\mathcal{P}$ is the set of all geodesics perpendicular to h.

Type $(-1,0)$. The line $S = P^{\perp}$ is isotropic. The pencil $\mathcal{P}$ is the set of geodesics with end S.

Let us examine the intersection of two pencils. It follows from the discussion above that two distinct pencils can't have more than one geodesic in common. The following is a list of true statements about the intersection of pencils.

2.11 Through two given points A and B passes a unique geodesic h.

2.12 Through a given point A passes a unique geodesic perpendicular to a given geodesic l.

2.13 Through a point A passes a unique geodesic with a given end.

2.14 There is a unique geodesic with a given end S perpendicular to a given geodesic h provided the ends of h are different from S.

2.15 There is a unique geodesic with given ends R and S.

Proof Let A be a point of H^2 and $\mathcal{A}$ the pencil of geodesics through A. Observe that the plane $A^{\perp}$ has type $(-2,0)$ and conclude that the intersection between $A^{\perp}$ and any other plane P is a line of type -1. This proves the first three statements.

To prove 2.14, pick a normal vector N to h and observe that $N^{\perp}$ has type $(-1,1)$ while $S^{\perp}$ has type $(-1,0)$. It follows that the intersection of these two planes has type -1 or 0. Type 0 can be ruled out since S is not an end of h.

To prove 2.15, observe that the planes $R^{\perp}$ and $S^{\perp}$ have type $(-1,0)$. The intersection is a line of type -1 or 0. Use the fact that our form is non-singular to rule out 0. □

III.3 CLASSIFICATION OF ISOMETRIES

An isometry of the hyperbolic plane H^2 is the product of at most three reflections in geodesics in H^2 as we have seen in II.4. In particular we can write an even reflection as a product $\alpha\beta$ of reflections in geodesics a and b. A finer classification will depend on the relative position of the two geodesics a and b. In other words, the classification of even isometries is tied to the three types of pencils of geodesics. The classification of odd isometries turns out to be simpler. An odd isometry is a glide reflection, i.e. the composite of a translation along a geodesic and a reflection in the same geodesic.

Theorem on three reflections 3.1 Let α,β,γ be reflections in three geodesics a, b, c belonging to the same pencil $\mathcal{P}$. The product $\alpha\beta\gamma$ is itself a reflection in a geodesic from the pencil $\mathcal{P}$.

Proof Let $S \neq 0$ be a vector orthogonal to the plane generated by the normal vectors of the geodesics from $\mathcal{P}$. Let us show that a Lorentz transformation σ of $sl_2(\mathbb{R})$ which fixes S is the product of one or two reflections in geodesics from $\mathcal{P}$. Note that σ induces an orthogonal transformation of the plane $P = S^{\perp}$. When $<S,S> > 0$, the plane P has type $(-2,0)$ and the result follows from I.4.5. When $<S,S> = 0$, the plane P is parabolic and the result follows from I.5.3 and I.6.16. When $<S,S> < 0$, the plane P is hyperbolic and the result follows from I.6.11.

Observe that the vector S is fixed by the three reflections α, β and γ and conclude that $\alpha\beta\gamma$ can be written as a product of at most two reflections. On the basis of parity II.4.10, we conclude that $\alpha\beta\gamma$ is a reflection. □

The theorem on three reflections has an application to the geometry of a triangle ΔABC. Let m_a, m_b, m_c be the perpendicular bisectors for AB, BC and CA, compare the proof of II.4.8.

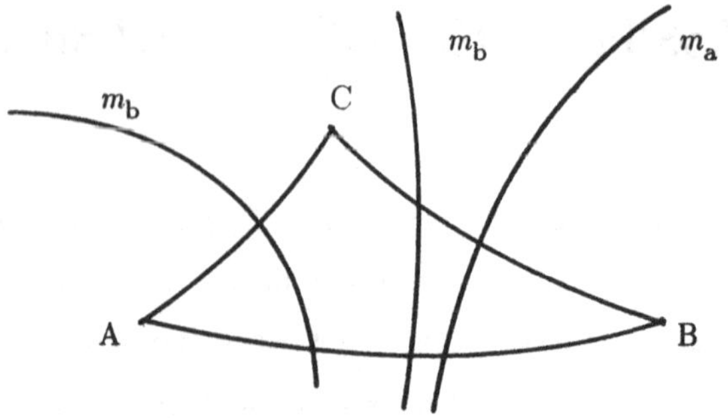

Corollary 3.2 The perpendicular bisectors m_a, m_b, m_c for triangle ΔABC lie in a pencil.

Proof Let $\mathcal{P}$ denote the pencil containing m_b and m_a and let $a \in \mathcal{P}$ be the geodesic through A. Reflections in the geodesics m_a, m_b, a are denoted μ_a, μ_b and α. It follows from 3.1 that $\mu_a\mu_b\alpha$ is a reflection in a geodesic $h \in \mathcal{P}$. Observe that $\mu_a\mu_b\alpha$ transforms A into B and conclude that $h = m_c$. Thus $m_c \in \mathcal{P}$. □

Horolations Let S be an isotropic line in $sl_2(\mathbb{R})$. A <u>horolation</u> with centre S is an isometry of the form $\alpha\beta$ where α and β are reflections in geodesics a and b with end S. The set of horolations with centre S form a group, 3.1.

Corollary 3.3 The group of horolations with centre S is abelian.

Proof For three reflections ρ,σ,τ in geodesics with end S we have $\rho\sigma\tau = \tau\sigma\rho$, as follows from $\rho\sigma\tau = (\rho\sigma\tau)^{-1} = \tau^{-1}\sigma^{-1}\rho^{-1}$. For four such reflections, we get that

$$(\alpha\beta)(\gamma\delta) = (\alpha\beta\gamma)\delta = (\gamma\beta\alpha)\delta = \gamma(\beta\alpha\delta) = \gamma(\delta\alpha\beta) = (\gamma\delta)(\alpha\beta)$$

and the result follows. □

A closer look at the proof of 3.1 reveals that the group of horolations with centre S is isomorphic to the additive group of $\mathbb{R}$, see also IV.2 The orbits on H^2 for the group of horolations with centre S are called <u>horocycles</u> with centre S.

Corollary 3.4 Two distinct points A and B belong to the same horocycle with centre S if and only if the perpendicular bisector m for A and B has end S.

Proof Suppose that the perpendicular bisector m for A and B has end S. Let μ be a reflection in m and let α be a reflection in the geodesic through A with end S. With this notation, the horolation $\mu\alpha$ transforms A into B. Conversely, let σ be a horolation with $\sigma(\mathrm{A}) = \mathrm{B}$. It follows from the theorem on three reflections that $\sigma\alpha$ is a reflection in a geodesic n with end S. Note that $\sigma\alpha(\mathrm{A}) = \mathrm{B}$ and conclude that n is the perpendicular bisector for A and B. □

Proposition 3.5 The group of horolations with centre S acts on the set of geodesics with end S in a simply transitive manner.

Proof Let a and b be geodesics with end S and let M and N be normal vectors for a and b respectively. The linear automorphism σ of the plane $\mathrm{P} = \mathrm{S}^{\perp}$ with $\sigma(\mathrm{S}) = \mathrm{S}$ and $\sigma(\mathrm{M}) = \mathrm{N}$ is orthogonal. It follows from I.5.3 that σ can be written as a product of one or two reflections. Upon replacing N by $-\mathrm{N}$ we can assume that σ is the product of two reflections. It is immediate to extend σ to a horolation with centre S.

Let us assume that the horolation σ fixes the geodesic a. For $\mathrm{A} \in a$ we must have $\sigma(\mathrm{A}) = \mathrm{A}$: otherwise, the perpendicular bisector for A and $\sigma(\mathrm{A})$ has end S, contradicting the fact that two perpendicular lines can't have the same end. □

Corollary 3.6 A horocycle $\mathcal{H}$ with centre S and a geodesic a with end S intersect in precisely one point.

Proof If $\mathcal{H}$ and a have two distinct points A and B in common, then the perpendicular bisector m for A,B has end S, contradicting the fact that two perpendicular geodesics can't have a common end. Let us choose a point $\mathrm{B} \in \mathcal{H}$ and a horolation σ which transforms the geodesic b through B with end S into the geodesic a. It follows that $\sigma(\mathrm{B})$ is a point of intersection between $\mathcal{H}$ and a. □

We shall present the remaining two pencils leaving the detailed treatment to the reader. This can be done by algebra as above, but elementary geometry may be applied as well. See also II.8.

Rotations By a rotation around a point A of H^2 we understand an isometry of the form $\alpha\beta$ where α and β are reflections in geodesics a and b passing through A. The rotations form a group as it follows from the theorem on three reflections. The orbits for the action of the rotation group on H^2 are the circles with centre A. The group of rotations around A is isomorphic to $\mathbb{R}/2\pi\mathbb{Z}$: The rotation angle θ of a rotation σ is given by

3.7
$$\theta = \angle_{\text{or}}(\mathbf{t},\sigma(\mathbf{t})) \qquad ; \mathbf{t} \in T_{\mathrm{A}}(H^2)\,,\ \langle\mathbf{t},\mathbf{t}\rangle = -1$$

The rotation with angle π is called a half-turn or symmetry with respect to A. A half-turn can be expressed $\kappa\lambda$ where κ and λ are reflections in a pair of perpendicular geodesics through A.

Translations Let k denote a geodesic in H^2. By a translation along k we understand an isometry of the form $\alpha\beta$ where α and β are reflections in geodesics a and b perpendicular to k. The translations along k form a group as follows from the theorem on three reflections 3.1. The group of translations along k acts on k in a simply transitive manner. The orbits for the group of translations along k are called hypercycles .

For a translation σ along k the geodesic k is called the translation axis of σ. The translation length T of σ is given by $\mathrm{T} = d(\mathrm{A},\sigma(\mathrm{A}))$, $\mathrm{A} \in k$. It is worth noticing that the translation axis has a natural orientation.

If we let κ denote reflection in k then we can decompose the translation above as $\alpha\beta = (\alpha\kappa)(\kappa\beta)$. In other words, a translation along k can be written as the composite of two half-turns with respect to points of k. Conversely, the composite of two half-turns around points of k is a translation along k.

Odd isometries The odd isometries of H^2 are easy to classify by virtue of the following proposition.

Proposition 3.8 An odd isometry ϕ of H^2 is a glide reflection , i.e. of the form $\phi = \tau\kappa$ where τ is translation along a geodesic k and κ is a reflection in k.

Proof Let us fix $A \in H^2$ and focus on the point $\phi(A)$. If $A = \phi(A)$, we find that ϕ is a reflection in a geodesic through A. If $A \neq \phi(A)$, let m denote the geodesic through A and $\phi(A)$ and let l denote the perpendicular bisector for A and $\phi(A)$. The point of intersection between m and l is denoted M while μ and λ denote reflections in these geodesics. Observe that $\lambda\phi$ is even and fixes the point A. Thus $\lambda\phi$ is a rotation around A, and we can use the theorem on three reflections to write $\lambda\phi = \mu\nu$ where ν is a reflection in a geodesic n through A. Let us introduce reflection κ in the geodesic k through M perpendicular to n and write

$$\phi = \lambda(\mu\nu) = (\lambda\mu)(\nu\kappa)\kappa$$

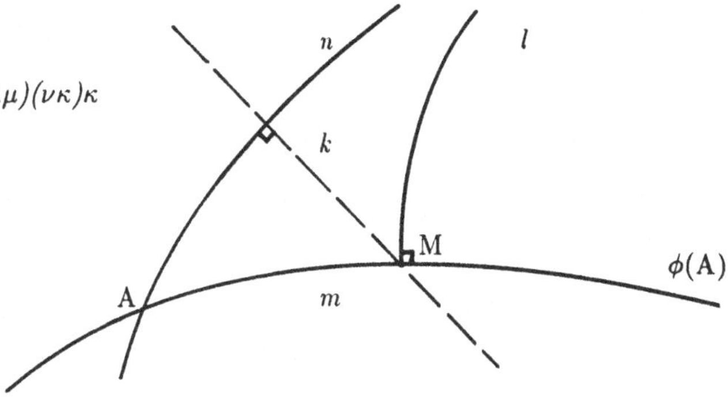

Note that $\lambda\mu$ and $\nu\kappa$ are half-turns around points of k. It follows that the isometry $\tau = (\lambda\mu)(\nu\kappa)$ is a translation along k. □

III.4 THE SPECIAL LINEAR GROUP

The group of isometries of the hyperbolic plane H^2 is the Lorentz group of the ambient space $sl_2(\mathbb{R})$. We shall use this to identify the group of isometries of H^2 with the group $PGl_2(\mathbb{R})$ given by

4.1
$$PGl_2(\mathbb{R}) = Gl_2(\mathbb{R})/\mathbb{R}^*$$

Let us first bring the whole group $Gl_2(\mathbb{R})$ to act on $sl_2(\mathbb{R})$. To this end we shall make use of the sign of the determinant

4.2
$$sign: Gl_2(\mathbb{R}) \to \{-1,1\}$$

With this notation, the group $Gl_2(\mathbb{R})$ acts on $sl_2(\mathbb{R})$ through the formula

4.3
$$\sigma\, X = sign(\sigma)\, \sigma X \sigma^{-1} \qquad ;\ \sigma \in Gl_2(\mathbb{R}),\ X \in sl_2(\mathbb{R})$$

It is clear that σ acts as an orthogonal transformation of $sl_2(\mathbb{R})$. We proceed to verify that this is a Lorentz transformation. To this end we ask the reader to verify the following formula

$$< \sigma \begin{bmatrix} 0 & -1 \\ 1 & 0 \end{bmatrix} \sigma^{-1}, \begin{bmatrix} 0 & -1 \\ 1 & 0 \end{bmatrix} > \; = \tfrac{1}{2}(det\, \sigma)^{-1}(a^2 + b^2 + c^2 + d^2)$$

where a,b,c,d are the entries of σ. It follows from this and I.6.4 that conjugation with σ interchanges the two sheets of the hyperbola if and only if $sign(\sigma) < 0$. It follows that the formula 4.3 defines a Lorentz transformation of $sl_2(\mathbb{R})$.

Theorem 4.4 The action of $Gl_2(\mathbb{R})$ on $sl_2(\mathbb{R})$ induces an isomorphism
$$PGl_2(\mathbb{R}) \xrightarrow{\sim} Lor(sl_2(\mathbb{R}))$$

Proof The main point is to realise any Lorentz transformation of $sl_2(\mathbb{R})$ by an element of $Gl_2(\mathbb{R})$. Let us start with reflection τ_K along a vector $K \in sl_2(\mathbb{R})$ of norm -1. We shall involve the "strange formula" 1.4

$$KX + XK = -2<K,X>\iota \qquad ; X \in sl_2(\mathbb{R})$$

Multiply this identity by the matrix $K = K^{-1}$ and deduce the formula

4.5
$$\boxed{-KXK^{-1} = X + 2<K,X>K} \qquad ; \ X \in sl_2(\mathbb{R})$$

From the fact I.6.11, that $Lor(sl_2(\mathbb{R}))$ is generated by reflections, it follows that the map $Gl_2(\mathbb{R}) \to Lor(sl_2(\mathbb{R}))$ is surjective.

Let us show that the kernel of the map $Gl_2(\mathbb{R}) \to Lor(sl_2(\mathbb{R}))$ consists of scalar matrices. Consider a $\sigma \in Gl_2(\mathbb{R})$ with $det(\sigma) > 0$ which acts trivially on $sl_2(\mathbb{R})$. Observe that the vector space $M_2(\mathbb{R})$ is generated by ι and $sl_2(\mathbb{R})$ and conclude that σ commutes with all matrices $X \in M_2(\mathbb{R})$. From this it follows that σ is a scalar matrix. Next, consider a $\sigma \in Gl_2(\mathbb{R})$ with $det(\sigma) < 0$ which acts trivially on $sl_2(\mathbb{R})$. This means that σ anti-commutes with $sl_2(\mathbb{R})$. We ask the reader to verify the general fact that a matrix $\sigma \in M_2(\mathbb{R})$ which anti-commutes with $sl_2(\mathbb{R})$ is the zero matrix. □

Corollary 4.6 The action of $Sl_2(\mathbb{R})$ on $sl_2(\mathbb{R})$ identifies the group $PSl_2(\mathbb{R})$ with the special Lorentz group $Lor^+(sl_2(\mathbb{R}))$.

Proof We shall show that the action of a $\sigma \in Gl_2(\mathbb{R})$ defined by 4.3 is an even Lorentz transformation on $sl_2(\mathbb{R})$ if and only if $det(\sigma) > 0$. To this end, we pick a decomposition of the action of σ on $sl_2(\mathbb{R})$ into a product $\tau_1 ... \tau_s$ of reflections along vectors $K_1,...,K_s$ of norm -1. It follows from 4.5 that the matrices σ and $K_1...K_s \in Gl_2(\mathbb{R})$ have the same action on $sl_2(\mathbb{R})$. Thus we can use 4.4 to pick a scalar $r \in \mathbb{R}^*$ such that

$$\sigma = r K_1 ... K_s$$

Calculate the determinant of both sides to get that $det(\sigma) = r^2(-1)^s$. This shows that $det(\sigma) > 0$ if and only if s is even. □

Theorem 4.7 Let the action of $\sigma \in Sl_2(\mathbb{R})$ on H^2 be decomposed as $\alpha\beta$ where α and β are reflections in geodesics h and k. If H and K denote normal vectors for h and k, then

$$tr^2(\sigma) = 4<\mathrm{H},\mathrm{K}>^2$$

Proof From the end of the proof of the previous proposition follows that we can write $\sigma = \epsilon \mathrm{HK}$, where $\epsilon = \pm 1$. This gives us

$$tr\,\sigma = \epsilon\, tr(\mathrm{HK}) = -2\epsilon<\mathrm{H},\mathrm{K}>$$

The result follows by taking the square of this relation. □

Let us summarise the discussion in a small table, compare also II.8.

4.8

Position of h and k	$tr^2(\sigma)$	Klein notation	geometric notation
intersecting	$[0,4[$	elliptic	rotation
common perpendicular	$]4,+\infty[$	hyperbolic	translation
common end	$[4]$	parabolic	horolation

For a general $\sigma \in Gl_2(\mathbb{R})$ we shall make use of the invariant $tr^2(\sigma) = \dfrac{(tr\,\sigma)^2}{det\,\sigma}$ introduced in I.9.14.

Proposition 4.9 Let $\sigma \in Gl_2(\mathbb{R})$ be a non-scalar matrix. The conjugacy class of σ in $PGl_2(\mathbb{R})$ is entirely determined by $tr^2(\sigma)$ and $sign\ det\sigma$.

Proof It is known from linear algebra that any conjugacy class in $Gl_2(\mathbb{R})$ contains a matrix from the following list

$$\begin{bmatrix} \lambda & 0 \\ 0 & \mu \end{bmatrix}, \begin{bmatrix} \alpha & -\beta \\ \beta & \alpha \end{bmatrix}, \begin{bmatrix} \lambda & 1 \\ 0 & \lambda \end{bmatrix}$$

The remaining details are left to the reader. □

III.5 TRIGONOMETRY

In the next four sections we shall study the trigonometry of the hyperbolic plane and apply this to a number of existence and classification problems.

Our treatment is based on the vector calculus from section III.1 and the concept of normal vectors from III.2. Let us use the convention that the *natural orientation of a closed polygon* is such that the polygon lies on the positive side of its edges. In other words, the corresponding normal vectors are *inward directed*, see the remarks made after 2.3. Let us start with the hyperbolic triangle ΔABC.

Cosine and sine relations 5.1

$$\boxed{\cosh a = \cosh b \cosh c - \sinh b \sinh c \cos \alpha}$$

$$\boxed{\frac{\sinh a}{\sin \alpha} = \frac{\sinh b}{\sin \beta} = \frac{\sinh c}{\sin \gamma}}$$

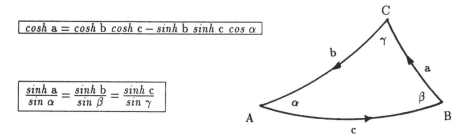

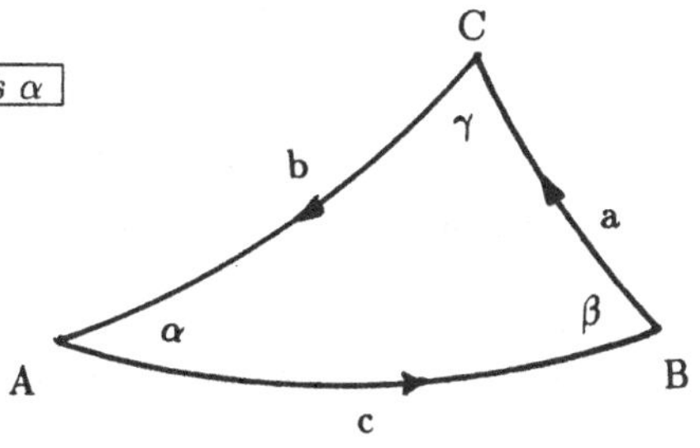

Proof Let us first prove the cosine relation. Recall from 2.5 that

$$B = A \cosh c + V \sinh c\ , \qquad C = A \cosh b - U \sinh b$$

Let us use this and $\langle U,V\rangle = \cos\alpha$ to calculate the inner product of B and C

$$\langle B,C\rangle = \cosh b \cosh c - \sinh b \sinh c \cos\alpha$$

Observe that $\langle B,C\rangle = \cosh a$ and the cosine relation follows.

In order to prove the sine realations we observe that $|vol(A,B,C)|$ is symmetrical in A,B,C. Let us evaluate this in an asymmetrical manner: with the notation from 2.5 we have

$$vol(A,B,C) = vol(A,U,V) \sinh b \sinh c$$

A few lines before 2.5 we found that $U \wedge V = A \sin \alpha$, thus

$$vol(A,B,C) = \sin \alpha \sinh b \sinh c = \frac{\sin \alpha}{\sinh a} \sinh a \sinh b \sinh c$$

and the result follows since $|\, vol(A,B,C)\, |$ is symmetrical in A,B,C. □

Alternative cosine relation 5.2

$$\cosh a = \frac{\cos\beta\ \cos\gamma + \cos\alpha}{\sin\beta\ \sin\gamma}$$

Proof Let H,K,L be the inward directed normal vectors of the edges of the triangle

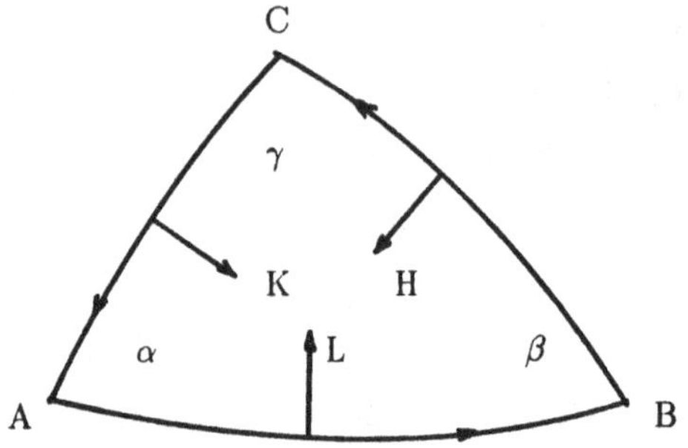

Formula 1.9 in combination with formula 2.5 gives us that

$$<H \wedge K, L \wedge H> = <K,H><H,L> - <H,H><K,L> = \cos\beta\ \cos\gamma + \cos\alpha$$

From the second formula in 2.5 we get that

$$H \wedge K = C \sin\gamma \ , \qquad L \wedge H = B \sin\beta$$

Combine this with the definition of hyperbolic distance to get that

$$<H \wedge K, L \wedge H> = <B,C> \sin\gamma\ \sin\beta = \cosh a\ \sin\gamma\ \sin\beta$$

and the alternative cosine relation follows. □

Corollary 5.3

The sum of the angles in ΔABC satisfies

$$\alpha + \beta + \gamma < \pi$$

Proof Let us assume that $\alpha \geq \beta$ and $\alpha \geq \gamma$. We ask the reader to use formula 5.2 to rewrite the inequality $\cosh a > 1$ in the form

$$\cos(\pi - \alpha) < \cos(\beta + \gamma)$$

When $\beta + \gamma \leq \pi$ we conclude that $\pi - \alpha > \beta + \gamma$ or $\alpha + \beta + \gamma < \pi$ as required. When $\pi \leq \beta + \gamma$ we can rewrite the inequality above as $\cos(\alpha+\pi) < \cos(\beta+\gamma)$; this gives $\alpha + \pi < \beta + \gamma$, which contradicts $\alpha \geq \beta$ and $\alpha \geq \gamma$. □

Theorem 5.4 Let $\alpha,\beta,\gamma \in]0,\pi[$ be real numbers with $\alpha+\beta+\gamma<\pi$, then there exists a triangle ΔABC in H^2 with $\angle A=\alpha$, $\angle B=\beta$, $\angle C=\gamma$.

Proof Let us rewrite the basic assumption $\beta+\gamma<\pi-\alpha$ and conclude that

$$\cos(\beta+\gamma) > \cos(\pi-\alpha)$$

Using the trigonometric addition formulas we find that

$$\sin\beta\ \sin\gamma < \cos\beta\ \cos\gamma + \cos\alpha$$

It follows that we can determine $a \in]0,+\infty[$ such that

$$\sin\beta\ \sin\gamma\ \cosh a = \cos\beta\ \cos\gamma + \cos\alpha$$

Let us now pick two points B and C with $d(B,C)=a$, and draw a geodesic l through B forming an angle β with BC and a geodesic k through C forming an angle γ with the geodesic h through BC. We let H,K,L denote inward directed normal vectors for these geodesics

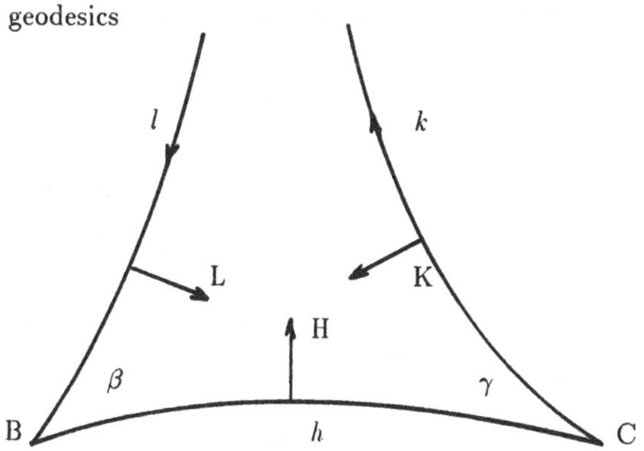

Proceeding as in the proof of 5.2 we find that

5.5
$$\cosh a\ \sin\beta\ \sin\gamma = \cos\beta\ \cos\gamma + \langle K,L\rangle$$

This gives us $\cos\alpha = \langle K,L\rangle$, which shows that $|\langle K,L\rangle| < 1$. We can use 2.6 to conclude that the geodesics l and k intersect. It follows from 5.2 that the angle of intersection is α. □

Let us investigate a quadrangle □ABCD with three right angles. Such a quadrangle is called a Lambert quadrangle after J.H.Lambert (1728–74). It is known since the early days of hyperbolic geometry that the fourth angle of a Lambert quadrangle is acute. We can deduce this from the second of the two trigonometric formulas to follow.

Lambert quadrangle 5.6

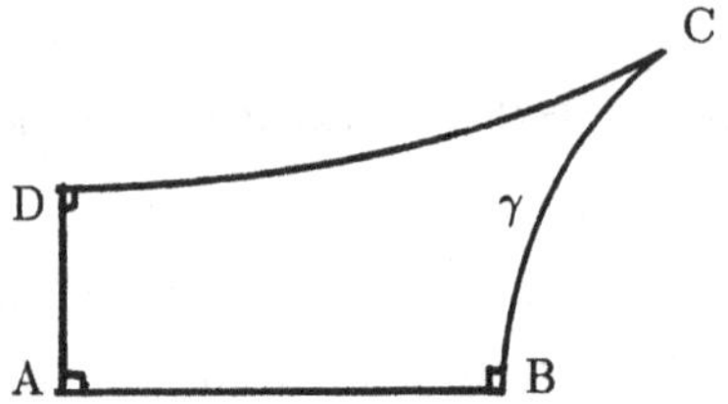

$$\cosh d(A,D) = \cosh d(B,C) \sin \gamma$$

$$\sinh d(A,B) \sinh d(D,A) = \cos \gamma$$

Proof Let H,K,L,M be inward directed normal vectors as indicated in the illustration below

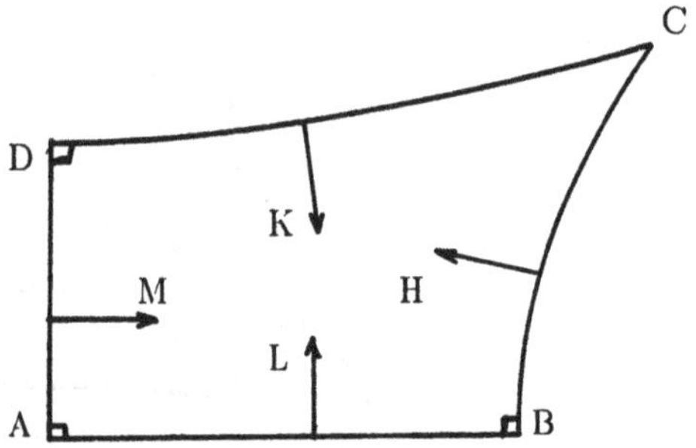

Let us substitute $\beta = \frac{\pi}{2}$ in formula 5.5 and deduce that

$$\cosh d(B,C) \sin \gamma = \langle K,L\rangle$$

Formula 2.8 gives us $\langle K,L\rangle = \cosh d(A,D)$, which takes care of the first formula. To prove the second formula, use 1.10 to get that

5.7 $$vol(K,M,L)\, vol(M,L,H) = det \begin{bmatrix} 0 & \langle K,L\rangle & \langle K,H\rangle \\ -1 & 0 & \langle M,H\rangle \\ 0 & -1 & 0 \end{bmatrix} = \langle K,H\rangle$$

and the result follows from formulas 2.8 and 2.5. □

III.6 ANGLE OF PARALLELISM

In this section we shall study the trigonometry of a triangle ΔABC in which the vertices B and C are ordinary points of H^2 while A is a point at infinity (an end). For the history of this problem and its relations to astronomy, the reader is referred to "Hyperbolic geometry : The first 150 years" [Milnor].

We are going to prove a relation for ΔABC which formally looks like the alternative cosine relation 5.2 of an ordinary triangle with $\angle A = \alpha = 0$.

6.1 $\cosh a \sin\beta \sin\gamma = \cos\beta \cos\gamma + 1$

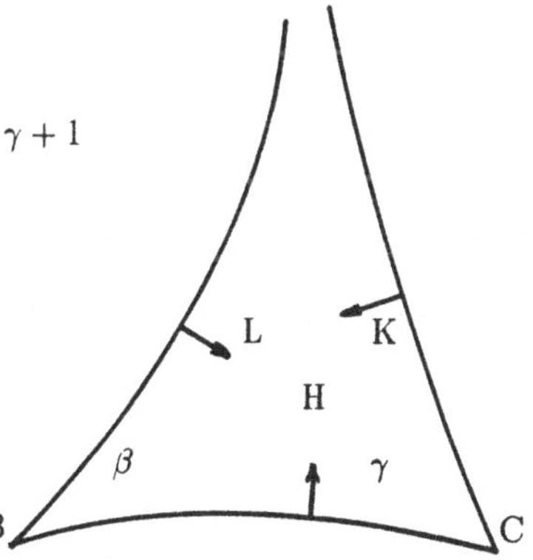

Proof Let us first observe that formula 5.5 remains valid in this context. It follows from 2.10 that $\langle K,L\rangle = \pm 1$. In order to decide between $+1$ and -1, observe that the left hand side of formula 5.5 is positive. □

In particular, if $\angle C$ is a right angle we find that

6.2
$$\cosh a \sin\beta = 1$$
$$\sinh a \tan\beta = 1$$
$$\tanh a \sec\beta = 1$$

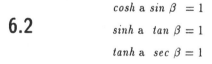

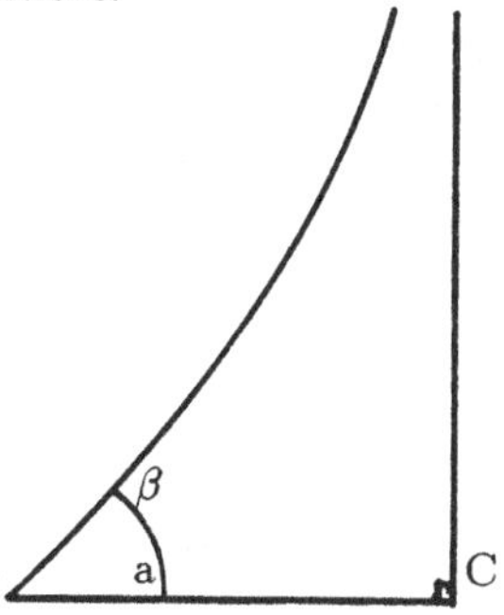

The last two formulas are elementary consequences of the first. In the third formula, we are using the old convention $\sec\beta = 1/\cos\beta$.

III.7 RIGHT ANGLED PENTAGONS

In this section we shall investigate a number of pentagons on which there are no analogues in Euclidean geometry. Let us start with the trigonometry of the right angled pentagon. The existence of such a thing will be demonstrated in 7.6.

7.1

$$\cosh a_5 = \sinh a_2 \, \sinh a_3$$
$$\cosh a_3 = \coth a_2 \, \coth a_4$$

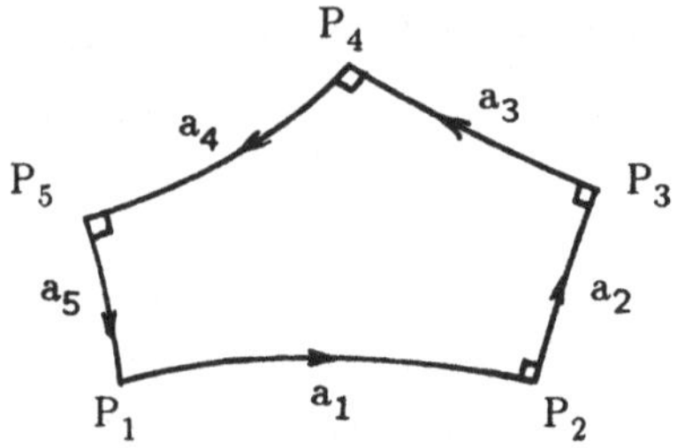

Proof We shall assume that the enumeration of the vertices is consistent with the natural orientation of the sides of the polygon, II.5. For $i = 1,\dots,5$ we let N_i be the inward directed normal vector for the edge P_iP_{i+1}. Let us at once record that

$$\langle N_i, N_{i+1} \rangle = 0 \qquad ; \; i = 1,\dots,5 \, , \; N_6 = N_1$$

From the general formula 1.10 (or the more specific formula 5.7), we get that

7.2

$$vol(N_1,N_2,N_3) \; vol(N_2,N_3,N_4) = \langle N_1,N_4 \rangle$$

Evaluate this by means of 2.8 to get the first of the two formulas in 7.1.

Let us now turn to the second formula, which we, by the way, intend to generalise beyond the case where $\angle P_1$ is a right angle. We shall first establish the following formula

7.3

$$vol(N_1,N_3,R) = vol(N_1,N_2,N_3) \; \langle N_2,R \rangle$$

Observe that this formula is linear in R and check the formula directly in the case where $R=N_1,N_2,N_3$. We ask the reader to use 1.10 to get that

$$vol(N_1,N_3,N_4) \; vol(N_3,N_4,N_5) = \langle N_1,N_5 \rangle + \langle N_1,N_3 \rangle \langle N_3,N_5 \rangle$$

Combine this formula with 7.3 with the specification $R = N_4$ to get that

7.4 $$vol(N_1,N_2,N_3)<N_2,N_4>vol(N_3,N_4,N_5) = <N_1,N_5> + <N_1,N_3><N_3,N_5>$$

Let us take advantage of formula 2.8 and deduce that

7.5 $$\sinh a_2 \sinh a_4 \cosh a_3 = <N_1,N_5> + \cosh a_2 \cosh a_4$$

In case $\angle P_1 = \frac{\pi}{2}$ we have $<N_1,N_5> = 0$, which concludes the proof. □

Proposition 7.6 Let $a_1 > 0$ and $a_2 > 0$ be real numbers. There exists a right angled pentagon with two consecutive sides of length a_1 and a_2 if and only if

$$\sinh a_1 \sinh a_2 > 1$$

Proof The condition is necessary as follows from the first formula in 7.1. To see that the condition is sufficient, start from a right angle $\angle P_1P_2P_3$ with $d(P_1,P_2) = a_1$ and $d(P_2,P_3) = a_2$. Draw the geodesic k_3 through P_3 perpendicular to P_2P_3 and the geodesic k_5 through P_1 perpendicular to P_1P_2

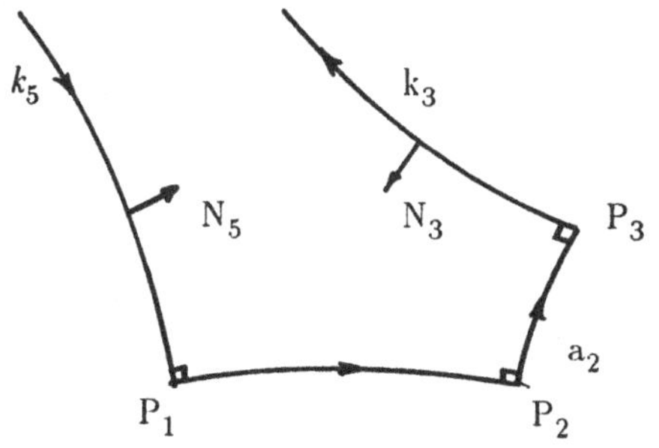

Formula 5.7, used in the analysis of Lambert quadrangles, at once gives that

$$\sinh a_1 \sinh a_2 = <N_3,N_5>$$

It follows that $<N_3,N_5> > 1$ which implies that k_3 and k_5 have a common perpendicular geodesic, compare 2.9. □

III.8 RIGHT ANGLED HEXAGONS

Let us now turn to the right angled hexagon. The method we have used in the analysis of the right angled pentagon applies to the hexagon as well. We shall show that right angled hexagons are classified by the length of three alternating sides a_2, a_4, a_6.

For geometric applications see [Fathi, Laudenbach, Poenaru], in particular, "exposé 3 et appendice aux exposé 8". The point is that the basic building blocks of hyperbolic geometry, hyperbolic pants, can be sewed out of two isometric right angled hexagons, see VII.2.6.

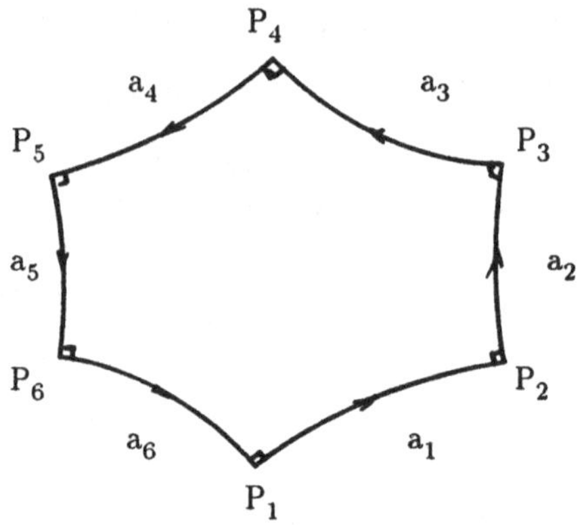

Sine relations 8.1

$$\frac{\sinh a_2}{\sinh a_5} = \frac{\sinh a_6}{\sinh a_3} = \frac{\sinh a_4}{\sinh a_1}$$

Proof With a notation similar to the one applied in the previous section we get from formula 7.2 that

$$vol(N_4,N_5,N_6)\ vol(N_5,N_6,N_1) = \langle N_4,N_1\rangle$$
$$vol(N_1,N_2,N_3)\ vol(N_2,N_3,N_4) = \langle N_1,N_4\rangle$$

Using 2.8 we deduce equality between the two first ratios. The remaining formulas follow by symmetry. □

Cosine relations 8.2

$$\sinh a_2 \sinh a_4 \cosh a_3 = \cosh a_6 + \cosh a_2 \cosh a_4$$

Proof Let us observe that formula 7.5 remains valid in this context. Thus the result follows after the specification $<N_1,N_5> = \cosh a_6$, compare 2.8. □

Classification theorem 8.3 Let a_2,a_4,a_6 be three strictly positive real numbers. Then, there exists a right angled hexagon in H^2 such that three alternating sides have lengths a_2,a_4,a_6. The hexagon is unique up to isometry.

Proof Let the number $a_3 > 0$ be determined by the cosine relation 8.2, and start the construction of the hexagon on the basis of the numbers a_2,a_3,a_4

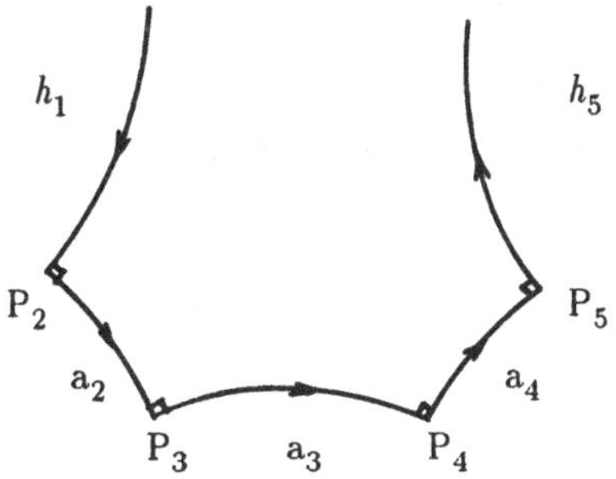

Let N_1 and N_5 denote the inward directed normal vectors for h_1 and h_5 and deduce from the relations 8.2 and 7.5 that

$$<N_1,N_5> = \cosh a_6$$

We conclude from this that the geodesics h_1 and h_5 have a common perpendicular h_6 say. We leave the remaining details to the reader. □

I would like to conclude this chapter with a reference to [Fenchel] which contains a wealth of trigonometric formulas. In particular, [Fenchel] p.84 gives information on "improper" right angled hexagons.

III.9 FROM POINCARÉ TO KLEIN

In this section we shall describe an isometry from the Poincaré half-plane H^2 to the $sl_2(\mathbb{R})$-model of the hyperbolic plane discussed in this chapter. Let us recall that the metric on the Poincaré half-plane is given by

9.1
$$\cosh d(z,w) = 1 + \frac{|z-w|^2}{2Im[z]Im[w]} \qquad ; \ z,w \in H^2$$

To a point $z \in H^2$ we assign the matrix $F(z) \in sl_2(\mathbb{R})$ given by

9.2
$$F(z) = \frac{1}{y}\begin{bmatrix} x & -|z|^2 \\ 1 & -x \end{bmatrix} \qquad ; \ z = x + iy \in H^2$$

We ask the reader to verify that F is an isometry of hyperbolic planes. We also leave it to the reader to show that F is $Gl_2(\mathbb{R})$-equivariant in the sense that it relates the action from II.8.4 to the action of III.4.3 as follows

9.3
$$F(\sigma(z)) = sign(\sigma)\ \sigma F(z)\sigma^{-1} \qquad ; \ z \in H^2, \ \ \sigma \in Gl_2(\mathbb{R})$$

Proposition 9.4 The transformation F from the Poincaré half-plane H^2 to the $sl_2(\mathbb{R})$-model preserves oriented angles.

Proof Let z vary along a smooth curve in H^2 and differentiate F(z) to get

$$F(z)' = -\frac{y'}{y^2}\begin{bmatrix} x & -|z|^2 \\ 1 & -x \end{bmatrix} + \frac{1}{y}\begin{bmatrix} x' & -2(xx'+yy') \\ 0 & -x' \end{bmatrix}$$

Let us rewrite this in terms of the tangent map of F

$$T_zF\binom{u}{v} = \frac{1}{y}\begin{bmatrix} 1 & -2x \\ 0 & -1 \end{bmatrix}u - \frac{1}{y^2}\begin{bmatrix} x & -x^2+y^2 \\ 1 & -x \end{bmatrix}v \qquad ; \ u,v \in \mathbb{R}$$

The norm of the tangent vector $T_zF\binom{u}{v}$ at the point F(z) is given by

9.5 $$< T_zF\binom{u}{v}, T_zF\binom{u}{v} > \; = \; det\; T_zF\binom{u}{v} = -\frac{1}{y^2}(u^2+v^2) \qquad ;\ u,v \in \mathbb{R}$$

From this follows that $T_zF\binom{u}{v}$ preserves angles. □

It is also worth making the calculation

9.6 $$vol(\,F(z), T_zF\binom{r}{s}, T_zF\binom{u}{v}) = \frac{1}{y^2}(rv-su) \qquad ;\ u,v,r,s \in \mathbb{R}$$

Let us end this section with some remarks for readers familiar with differential forms. The formula 9.6 may be written

9.7 $$F^*vol = \frac{1}{y^2}\, dx \wedge dy$$

Formula 9.5 gives us the Riemannian metric on the Poincaré half-plane

9.8 $$\frac{1}{y^2}(dx^2 + dy^2) \qquad x+iy \in \mathbb{C},\ y>0$$

Hyperbolic area

Let us define the hyperbolic area of an ordinary hyperbolic polygon Δ on n vertices by the formula

9.9 $$Area\;\Delta = (n-2)\pi - \sum_{P} \angle_{int}P$$

where the sum is over all vertices P of Δ. We ask the reader to check that the area is additive under simple subdivision of polygons. Once this is done, it follows from 5.3 that $Area\;\Delta > 0$. In the Poincaré half-plane model the area can be calculated from the area-form 9.7

9.10 $$Area\;\Delta = \int_{\Delta} \frac{1}{y^2}\, dx\, dy$$

There are hints on how to verify this in the exercises. In general, formula 9.8 may be taken as the definition of area.

III.10 LIGHT CONE

Let us investigate the space PC of isotropic lines in $sl_2(\mathbb{R})$. This means that we are investigating lines generated by matrices of determinant 0. To a non-zero vector $\mathbf{e} \in \mathbb{R}^2$ we shall assign the matrix

10.1
$$L(\mathbf{e}) = \begin{bmatrix} zw & -z^2 \\ w^2 & -zw \end{bmatrix} \quad ; \ \mathbf{e} = \begin{bmatrix} z \\ w \end{bmatrix} \in \mathbb{R}^2$$

The matrix $L(\mathbf{e})$ has determinant 0, more precisely the vector $\mathbf{e}$ belongs to the kernel of the linear map with matrix $L(\mathbf{e})$. From this it easily follows that our construction defines a bijection $L: \hat{\mathbb{R}} \xrightarrow{\sim} PC(sl_2(\mathbb{R}))$. This map obeys the transformation rule (see I.9.2 for notation)

10.2
$$L(\sigma(\mathbf{e})) = \sigma L(\mathbf{e}) \sigma^{\check{}} \quad ; \ \sigma \in Gl_2(\mathbb{R}), \ \mathbf{e} \in \mathbb{R}^2$$

Proof Let us first perform the decomposition

$$L(\mathbf{e}) = \begin{bmatrix} z \\ w \end{bmatrix} \begin{bmatrix} z \\ w \end{bmatrix}^T \begin{bmatrix} 0 & -1 \\ 1 & 0 \end{bmatrix} \quad ; \ \mathbf{e} = \begin{bmatrix} z \\ w \end{bmatrix} \in \mathbb{R}^2$$

Next, observe the matrix formula

$$\begin{bmatrix} a & b \\ c & d \end{bmatrix}^T \begin{bmatrix} 0 & -1 \\ 1 & 0 \end{bmatrix} = \begin{bmatrix} 0 & -1 \\ 1 & 0 \end{bmatrix} \begin{bmatrix} a & b \\ c & d \end{bmatrix}^{\check{}}$$

and the transformation rule follows rather easily. □

To an ordered pair (Z,U) of distinct points of $\hat{\mathbb{R}}$ we assign the vector $L(Z,U) \in sl_2(\mathbb{R})$ of determinant -1 given by

10.3 $$L(Z,U) = \frac{1}{wu - zv}\begin{bmatrix} zv+wu & -2zu \\ 2wv & -zv-wu \end{bmatrix} \qquad ; Z = \begin{bmatrix} z \\ w \end{bmatrix},\ U = \begin{bmatrix} u \\ v \end{bmatrix}$$

Alternatively, this can be described by means of the wedge product 1.5

10.4 $$L(Z,U) = \frac{2}{D^2}\, L\left(\begin{smallmatrix} z \\ w \end{smallmatrix}\right) \wedge L\left(\begin{smallmatrix} u \\ v \end{smallmatrix}\right) \qquad ; D = det\begin{bmatrix} z & u \\ w & v \end{bmatrix}$$

It follows from 1.6 that the geodesic in H^2 with normal vector $L(Z,U)$ has ends (Z,U). Moreover, we conclude from 10.2 and 10.4 that

10.5 $$\sigma L(Z,U)\sigma^{-1} = L(\sigma(Z),\sigma(U)) \qquad ; \sigma \in Gl_2(\mathbb{R})$$

We shall now evaluate the inner product of these vectors in terms of the cross ratio on $\hat{\mathbb{R}}$. For four distinct points A,B,P,Q we shall prove that

10.6 $$<L(A,B),L(P,Q)> = \frac{[A,B,P,Q] + 1}{[A,B,P,Q] - 1}$$

Proof According to 10.5 and I.9.7 both sides of the formula are invariant under $Sl_2(\mathbb{R})$. From the fact, I.9.3, that the action of $Sl_2(\mathbb{R})$ on $\hat{\mathbb{R}}$ is triply transitive it follows that it suffices to verify the formula in the case $A = \infty$, $B = 0$, $P = 1$, $Q = q$, $q \in \mathbb{R}$. Let us recall from I.9.9 that $[\infty,0,1,q] = q$. Thus it suffices to prove that

$$<L(\infty,0),L(1,q)> = \frac{q+1}{q-1} \qquad ; q \in \mathbb{R}$$

This is a straightforward verification, which we leave to the reader. □

We ask the reader to establish the following supplement to the general formula displayed in 10.6

10.7 $$<L(A,B),L(B,C)> = 1 \qquad ; A \neq B \neq C$$

Let us consider two distinct geodesics h and k in H^2 with normal vectors H and K and ends (A,B) and (P,Q). We can summarise our investigations in the following table

10.8

Relative position of h and k	$<H,K>$	Sylvester type	[A,B,P,Q]
intersecting	$]-1,1[$	$(-2,0)$	$]-\infty,0[$
perpendicular	0	$(-2,0)$	-1
common perpendicular	$\mathbb{R}-[-1,1]$	$(-1,1)$	$]0,1[\cup]1,+\infty[$
common end	$1,-1$	$(-1,1)$	1

The column entitled Sylvester type refers to the pencil generated by h and k, see the end of III.2.

III EXERCISES

EXERCISE 1.1 Determine the eigenspace decomposition for the operator $X \mapsto X^{\check{}}$ on $M_2(\mathbb{R})$. For notation see I.9.2.

EXERCISE 1.2 Let E denote a two-dimensional real vector space. By a complex structure on E we understand a linear endomorphism J of E with $J^2 = -I$.

1° Show that $J \in End(E)$ is a complex structure $\Leftrightarrow$ $tr\, J = 0$ and $det\, J = 1$.

2° Show that two complex structures on E belong to the same sheet of the hyperbola $det = 1$ in $sl(E) = \{ R \in End(E) \mid tr\, R = 0 \}$ if and only if they define the same orientation of E.

EXERCISE 2.1 1° Show that the vertices of an equilateral triangle lie on a circle.

2° Show that this is not true for a general triangle in H^2.

EXERCISE 2.2 Let h be a geodesic in H^2. For a point $P \in H^2$ let P^h denote the perpendicular projection of P onto h. Show that the map $P \mapsto P^h$ decreases distance, in the sense that $d(P,Q) \geq d(P^h,Q^h)$ for all points P and Q.

EXERCISE 3.1 Let $\square$ABCD be a parallelogram, i.e. a convex quadrangle in which opposite sides have equal length. Show that the diagonals bisect each other and that the geodesics through opposite sides have a common perpendicular. Hint: Show that half-turn μ with respect to the midpoint M of the diagonal AC takes the quadrangle into itself.

EXERCISE 3.2 Let S be an isotropic line in $sl_2(\mathbb{R})$. Show that the horocycles with centre S are cut by the affine planes in $sl_2(\mathbb{R})$ parallel to $T_S(H^2) = S^{\perp}$.

EXERCISE 3.3 1° Given a triangle ΔABC. Let ρ_A denote the rotation with centre A which transforms the ray AB into the ray AC. Show that $\rho_A{}^2 = \gamma\beta$, where γ is reflection in the geodesic AB and β is reflection in the line AC.

2° Define rotations ρ_B and ρ_C as above and prove that $\rho_A{}^2 \circ \rho_B{}^2 \circ \rho_C{}^2 = \iota$.

3° If ΔABC is equilateral show that ρ_A, ρ_B, ρ_C are conjugated in $Isom^+$.

EXERCISE 4.1 Show that an involution in $PGl_2(\mathbb{R})$ can be represented by a matrix $R \in Gl_2(\mathbb{R})$ of trace 0 and determinant 1 or -1. Hint: Show that a matrix $R \in Gl_2(R)$ which represents an involution in $PGl_2(\mathbb{R})$ is an eigenvector for the operator $X \mapsto X^{\check{}}$.

EXERCISE 4.2 Let $A \in sl_2(\mathbb{R})$ represent a point of H^2. Show that the transformation $X \mapsto AXA^{-1}$ of $sl_2(\mathbb{R})$ is half-turn with respect to the point A.

EXERCISE 4.3 1° Show that the set of vectors in $sl_2(\mathbb{R})$ of norm -1 is connected.

2° Prove that $Sl_2(\mathbb{R})$ is connected. Hint: Any $\sigma \in Sl_2(\mathbb{R})$ can be written in the form $\sigma = XY$, where $X, Y \in sl_2(\mathbb{R})$ are vectors of norm -1.

EXERCISE 4.4 1° Show that an even Möbius transformation α of $\hat{\mathbb{C}}$ with matrix

$$A = \begin{bmatrix} a & b \\ \bar{b} & \bar{a} \end{bmatrix} \qquad ; \; a,b \in \mathbb{C} \, , \; |a|^2 - |b|^2 > 0$$

leaves the unit disc D invariant.

2° Show that any even isometry of D has this form. Hint: Find all even Möbius transformations which commute with reflection in ∂D.

3° Find the connection between $tr^2 A$ and the "type" of the isometry α.

4° Show that an even isometry of D can be written

$$z \mapsto exp\, 2\pi i\theta \, \frac{z + c}{1 + \bar{c}z} \qquad ; \; c \in D \, , \; \theta \in \mathbb{R}$$

EXERCISE 5.1 Consider a triangle ΔABC with $\angle C$ a right angle. Show that

$$sin\, A = \lim_{a \to 0} \frac{a}{c} \qquad cos\, A = \lim_{b \to 0} \frac{b}{c}$$

EXERCISE 5.2 Show that a triangle ΔABC has an inscribed circle with radius

$$r = \frac{\cos^2 A + \cos^2 B + \cos^2 C + 2\cos A \cos B \cos C}{2(1+\sin A)(1+\sin B)(1+\sin C)}$$

EXERCISE 6.1 Let S be an isotropic line in $sl_2(\mathbb{R})$. Show that the set of vectors of norm -1 in $S^\perp$ has two connected components. Use this to replace the second half of the proof of 6.1 by a "deformation argument".

EXERCISE 8.1 1° Show that the normal vectors of a right angled hexagon satisfy the formula

$$vol(N_6,N_1,N_2) <N_1,N_3> vol(N_3,N_4,N_5) <N_2,N_4> =$$
$$<N_2,N_5><N_2,N_6> + <N_3,N_5><N_3,N_6>$$

2° Prove the following <u>cotangent</u> <u>relation</u> for a right angled hexagon

$$\cosh a_2 \cosh a_3 = \coth a_1 \sinh a_3 + \coth a_4 \sinh a_2$$

3° Investigate an improper hexagon with five right angles in which the sixth vertex P_6 lies at infinity. Hint: Derive a formula for the normal vectors similar to the formula above (the term $<N_5,N_6>$ must be added to the right hand side).

EXERCISE 9.1 Let ΔABC be a triangle in the hyperbolic plane H^2, and let α,β,γ be translation from B to C, C to A resp. A to B.

1° Show that $\beta\alpha\gamma$ is a rotation around A of angle $-Area\ \Delta$ relative to the orientation defined by the triangle. Hint: For two distinct points P and Q we let $\mathbf{t}_{PQ}$ denote the unit tangent vector at P to the geodesic from P to Q. Justify the following "$mod\ 2\pi$" calculation

$$\angle_{or}(\mathbf{t}_{CA},\alpha\gamma\mathbf{t}_{AB}) = \angle_{or}(\mathbf{t}_{CA},\mathbf{t}_{CB}) + \angle_{or}(\mathbf{t}_{CB},\alpha\gamma\mathbf{t}_{AB}) \equiv \angle_{int}A - \angle_{or}(\mathbf{t}_{CB},\alpha\mathbf{t}_{BA}) \equiv$$
$$\angle_{int}A - \angle_{or}(\alpha^{-1}\mathbf{t}_{CB},\mathbf{t}_{BA}) \equiv \angle_{int}A + \angle_{or}(\mathbf{t}_{BC},\mathbf{t}_{BA}) \equiv \angle_{int}A + \angle_{int}B$$

Let θ denote the rotation angle in accordance with 3.7. Justify the calculation

$$\angle_{or}(\beta\mathbf{t}_{CA},\beta\alpha\gamma\mathbf{t}_{AB}) \equiv \angle_{or}(\beta\mathbf{t}_{CA},\mathbf{t}_{AB}) + \angle_{or}(\mathbf{t}_{AB},\beta\alpha\gamma\mathbf{t}_{AB}) \equiv \pi - \angle_{int}A + \theta$$

Finally, combine the two results.

2° Generalise the result to an arbitrary polygon in H^2.

3° Prove a similar result for the sphere S^2.

EXERCISE 9.2 Show that a disc D in H^2 with radius r has

$$Area(D) = 2\pi\ (\cosh r - 1)$$

EXERCISE 9.3 Let the hyperbolic area of say a compact subset Δ of the Poincaré half-plane H^2 be defined by formula 9.8. Show that

$$Area(\sigma(\Delta)) = Area(\Delta) \qquad ;\ \sigma \in Isom(H^2)$$

2° Given r>0 and $v \in [0,\frac{\pi}{2}]$. For s>r let E(s) denote the subset of H^2 bounded by the y-axis, $x = r\cos v$, $y = s$ and the circle with centre 0 and radius r. Show that

$$\lim_{s\to\infty} Area\ E(s) = \frac{\pi}{2} - v$$

3° Prove that the area of ΔABC is $\pi - \angle A - \angle B - \angle C$. Hint: Move the triangle such that two vertices are located on the y-axis and apply 2°.

IV FUCHSIAN GROUPS

We shall introduce the central topic of this book, discrete groups of isometries of the hyperbolic plane H^2. A discrete group of even isometries is called a Fuchsian group. The set of fixed points for the elliptic elements of a discrete group Γ form a discrete subset of H^2. We shall prove after Jacob Nielsen a converse to this: A non-elementary group Γ with discrete elliptic fixed point set is itself discrete.

It is a common feature of the Poincaré half-plane and the $sl_2(\mathbb{R})$-model of H^2 that these models provide natural isomorphisms between $Isom(H^2)$ and $PGl_2(\mathbb{R})$, compare III.9.3. The results of this chapter can be interpreted in either of the two models. The $sl_2(\mathbb{R})$-model is used throughout in the more technical parts of the proofs.

IV.1 DISCRETE SUBGROUPS

The group $PGl_2(\mathbb{R})$, being the factor group of the topological group $Gl_2(\mathbb{R})$ with respect to a its centre $\mathbb{R}^*$, carries the structure of a topological group: the open subsets of $PGl_2(\mathbb{R})$ correspond to open subsets of $Gl_2(\mathbb{R})$ stable under $\mathbb{R}^*$. We shall use the action of $PGl_2(\mathbb{R})$ on H^2 to study discrete[1] subgroups of $PGl_2(\mathbb{R})$. But first, we shall exhibit some convenient compact neighbourhoods of the identity ι in $PGl_2(\mathbb{R})$.

Theorem 1.1 A point $z \in H^2$ and an $\epsilon > 0$ determine a subset of $PGl_2(\mathbb{R})$

$$\mathcal{R}(z;\epsilon) = \{\sigma \in PGl_2(\mathbb{R}) \mid d(z,\sigma z) \leq \epsilon\}$$

which is a compact neighbourhood of the identity ι in $PGl_2(\mathbb{R})$.

[1]The notion of a discrete subset is discussed in an appendix to this section.

Proof Let us introduce the norm $\|\sigma\|$ of a matrix $\sigma \in Gl_2(\mathbb{R})$ as

1.2 $$\|\sigma\| = \sqrt{\frac{a^2+b^2+c^2+d^2}{|\det \sigma|}} \qquad ; \ \sigma = \begin{bmatrix} a & b \\ c & d \end{bmatrix}$$

Observe the formula $\|a\sigma\| = \|\sigma\|$, $a \in \mathbb{R}^*$, and conclude that the norm induces a continuous function on $PGl_2(\mathbb{R})$. It follows that a set of the form

1.3 $$\{ \sigma \in PGl_2(\mathbb{R}) \mid \|\sigma\| \leq r \} \qquad ; \ r > \sqrt{2}$$

is a neighbourhood of ι in $PGl_2(\mathbb{R})$. Let us show that a set of the form 1.3 is compact. Observe first that the set 1.3 is the image by a continuous function of the set

$$\{ \sigma \in Gl_2(\mathbb{R}) \mid \|\sigma\| \leq r \, , |\det\sigma| = 1\}$$

It follows from 1.2 that this set is a closed and bounded subset of $M_2(\mathbb{R})$. But a closed and bounded subset of $M_2(\mathbb{R}) \xrightarrow{\sim} \mathbb{R}^4$ is compact.

Let us recall that the action of $PGl_2(\mathbb{R})$ on H^2 is transitive. It follows that it suffices to treat the point $i \in H^2$ of the $sl_2(\mathbb{R})$-model given by

$$i = \begin{bmatrix} 0 & -1 \\ 1 & 0 \end{bmatrix}$$

Let us recall from III.4.3 that the action of $\sigma \in PGl_2(\mathbb{R})$ on i is given by

$$\sigma(i) = \frac{1}{|ad-bc|} \begin{bmatrix} a & b \\ c & d \end{bmatrix} \begin{bmatrix} 0 & -1 \\ 1 & 0 \end{bmatrix} \begin{bmatrix} d & -b \\ -c & a \end{bmatrix}$$

Simple calculation gives us

$$2\langle\sigma(i),i\rangle = -tr[\sigma(i)\ i] = |ad-bc|^{-1}(a^2+b^2+c^2+d^2)$$

With the notation above this may be written

1.4 $$2 \cosh d(i,\sigma i) = \|\sigma\| \qquad ; \ \sigma \in PGl_2(\mathbb{R})$$

The result follows from the fact that sets of the form 1.3 are compact neighbourhoods of the identity ι in $PGl_2(\mathbb{R})$. □

Corollary 1.5 Let Γ denote a subgroup of $PGl_2(\mathbb{R})$. If Γ is discrete, then for all $z \in H^2$ and $\epsilon > 0$ the following set is finite

$$\{ \sigma \in \Gamma \mid d(\sigma(z),z) \leq \epsilon \}$$

Conversely, if for one $\epsilon > 0$ and one point $z \in H^2$ this set is finite, then Γ is a discrete subgroup of $PGl_2(\mathbb{R})$.

Proof With the notation above observe that

$$\{ \sigma \in \Gamma \mid d(\sigma(z),z) \leq \epsilon \} = \mathfrak{R}(z;\epsilon) \cap \Gamma$$

and the result follows from 1.1 and the general lemma 1.13. □

The modular group 1.6 The group $Gl_2(\mathbb{Z})$ of 2×2 matrices with entries from $\mathbb{Z}$ and determinant ± 1 is a discrete subgroup of $Gl_2(\mathbb{R})$ as follows from the elementary fact that any of the subsets

$$\{\sigma \in Gl_2(\mathbb{Z}) \mid |\sigma| \leq n \} \qquad ; \; n > \sqrt{2}$$

is finite. The image of $Gl_2(\mathbb{Z})$ in $PGl_2(\mathbb{R})$ is discrete: in fact we ask the reader to show that any discrete subgroup $\Gamma \subset Gl_2(\mathbb{R})$ consisting of matrices with determinant ± 1 maps onto a discrete subgroup of $PGl_2(\mathbb{R})$.

Proposition 1.7 A discrete subgroup Γ of $PGl_2(\mathbb{R})$ operates on the hyperbolic plane H^2 with discrete orbits.

Proof Consider an orbit $\mathcal{A}$ for Γ and pick a base point $z \in \mathcal{A}$. A disc D with centre $z \in \mathcal{A}$ will contain finitely many points only from $\mathcal{A}$ as follows from 1.5. Thus we can find $\epsilon > 0$ such that $D(z;\epsilon) \cap \mathcal{A} = \{z\}$. Let us show that in fact

1.8 $$d(u,v) \geq \epsilon \qquad ; \; u,v \in \mathcal{A}, \; u \neq v$$

To see this choose $\sigma \in \Gamma$ with $\sigma(z) = u$. For a second point $v \neq u$ of the orbit $\mathcal{A}$ we have $\sigma(v) \notin D(z;\epsilon)$ and consequently $d(u,v) = d(\sigma(u),\sigma(v)) \geq \epsilon$. In order to see that $\mathcal{A}$ is closed, consider a point $w \notin \mathcal{A}$. The open disc $D(w;\epsilon/2)$ contains at most two points of $\mathcal{A}$ as follows from 1.8. It is now easy to find a disc with centre w not meeting $\mathcal{A}$. □

Theorem 1.9 Let Γ be a discrete subgroup of $PGl_2(\mathbb{R})$. For compact subsets K and L of the hyperbolic plane H^2, the sets $\sigma(K)$ and L are disjoint for all but finitely many $\sigma \in \Gamma$.

Proof Given a compact set K, a point $z \in H^2$ and a number $r > 0$, let us show that D(z;r) and $\sigma(K)$ are disjoint for all but finitely many $\sigma \in \Gamma$. Choose $s > 0$ with $K \subset D(z;r+s)$ and conclude from 1.5 that $S = \{\gamma \in \Gamma \mid \gamma(z) \in D(z;r+s)\}$ is finite. Observe that $\gamma D(z;r) = D(\gamma(z);r)$ and conclude that $K \cap \gamma D(z;r) = \emptyset$ for all $\gamma \in \Gamma - S$. Given a second compact set L, choose $z \in L$ and $r > 0$ such that $L \subset D(z;r)$. It follows that $\sigma(K)$ and L meet for at most finitely many $\sigma \in \Gamma$. □

Corollary 1.10 Let Γ be a discrete subgroup of $PGl_2(\mathbb{R})$ and K a compact subset of H^2. Only finitely many elements of Γ have fixed points on K.

Proof For all but finitely many $\sigma \in \Gamma$ the sets K and $\sigma(K)$ are disjoint. It follows that only a finite number of elements from Γ have fixed points on K. □

By an <u>elliptic fixed point</u> for Γ on H^2 we understand a point $z \in H^2$ which is fixed by some proper elliptic element $\sigma \in \Gamma$.

Corollary 1.11 Let Γ be a discrete subgroup of $PGl_2(\mathbb{R})$. The set P of elliptic fixed points for Γ on H^2 is a discrete subset of H^2.

APPENDIX: DISCRETE SUBSETS

Let us consider a locally compact space X, i.e. a topological space X which is Hausdorff and in which every point has a compact neighbourhood. We say that a subset S of X is discrete if S is locally finite in X, meaning that every point $x \in X$ has a neighbourhood which meets at most finitely many points from S.

Observe that a discrete subset S of X is closed and that every point $v \in S$ is an isolated point of S in the sense that v has a neighbourhood V in X with $V \cap S = \{v\}$. Conversely, a closed subset S of X consisting entirely of isolated points is discrete.

Let us say that a point $w \in X$ is an accumulation point for S if every neighbourhood of w contains infinitely many points of S. A discrete set in S has no accumulation points in X and conversely, a subset S of X without accumulation points in X is a discrete subset of X. Discreteness of a subset S can be expressed geometrically as follows

Proposition 1.12 A subset S of a locally compact space X is discrete if and only if S intersects every compact subset of X in a finite set.

Proof This follows immediately from the Borel–Heine theorem. □

By a discrete subgroup Γ of a locally compact group G we understand a subgroup Γ of G which is a discrete subset of G. Discreteness of subgroups can be detected as follows

Proposition 1.13 Let Γ be a subgroup of the locally compact topological group G. If there exists a neighbourhood V of ι in G such that $V \cap \Gamma = \{\iota\}$ then Γ is a discrete subgroup of G.

Proof Let V be an open neighbourhood of ι in G with $V \cap \Gamma = \{\iota\}$. Let us consider a point $\sigma \in G$ with the intention of constructing an open neighbourhood of σ in G which meets Γ in at most one point. Consider the alternatives

$$\sigma V^{-1} \cap \Gamma = \emptyset \ , \qquad \sigma V^{-1} \cap \Gamma \neq \emptyset$$

The first alternative gives us the open neighbourhood σV^{-1} of σ disjoint from Γ. The second alternative allows us to pick a $\gamma \in \sigma V^{-1} \cap \Gamma$. From $\sigma \in \gamma V$ and $\gamma \in \Gamma$ we conclude that $\gamma V \cap \Gamma = \{\gamma\}$:

$$\gamma^{-1}(\gamma V \cap \Gamma) = V \cap \gamma^{-1}\Gamma = V \cap \Gamma = \{\iota\}$$

In conclusion the open neighbourhood γV of σ meets Γ in the point γ only. □

Corollary 1.14 Let the locally compact group G act continuously on the topological space X, and let Γ be subgroup of G. If there exists a point $x \in X$, isolated in its orbit Γx, and such that the stabiliser Γ_x is discrete in G, then Γ is a discrete subgroup of G.

Proof Let U be an open neighbourhood of x in X such that $U \cap \Gamma x = \{x\}$. Then the inverse image of U along the map $\tau_x{:}G \rightarrow X$, $g \mapsto gx$, satisfies

$$\tau_x^{-1}(U) \cap \Gamma = \Gamma_x$$

Let us pick an open neighbourhood of V of ι in G such that $V \cap \Gamma_x = \{\iota\}$. The open neighbourhood $V \cap \tau_x^{-1}(U)$ isolates ι in Γ, and 1.13 applies. □

IV.2 ELEMENTARY SUBGROUPS

Let us start the study of subgroups of $PGl_2(\mathbb{R})$ with a list of its so called elementary subgroups. By this we mean a subgroup G which fixes a point of H^2, fixes a point of ∂H^2 or stabilises a geodesic in H^2. We shall see that a subgroup G of $PGl_2(\mathbb{R})$ is elementary if and only if the group G is solvable.

Proposition 2.1 The stabiliser of a point of H^2 under the action of the group $Gl_2(\mathbb{R})$ is conjugated to a group of matrices of the form

$$\begin{bmatrix} a & b \\ -b & a \end{bmatrix}, \begin{bmatrix} a & b \\ b & -a \end{bmatrix} \qquad ; a^2+b^2 \neq 0$$

$PGl_2(\mathbb{R})$ acts on H^2 with stabilisers conjugated to $PO_2(\mathbb{R}) = O_2(\mathbb{R})/\{-\iota\}$.

Proof It is easily seen that the matrices from the list above stabilises the point $i \in H^2$ considered in the proof of 1.1. From formula 1.4 follows that matrix $\begin{bmatrix} a & b \\ c & d \end{bmatrix}$ stabilises i if and only if

$$a^2+b^2+c^2+d^2 = 2|ad-bc|$$

When $ad-bc>0$, this can be untangled via the identity

$$(a-d)^2+(b+c)^2 = a^2+b^2+c^2+d^2-2(ad-bc)$$

The case $ad-bc<0$ is handled in a similar manner. It follows that the stabiliser of i in $Gl_2{}^*(\mathbb{R}) = \{\sigma \in Gl_2(\mathbb{R}) \mid det\sigma = \pm 1\}$ is $O_2(\mathbb{R})$. This shows that the stabiliser of i in $PGl_2(\mathbb{R})$ is $O_2(\mathbb{R})/\{-\iota\}$. □

Proposition 2.2 The stabiliser in $Gl_2(\mathbb{R})$ of a given geodesic k in H^2 is conjugated to the subgroup of $Gl_2(\mathbb{R})$ represented by matrices

$$\begin{bmatrix} a & 0 \\ 0 & d \end{bmatrix}, \begin{bmatrix} 0 & b \\ c & 0 \end{bmatrix} \qquad ; ad \neq 0,\ bc \neq 0$$

Proof Let y be the geodesic with normal vector $N = \begin{bmatrix} 1 & 0 \\ 0 & -1 \end{bmatrix}$. Let us show that the matrices we have listed above make up the set of transformations which acts on $sl_2(\mathbb{R})$ with N as eigenvector. Direct calculation gives us

$$\begin{bmatrix} a & b \\ c & d \end{bmatrix}\begin{bmatrix} 1 & 0 \\ 0 & -1 \end{bmatrix}\begin{bmatrix} d & -b \\ -c & a \end{bmatrix} = \begin{bmatrix} ad+bc & -2ab \\ 2cd & -ad-bc \end{bmatrix}$$

The details are left with the reader. □

Proposition 2.3 The stabiliser in $Gl_2(\mathbb{R})$ of an end of H^2 is conjugated to the subgroup

$$\begin{bmatrix} a & b \\ 0 & d \end{bmatrix} \quad ; \text{ad} \neq 0$$

Proof It is cleare that a matrix of the above form stabilises the special end $\infty = \begin{bmatrix} 0 & 1 \\ 0 & 0 \end{bmatrix}$. Conversely, let the matrix $\begin{bmatrix} a & b \\ c & d \end{bmatrix} \in Gl_2(\mathbb{R})$ stabilise the end ∞. Direct calculation gives us

$$\begin{bmatrix} a & b \\ c & d \end{bmatrix}\begin{bmatrix} 0 & 1 \\ 0 & 0 \end{bmatrix}\begin{bmatrix} d & -b \\ -c & a \end{bmatrix} = \begin{bmatrix} -ac & a^2 \\ -c^2 & ac \end{bmatrix}$$

From this we conclude that $c = 0$, compare III.4.3. □

Discrete groups Let us find the discrete elementary groups. From 2.1 we find that a group Γ which fixes a point is discrete if and only if Γ is finite. From 2.2 we find that a discrete group Γ which stabilises a geodesic k is discrete if and only if the translation subgroup is cyclic. In the case of a fixed point on ∂H^2 we have the following result.

Proposition 2.4 Let Γ be a subgroup of $PSl_2(\mathbb{R})$ which fixes a point of ∂H^2. The group is discrete if and only if it is cyclic.

Proof Let us consider a group generated by the two matrices

$$\sigma = \begin{bmatrix} u & r \\ 0 & u^{-1} \end{bmatrix}, \quad \tau = \begin{bmatrix} v & s \\ 0 & v^{-1} \end{bmatrix} \qquad ; u > 1, v > 1$$

We ask the reader to verify that after conjugation by a matrix of the form $\begin{bmatrix} 1 & x \\ 0 & 1 \end{bmatrix}$ we may assume that $r = 0$. Let us study the group generated by σ and the commutator $\gamma = \sigma\tau\sigma^{-1}\tau^{-1}$

$$\sigma = \begin{bmatrix} u & 0 \\ 0 & u^{-1} \end{bmatrix}, \quad \gamma = \begin{bmatrix} 1 & c \\ 0 & 1 \end{bmatrix} \qquad ; u > 1$$

Direct computation yields the following family of elements of Γ

$$\sigma^n \gamma \sigma^{-n} = \begin{bmatrix} 1 & c\, u^{2n} \\ 0 & 1 \end{bmatrix} \qquad ; \; n \in \mathbb{Z}$$

which shows that Γ is a non-discrete group unless $u = 1$ or $c = 0$. □

Let us present an important class of examples of non-discrete groups. The material is presented as a lemma and will serve as such in section IV.4.

Lemma 2.5 Let Γ be a non-elementary subgroup of $PSl_2(\mathbb{R})$ containing translations σ and τ whose translation axes h and k are distinct but have a common end ∞. Then the rotations in Γ accumulate at ι.

Proof Let us first show that Γ contains a translation ρ which doesn't fix ∞: pick $\mu \in \Gamma$ which doesn't fix ∞. The geodesics $\mu(h)$ and $\mu(k)$ have $\mu(\infty)$ as an end. The second end of one of these geodesics must be different from ∞. It follows that one of the translations $\mu\sigma\mu^{-1}$ and $\mu\tau\mu^{-1}$ doesn't fix ∞.

The group Γ_∞ of horolations in Γ with centre ∞ is non-discrete as we saw during the proof of 2.4. Let us investigate the commutator $\eta\rho\eta^{-1}\rho^{-1}$ as η varies through Γ_∞. To this end we put

$$\rho = \begin{bmatrix} a & b \\ c & d \end{bmatrix}, \quad \eta = \begin{bmatrix} 1 & e \\ 0 & 1 \end{bmatrix}$$

We ask the reader to verify that

$$tr(\eta\rho\eta^{-1}\rho^{-1}) = 2 - c^2e^2 \qquad ; e \in \mathbb{R}$$

By assumption we have $c \neq 0$ and we conclude that $tr^2(\eta\rho\eta^{-1}\rho^{-1}) < 4$ for $e \neq 0$ sufficiently small. □

Solvable groups We proceed to make some investigations of group theoretical nature based on the following fundamental principle.

Normaliser Principle 2.6 Let X be a set on which a group G acts. For any normal subgroup N of G the fixed point set X^N is stable under G.

Proof For $x \in X^N$ and $\gamma \in G$ we have that

$$\nu(\gamma x) = \gamma(\gamma^{-1}\nu\gamma)x = \gamma x \qquad ; \nu \in N$$

which shows that γx is a fixed point for N. □

Proposition 2.7 A subgroup of $PGl_2(\mathbb{R})$ is elementary if and only if it is a solvable group.

Proof The elementary groups we have displayed in the three first propositions of this section are easily seen to be solvable. Conversely, let us consider a solvable subgroup $G \neq 0$ and form the derived series of commutator groups

$$G = D^0G \supseteq D^1G \supseteq \ \dots\ D^{n-1}G \supseteq D^nG \supseteq D^{n+1}G = \{\iota\} \ ; \ D^nG \neq \{\iota\}$$

Let us first assume that G consists of even transformations. We shall make use of the classification of even isometries and proceed accordingly.

1° D^nG contains a rotation σ with centre A. This means that the fixed point set for σ on H^2 consists of A only. Let us prove by decreasing induction that the fixed point set for D^iG on H^2 is {A}, i = 0,...,n. For i = n observe that the group D^nG is abelian and use the "normaliser principle" 2.6. The induction step is likewise conducted by means of 2.6.

2° D^nG contains a horolation σ with centre $P \in \partial H^2$. It follows that the fixed point set for σ on ∂H^2 is {P}. We can proceed as above and conclude from 2.6 that the fixed point set for G on ∂H^2 is {P}.

3° D^nG contains a translation $\sigma \neq \iota$ along the geodesic k. We shall utilise the action of G on the set $\mathfrak{G}$ of geodesics in H^2; let us show that $\mathfrak{G}^\sigma = \{k\}$. Let σ stabilise the geodesic h which means that the set {A,B} of ends for h is stabilised by σ. It follows that σ^2 fixes A and B. Observe that $\sigma^2 \neq \iota$ has precisely two fixed points on ∂H^2 and conclude that A and B must be the ends for k, thus $h = k$. We conclude as above that the fixed point set for G on $\mathfrak{G}$ is {k}.

Let us now return to the general solvable group G and observe that G^+ is solvable as well. When $G^+ \neq \{\iota\}$ we can assume that G^+ has precisely one fixed point on either one of the following three sets H^2, ∂H^2 or $\mathfrak{G}$ and we can use the "normaliser principle" 2.6 to conclude that G has a fixed point as well. □

Let us investigate various types of groups which lack hyperbolic elements: groups of elliptic elements and parabolic elements.

Proposition 2.8 A non-elementary subgroup G of $PGl_2(\mathbb{R})$ contains a translation.

Proof Let us make the general remark that the commutator of two rotations σ and τ is always a translation. To see this, let their centres be A and B and let β denote reflection in the geodesic through A and B. Let us use the theorem on three reflections III.3.1 to write $\sigma = \alpha\beta$ and $\tau = \beta\gamma$ where α and γ are reflections in geodesics through A and B. Direct calculation gives us

$$\sigma\tau\sigma^{-1}\tau^{-1} = \alpha\beta\beta\gamma\beta\alpha\gamma\beta = (\alpha\gamma\beta)^2$$

Using that $\alpha\gamma\beta$ that is a glide reflection, compare III.3.8, we conclude that its square $\sigma\tau\sigma^{-1}\tau^{-1}$ is a translation. Thus $\sigma\tau\sigma^{-1}\tau^{-1} = \iota$.

The same argument shows that the commutator of two horolations is a translation and that the commutator of a rotation and a horolation is a translation.

Let us now assume that G doesn't contain translations. This implies that that G^+ is abelian and that G is solvable. It follows from proposition 2.7 that G is an elementary subgroup. □

Proposition 2.9 Let G be a non-elementary subgroup of $PGl_2(\mathbb{R})$. If $G \neq G^+$ then G contains a glide reflection.

Proof Let us assume that all odd transformations from G are reflections. Pick a non-trivial transformation $\sigma \in G^+$. If σ is a rotation around the point A, then we conclude from 2.10 that all reflection axes of elements of G pass through A, and we conclude that A is fixed by G. If σ is a horolation with centre S, then we conclude from 2.10 that all reflection axes from G pass through S, which implies that G fixes S. If σ is a translation then we conclude from 2.10 that all reflection axes from G are perpendicular to the translation axis s of σ. This implies that s is stabilised by the group G. □

Lemma 2.10 Let κ be a reflection in a geodesic k and σ an even isometry of H^2. If $\kappa\sigma$ is a reflection, then σ is a rotation around a point of k or a horolation with a centre at one of the ends of k or a translation with axis perpendicular to k.

Proof Observe that $tr^2(\kappa\sigma) = 0$, exercise II.8.1, and use explicit calculation in the Poincaré model taking for k the y-axis. See also exercise IV.3.1. □

IV.3 GEOMETRY OF COMMUTATORS

In this section we shall work out a number of trace formulas for the composition of two even isometries and for their commutator.

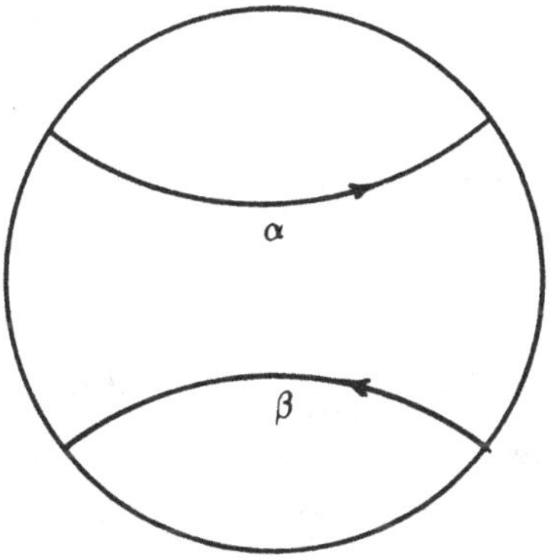

Proposition 3.1 Let α and β be translations whose axes a and b have a common perpendicular. If α and β have the same translation length T, but are oppositely directed, then with $\mathrm{d} = \mathrm{d}(a,b)$

$$tr^2(\alpha\beta) - 4 = 16\ sinh^2(\tfrac{1}{2}\mathrm{T})\ sinh^2(\tfrac{1}{2}\mathrm{d})\ [sinh^2(\tfrac{1}{2}\mathrm{T})\ sinh^2(\tfrac{1}{2}\mathrm{d}) - 1]$$

Proof Let us represent α by the matrix $\mathrm{A} \in Sl_2(\mathbb{R})$ with $tr\ \mathrm{A} > 0$. This allows us to assign to α the <u>transformation vector</u> $\mathrm{U} = \frac{1}{2}(\mathrm{A}^{\cdot} - \mathrm{A})$. Observe that $tr\ \mathrm{U} = 0$, otherwise expressed, $\mathrm{U} \in sl_2(\mathbb{R})$. The axis a of the translation α has a natural orientation and M is defined as the corresponding normal vector. We shall prove that U and M and the translation length T are related through

3.2 $$\mathrm{U} = sinh(\tfrac{1}{2}\mathrm{T})\ \mathrm{M}\ , \qquad \tfrac{1}{2} tr\ \mathrm{A} = cosh(\tfrac{1}{2}\mathrm{T})$$

To see this, let us pick points P and Q on a such that Q is the midpoint of P and α(P). Let H and K be positively directed unit tangent vectors to a at P and Q respectively

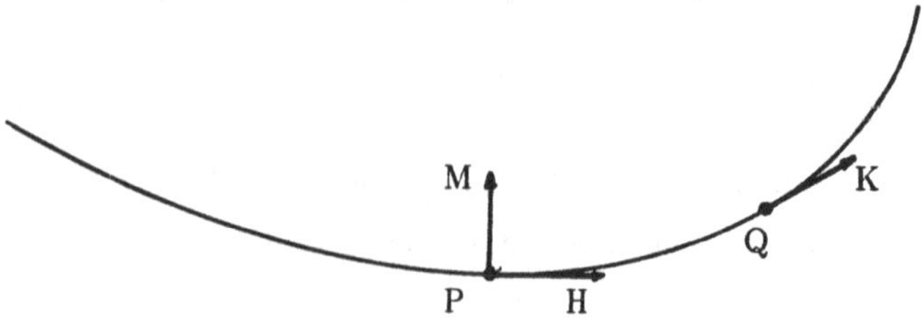

From the parametrisation of a given by $P\cosh(s) + H\sinh(s)$, $s \in \mathbb{R}$, we get that

$$Q = P\cosh(\tfrac{1}{2}T) + H\sinh(\tfrac{1}{2}T)\ , \qquad K = P\sinh(\tfrac{1}{2}T) + H\cosh(\tfrac{1}{2}T)$$

The second of these two formulas gives us

$$H \wedge K = H \wedge P\sinh(\tfrac{1}{2}T)\ , \qquad \langle H,K \rangle = -\cosh(\tfrac{1}{2}T)$$

Let us use III.4.5 to represent α by the the matrix $A = KH$. This gives us

$$\tfrac{1}{2}(A^{\vee} - A) = \tfrac{1}{2}(HK - KH) = H \wedge K\ ,\ \tfrac{1}{2} tr\ A = -\langle H,K \rangle = \cosh(\tfrac{1}{2}T)$$

Recall from III.2.1 that $M = H \wedge P$, which concludes the verification of 3.2.

Similarly, let us represent β by a $B \in Sl_2(\mathbb{R})$ with $\tfrac{1}{2} tr\ B = \cosh(\tfrac{1}{2}T)$ and $V = \tfrac{1}{2}(B^{\vee} - B)$. The normal vector for the natural orientation of the axis b for β is denoted N. The fact that M and N represent opposite directions can be expressed by $\langle M,N \rangle > 0$. As a consequence of III.2.8 we have that

$$\langle M,N \rangle = \cosh d$$

Let us substitute this information into lemma 3.3 below to get

$$\begin{aligned} &\tfrac{1}{2}\, tr(AB) = \cosh^2(\tfrac{1}{2}T) - \sinh^2(\tfrac{1}{2}T)\cosh d = \\ &1 + \sinh^2(\tfrac{1}{2}T) - \sinh^2(\tfrac{1}{2}T)(1 + 2\sinh^2(\tfrac{1}{2}d)) = \\ &1 - 2\sinh^2(\tfrac{1}{2}T)\sinh^2(\tfrac{1}{2}d) \end{aligned}$$

Finally, square this relation and the announced formula follows. □

Given matrices $A,B \in M_2(\mathbb{R})$. The <u>transformation</u> <u>vectors</u> $U = \tfrac{1}{2}(A^{\vee} - A)$ and $V = \tfrac{1}{2}(B^{\vee} - B)$ satisfy

3.3 $$\tfrac{1}{2} tr(AB) = \tfrac{1}{2}\ tr\ A\ \tfrac{1}{2} tr\ B - \langle U,V \rangle$$

Proof The elementary relation $A + A^{\check{}} = \iota\, trA$ gives us $U = \frac{1}{2}(\iota\ trA - 2A)$ and similarly $V = \frac{1}{2}(\iota\ trB - 2B)$. Straightforward calculation yields

$$-2\langle U,V\rangle = tr(UV) = \tfrac{1}{4}\, tr[(\iota\ trA - 2A)(\iota\ trB - 2B)] =$$
$$\tfrac{1}{4}\, tr\iota\ trA\ trB + trAB - \tfrac{1}{4} trA\ tr2B - \tfrac{1}{4} tr2A\ trB$$

from which formula 3.3 follows. □

Corollary 3.4 Let α and β be translations with the same translation length T whose axes a and b have a common perpendicular. If neither $\alpha\beta$ nor $\alpha\beta^{-1}$ is a rotation, then

$$\sinh \tfrac{1}{2}T \sinh \tfrac{1}{2}d \geq 1 \qquad ; \quad d = d(a,b)$$

Let us now turn to commutators. It is useful to remark that the symbol $tr(\alpha\beta\alpha^{-1}\beta^{-1}) \in \mathbb{R}$ is a well defined function of $\alpha,\beta \in PSl_2(\mathbb{R})$ since $tr(ABA^{-1}B^{-1})$ is independent of representatives A and B for α and β.

Lemma 3.5 Any two transformations $\alpha,\beta \in PSl_2(\mathbb{R})$ satisfy

$$\tfrac{1}{2}\, tr(\beta\alpha\beta^{-1}\alpha^{-1}) = \tfrac{1}{4}\, tr^2\alpha + \langle U,\beta(U)\rangle$$

where the vector U is given by $U = \frac{1}{2}(A^{\check{}} - A)$ for $A \in Sl_2(\mathbb{R})$ representing α.

Proof Let us represent α and β by matrices A and B. The transformation vector for A is $U = \frac{1}{2}(A^{\check{}} - A)$, while the transformation vector for BAB^{-1} is

$$\tfrac{1}{2}((BAB^{\check{}})^{\check{}} - BAB^{\check{}}) = \tfrac{1}{2}(\ BA^{\check{}}B^{\check{}} - BA^{\check{}}) = BUB^{-1}$$

and the transformation vector for A^{-1} is $-U$. Let us substitute this into 3.3

$$\tfrac{1}{2} tr(BAB^{-1}A^{-1}) = \tfrac{1}{2}\, trBAB^{-1}\ \tfrac{1}{2}\, trA^{-1} + \langle BUB^{-1},U\rangle$$

which concludes the proof. □

We shall now investigate the commutator of two translations α and β with the same translation length. This situation occours for example when α and β are conjugated.

Proposition 3.6 Let α and β be two translations with the same translation length T. If the axes a and b for α and β intersect at an angle $\theta \in]0,\frac{1}{2}\pi]$ then

$$tr^2(\beta\alpha\beta^{-1}\alpha^{-1}) - 4 = 16\ sinh^4(\tfrac{1}{2}T)\ sin^2\theta\ [sinh^4(\tfrac{1}{2}T)\ sin^2\theta - 1\ ]$$

Proof From the formulas 3.2 and lemma 3.5 we deduce that

$$\tfrac{1}{2}\ tr(\beta\alpha\beta^{-1}\alpha^{-1}) = cosh^2(\tfrac{1}{2}T) + sinh^2(\tfrac{1}{2}T) < M, \beta(M) >$$

where M is a normal vector for a. Observe that $\beta(M)$ is a normal vector for $\beta(a)$

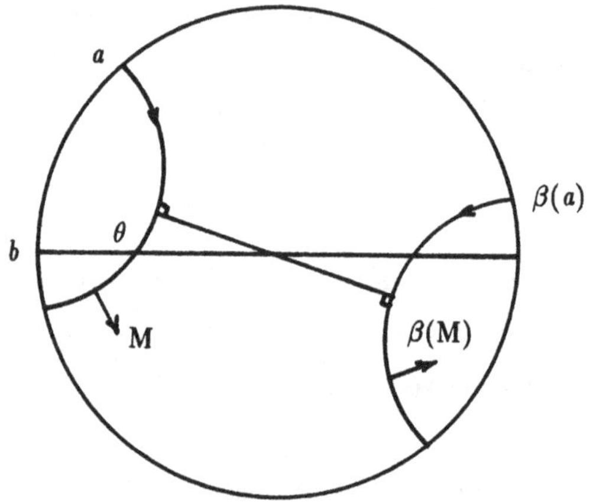

From III.2.8 we get with $d(a,\beta(a)) = d$ that

$$- < M, \beta(M) > = cosh\ d(a,\beta(a)) = 1 + 2\ sinh^2(\tfrac{1}{2}d)$$

The trigonometry of the right angled triangle, III.5.1, gives us

$$sinh(\tfrac{1}{2}d) = sin\ \theta\ sinh(\tfrac{1}{2}T)$$

Substitute this into the formula above to get that

$$\tfrac{1}{2}tr(\beta\alpha\beta^{-1}\alpha^{-1}) = cosh^2(\tfrac{1}{2}T) - sinh^2(\tfrac{1}{2}T)(1+2sin^2\theta\ sinh^2(\tfrac{1}{2}T)) = 1-2sin^2\theta\ sinh^4(\tfrac{1}{2}T)$$

Finally, square this formula to obtain the desired result. □

Corollary 3.7 Let α and β be two translations with the same translation length T, whose axes a and b intersect at an angle $\theta \in]0,\frac{1}{2}\pi]$. The commutator $\beta\alpha\beta^{-1}\alpha^{-1}$ is not a rotation if and only if

$$sinh^2(\tfrac{1}{2}T)\ sin\ \theta \geq 1$$

IV.4 JACOB NIELSEN'S THEOREM

We have seen that a discrete subgroup Γ of $PGl_2(\mathbb{R})$ acts on H^2 with a discrete set of elliptic fixed points. We shall prove the converse for a non-elementary group.

Jacob Nielsen's theorem 4.1 A non-elementary subgroup Γ of $PGl_2(\mathbb{R})$ is discrete if and only if the set P of elliptic fixed points on H^2 is discrete.

Proof It follows from 2.7 that a subgroup Γ of $PGl_2(\mathbb{R})$ is elementary if and only if Γ^+ is elementary. Thus we may assume that Γ is Fuchsian.

Let us assume that P is non empty. Pick a point $z \in P$ and observe that its orbit under Γ is discrete since it is a subset of the discrete set P. According to 1.5 it suffices to prove that Γ_z is finite. To see this, use that Γ is non-elementary to pick a second point $w \in P$, $w \neq z$. The orbit of w under Γ_z is finite since it is a discrete subset of the circle S^1 with centre z through w. We can use that the action of Γ_z on S^1 is faithful to conclude that Γ_z is finite as required.

In the rest of the proof we assume that Γ is non-elementary and contains no rotations. It follows from 2.8 that Γ must contain proper translations. Let us pick a geodesic s such that the subgroup Γ_s made up of translations along s is non-trivial. Since Γ is non-elementary we have $\Gamma \neq \Gamma_s$. For $\tau \notin \Gamma_s$ we have $\tau(s) \neq s$ since we can rule out half-turns. It follows from 2.5 that $\tau(s)$ and s will intersect or have a common perpendicular; we shall examine the two cases separately.

When s and $\tau(s)$ have a common perpendicular, let us introduce the distance d between s and $\tau(s)$. For a non-trivial $\sigma \in \Gamma_s$ let us apply corollary 3.4 to σ and $\tau\sigma\tau^{-1}$ and deduce the inequality

4.2 $$sinh(\tfrac{1}{2}T_\sigma)\, sinh(\tfrac{1}{2}d) \geq 1 \qquad ; \sigma \in \Gamma_s\,,\ \sigma \neq \iota$$

When s and $\tau(s)$ intersect at an angle $\theta \in]0,\frac{\pi}{2}]$ say, we can apply corollary 3.7 to

σ and $\tau\sigma\tau^{-1}$ and deduce the inequality

4.3 $$\sinh^2(\tfrac{1}{2}T_\sigma)\ \sin(\theta) \geq 1 \qquad ;\ \sigma \in \Gamma_s\,,\ \sigma \neq \iota$$

Let us fix τ for a moment. We can then use 4.2 and 4.3 to conclude that the numbers T_σ are bounded away from 0 as σ runs through $\Gamma_s - \{\iota\}$. It follows that Γ_s is discrete, in fact, Γ_s is cyclic generated by an element σ of translation length T, say. For any $\tau \in \Gamma$ such that s and $\tau(s)$ have a common perpendicular we get from 4.2 that

4.4 $$\sinh(\tfrac{1}{2}T)\ \sinh(\tfrac{1}{2}d) \geq 1 \qquad ;\ d = d(s,\tau(s))$$

For any $\tau \in \Gamma$ such that s and $\tau(s)$ intersect we get from 4.3 that

4.5 $$\sinh^2(\tfrac{1}{2}T)\ \sin(\theta) \geq 1 \qquad ;\ \theta = \angle(s,\tau(s))$$

We shall now study the action of Γ on $N = N(sl_2(\mathbb{R}))$, the space of vectors in $sl_2(\mathbb{R})$ of norm -1. Let us pick a normal vector N for s and observe that the stabiliser for N is Γ_s. Let us recall from III.2.5 and III.2.8 that for the inner product between N and τN we have respectively

$$|\langle N,\tau N\rangle| = \cosh d\ , \quad |\langle N,\tau N\rangle| = \cos\theta$$

We conclude from 4.4 that d is bounded away from 0 and from 4.5 that θ is bounded away from 0. It follows that we can find a constant c>0 such that

$$|\langle N,\tau N\rangle| \notin\ ]1-c,1+c[\qquad ;\ \tau \notin \Gamma_N$$

From this we conclude that N is isolated in its orbit ΓN in N. Recall once again that Γ_N is discrete and conclude from 1.14 that Γ is discrete. □

A slight modification of the proof of Jacob Nielsen's theorem yields the following variation due to C.L. Siegel.

Theorem 4.6 A non-elementary subgroup Γ of $PGl_2(\mathbb{R})$ is discrete if and only if the elliptic elements of Γ do not accumulate at ι.

Proof Let us assume that Γ is non-elementary and that the elliptic elements do not accumulate at ι. Let us use 2.8 to pick a proper translation $\sigma \in \Gamma$ with axis s, say, and introduce the group Γ_s made up of the translations in Γ with axis s. Let us now pick a neighbourhood V of ι in $PGl_2(\mathbb{R})$ such that

$$<N,\tau N> < 0 \quad \text{and} \quad \tau\sigma\tau^{-1}\sigma^{-1} \text{ is not a proper rotation} \qquad ; \sigma,\tau \in V$$

Let us analyse a pair of elements $\tau \in V - \Gamma_s$, $\sigma \in \Gamma_s \cap V$, $\sigma \neq \iota$. Observe that τ is not a half-turn around a point P of s: a half-turn π around P satisfies $<N,\pi N> = 1$. We conclude that $\tau(s) \neq s$. Let us remark that s and $\tau(s)$ cannot have a common end as follows from 2.5 applied to σ and $\tau\sigma\tau^{-1}$. If $\tau(s)$ intersects s, then we can observe that

$$\sigma(\tau\sigma^{-1}\tau^{-1})\sigma^{-1}(\tau\sigma^{-1}\tau^{-1})^{-1} = (\tau\sigma\tau^{-1}\sigma^{-1})^{-2}$$

is not a rotation and we conclude from 3.4 that inequality 4.3 holds. If $\tau(s)$ and s have a common perpendicular then we conclude from $<N,\tau N> < 0$ that σ^{-1} and $\tau\sigma\tau^{-1}$ translate in opposite directions, compare the proof of 3.2. We conclude from 3.1 that inequality 4.2 holds. This is sufficient to conclude that Γ_s is discrete, so let us pick a definite generator $\sigma \in \Gamma_s$ and change the neighbourhood V of ι such that

$$<N,\tau N> < 0 \quad \text{and} \quad \tau\sigma\tau^{-1}\sigma^{-1} \text{ is not a proper rotation} \qquad ; \tau \in V$$

At this point we leave the remaining details to the reader. □

IV.5 CUSPS

In this section we shall be concerned with the action of a discrete group Γ on the boundary of the hyperbolic plane H^2. We say that a point $S \in \partial H^2$ is a cusp for H^2 if it is fixed by some proper parabolic transformation $\beta \in \Gamma$. The stabiliser Γ_S of a cusp $S \in \partial H^2$ consists of horolations and reflections in geodesics through S, 2.4. By a horodisc we understand a region in H^2 bounded by a horocycle.

Proposition 5.1 Given a cusp $S \in \partial H^2$ for the discrete group Γ of isometries of H^2. There exists a horodisc D with centre $S \in H^2$ such that

$$\gamma(D) \cap D = \emptyset \qquad ; \gamma \in \Gamma - \Gamma_S$$

Proof We shall work with the point ∞ of the Poincaré upper half-plane. Moreover, we shall assume that the group $\Gamma_S{}^+$ is generated by $A = \begin{bmatrix} 1 & 1 \\ 0 & 1 \end{bmatrix}$. Let $B = \begin{bmatrix} a & b \\ c & d \end{bmatrix} \in Gl_2(\mathbb{R})$ with $det\, B = \pm 1$ represent $\beta \in \Gamma - \Gamma_S$. According to lemma 5.3 we have $|c| \geq 1$. An elementary estimate based on II.8.3 and II.8.4 gives us

$$Im[\beta(x + iy)] = y\,|c(x \pm iy) + d|^{-2} \leq y\,c^{-2}y^{-2} \leq y^{-1}$$

Otherwise expressed

5.2 $$Im[z]\; Im[\beta(z)] \leq 1 \qquad ; z \in H^2,\; \beta \in \Gamma - \Gamma_\infty$$

It follows that the horodisc $Im[z] > 1$ will serve our purpose. □

Lemma 5.3 Let us consider two matrices A and B from $Gl_2(\mathbb{R})$

$$A = \begin{bmatrix} 1 & 1 \\ 0 & 1 \end{bmatrix}, \quad B = \begin{bmatrix} a & b \\ c & d \end{bmatrix} \qquad c \neq 0,\ det B = \pm 1$$

If A and B generate a discrete subgroup of $PGl_2(\mathbb{R})$ then $|c| \geq 1$.

Proof Let us proceed under the assumption that $|c| < 0$. In order to derive a contradiction we consider the sequence of matrices $(B_n)_{n \in \mathbb{N}}$ given by $B_0 = B$ and

$$B_{n+1} = B_n A B_n^{-1} \qquad ; \ n \in \mathbb{N}$$

We ask the reader to establish the recursion formula

$$\begin{bmatrix} a_{n+1} & b_{n+1} \\ c_{n+1} & d_{n+1} \end{bmatrix} = \begin{bmatrix} 1-a_n c_n & a_n^2 \\ -c_n^2 & 1+a_n c_n \end{bmatrix} \qquad ; \ n \geq 1$$

A simple induction on $n \in \mathbb{N}$ gives us that

$$|c_n| = |c_0|^{2^n} \quad \text{and} \quad |a_n| \leq n + |a_0| \qquad ; \ n \in \mathbb{N}$$

Using $c_0 = c$ and $|c| < 1$ we conclude from this that $c_n \to 0$ and $a_n c_n \to 0$ for $n \to \infty$. Put this information into the right hand side of the recursion formula and conclude that $B_n \to A$ for $n \to \infty$. Let us assume that the corresponding classes α and β generate a discrete subgroup of $PGl_2(\mathbb{R})$. Then we get that $\beta_n = \alpha$ for n large and deduce, by descending induction on n, that ∞ is fixed by $\ldots, \beta_2, \beta_1, \beta_0$. Since $\beta = \beta_0$ this contradicts the assumption $c \neq 0$. □

With the notation of 5.1, let us observe that the orbit of a point $A \in \partial D$ does not meet D: for $\gamma \in \Gamma_S$ we have $\gamma(\partial D) = \partial D$ and for $\gamma \in \Gamma - \Gamma_S$ we find that $\gamma(\partial D)$ and D are disjoint.

Proposition 5.4 Let $S \in \partial H^2$ be a cusp for the discrete group Γ of isometries of H^2. For any compact subset K of H^2 there exists a horodisc D with centre $S \in H^2$ such that $\gamma(D) \cap K = \emptyset$ for all $\gamma \in \Gamma$.

Proof Let us return to the proof of 5.1 and choose $r>1$ such that K lies in the vertical strip of the Poincaré half-plane bounded by $Im[z] = r$ and $Im[z] = r^{-1}$. For $\gamma \in \Gamma_S$ we have $\gamma(D) = D$ and for $\gamma \in \Gamma - \Gamma_S$ we have $\gamma(D) \cap K = \emptyset$ as a consequence of the inequality 5.2. □

Corollary 5.5 Given cusps $S,T \in \partial H^2$ in different Γ–orbits, for any horodisc E with centre T there exists a horodisc D with centre S such that $\gamma(D) \cap E = \emptyset$ for all $\gamma \in \Gamma$.

Proof Let us pick a compact subset K of ∂E such that $\Gamma_T K = \partial E$. Use 5.4 to choose a horodisc D with centre S such that $D \cap \Gamma K = \emptyset$. This gives $\Gamma D \cap \partial E = \emptyset$. For $\gamma \in \Gamma$ we conclude from the connectedness of γD that $\gamma D \subset E$ or $\gamma D \cap E = \emptyset$. The first option can be ruled out on the grounds that $\gamma(S) \neq T$. □

Corollary 5.6 If H^2/Γ is compact, then the group Γ contains no proper horolations.

Proof Let us first observe that the projection $H^2 \to H^2/\Gamma$ maps open subsets to open subset. Next, cover H^2 with open discs and apply the Borel–Heine theorem to the images of the discs in the compact space H^2/Γ. The upshot is that we can find a finite set of discs in H^2 which meet all Γ-orbits. It follows that we can find a compact set K in H^2 which meets all Γ-orbits. We conclude from 5.4 that Γ contains no horolations. □

Horocyclic Topology For a discrete group Γ of isometries of H^2 let us consider the subset Y of $\overline{H}^2$ consisting of all ordinary points of H^2 and the set of cusps for Γ. Let us introduce the horocyclic topology on Y: a subset W of Y is open if for all points $S \in W$ there exists a disc/horodisc with centre S entirely contained in W. The horodiscs with centres at a cusp $S \in \partial H^2$ form a fundamental system of neighbourhoods for S, while the ordinary discs with a given centre $A \in H^2$ form a fundamental system of neighbourhoods of $A \in Y$.

The group Γ acts continuously on Y, which gives the orbit space $X = Y/\Gamma$ a structure of topological space. This new space is a two-dimensional <u>topological manifold with boundary</u>:

Proposition 5.7 With the notation above, the topological space $X = Y/\Gamma$ is a Hausdorff space in which every point has a neighbourhood homeomorphic to a neighbourhood of 0 in $\mathbb{R}^2$ or a neighbourhood of 0 in $\mathbb{R}\times[0,\infty[$.

Proof The space $X = Y/\Gamma$ is Hausdorff as follows from 5.4 and 5.5. In order to construct a neighbourhood of a cusp point, it suffices to treat the point ∞ of the Poincaré half-plane. It follows from 5.1 that it suffices to treat the case where $\Gamma = \Gamma_\infty$. Let us assume that Γ is generated by $z \mapsto z+k$, $k>0$. The space Y/Γ is homeomorphic to the open unit disc D in the complex plane

$$Y/\Gamma \xrightarrow{\sim} D \;,\quad z \mapsto exp(2\pi i k^{-1} z) \;,\; z \in H^2$$

In the general case we can arrange for Γ to be generated by $z \mapsto -\bar{z}$ and $z \mapsto z+k$ as above. In this case we find that the orbit space X is homeomorphic to the orbit space for the action of complex conjugation on the open unit disc D in $\mathbb{C}$.

In order to investigate a finite point, it suffices to treat the origin 0 in the Poincaré disc D. According to 1.5 (for more details see VI.5.8) we may assume that the full group Γ fixes 0. Let us assume that Γ^+ is generated by $z \mapsto \theta z$, where θ is a primitive nth root of unity. Using the well known fact that $z \mapsto z^n$ maps open set to open set, we can make the identification

$$H^2/\Gamma^+ \xrightarrow{\sim} D$$

A non-trivial element of Γ/Γ^+ corresponds to reflection of D in a line through 0. □

Corollary 5.8 The space $X = Y/\Gamma$ is compact if and only if there exists a compact subset K of H^2 and a finite number of horocycles $D_1,\dots,D_r$ whose centres $S_1,\dots,S_r$ are cusps for Γ such that any Γ-orbit in H^2 meets $K \cup D_1 \cup \dots \cup D_r$.

In the case of a Fuchsian group Γ, the proof of 5.7 provides a structure of Riemann surface [Farkas, Kra] for the space $X = Y/\Gamma$.

When the space X is compact, it is called the horocyclic compactification of H^2/Γ, compare VII.5. For general information see [Lang] and [Shimura].

IV EXERCISES

EXERCISE 1.1 Let Γ be a subgroup of $Gl_2(\mathbb{R})$ and put $\Delta = \Gamma \cap Sl_2(\mathbb{R})$. Let $P\Gamma$ and $P\Delta$ denote images of Γ and Δ in $PGl_2(\mathbb{R})$ and assume that $[P\Gamma : P\Delta]$ is finite. Show that $P\Gamma$ is discrete in $PGl_2(\mathbb{R})$ if and only if $P\Delta$ is discrete in $PGl_2(\mathbb{R})$.

EXERCISE 1.2 The additive group $\mathbb{R}$ of real numbers acts on $\mathbb{R}/2\pi\mathbb{Z}$. Show that the $\mathbb{Z}$–orbit of $1 \ mod \ 2\pi$ is not discrete.

EXERCISE 2.1 1° Show that the group of rotations around a point $A \in H^2$ equals its normaliser in $PSl_2(\mathbb{R})$.

2° Show that group of even isometries fixing a given geodesic line k equals its own normaliser in $PSl_2(\mathbb{R})$.

3° Show that the group of even isometries fixing a given end S equals its own normaliser in $PGl_2(\mathbb{R})$.

EXERCISE 2.2 Let α and β be translations whose axes have a common perpendicular. Show that the commutator $\alpha\beta\alpha^{-1}\beta^{-1}$ is a translation.

EXERCISE 2.3 1° Show that if a subgroup Γ of $PSl_2(\mathbb{R})$ has a finite orbit on H^2 of order n, then $n = 1$. Hint: If $n \geq 2$, show that $\gamma^{n!} = \iota$ for all $\gamma \in G$.

2° Show that if a subgroup Γ of $PSl_2(\mathbb{R})$ has a finite orbit of order n on ∂H^2, then $n = 1,2$. Hint: If $n \geq 3$ show that $\gamma^{n!} = \iota$ for all $\gamma \in \Gamma$.

EXERCISE 2.4 Let $\kappa \in PGl_2(\mathbb{R})$ be a proper translation with fixed points A,B.

1° Show that for any point $Q \notin \{A,B\}$ the set $\{\kappa^n(Q) \mid n \in \mathbb{Z}\}$ accumulates at A and B. Hint: Take A=0 and B=∞, in which case $\kappa = \begin{bmatrix} r & 0 \\ 0 & r^{-1} \end{bmatrix}$.

2° Show that a non-empty κ-stable closed subset Z of $\hat{\mathbb{R}}$ contains A and B.

EXERCISE 3.1 Let $S \in Gl_2(\mathbb{R})$ and put $U = \frac{1}{2}(S^{\check{}} - S)$. Show that $U \in sl_2(\mathbb{R})$ is an eigenvector for σ and use $det S = (\frac{1}{2}\, tr S)^2 - tr\, S^2$ (exercise I.3.1) to prove that

$$<U,U> = \tfrac{1}{4}\, det S\, (4 - tr^2 S)$$

2° For a second matrix $K \in sl_2(\mathbb{R})$ prove that

$$<U,K> = tr SK - \tfrac{1}{4}\, tr S\; tr K$$

3° Use this to give an alternative proof of lemma 2.10.

EXERCISE 4.1 1° For $A \in H^2$ and $r > 0$ fixed, prove that the set below is compact

$$\{\, N \in N(sl_2(\mathbb{R})) \mid |<N,A>| \le r \,\}$$

Hint: Pick a basis S,T for the tangent space of H^2 at A and use that any point of F can be written in the form $N = xA + yS + zT$, $x,y,z \in \mathbb{R}$.

2° Let $\mathfrak{D}$ be a discrete subset of $N(sl_2(\mathbb{R}))$. Given a point $A \in H^2$ and $r > 0$, show that only finitely many geodesics in H^2 with normal vectors in $\mathfrak{D}$ will meet the hyperbolic disc with centre A and radius r.

V FUNDAMENTAL DOMAINS

The most important way to get information about a discrete group Γ of isometries of H^2 is through fundamental domains. The art of reading information about discrete group from a fundamental domain dates from the last century. A good example for this is the "modular figure" which served Gauss as a fundamental domain for $Gl_2(\mathbb{Z})$. The modular figure can be constructed by a general method due to Dirichlet to be presented in V.4.

The finer analysis of a fundamental domain involves a decomposition of its boundary into linear pieces which are paired together through side transformations. A good fundamental domain leads to a representation of the group by generators and relations, compare Poincaré's theorem in chapter VII.

V.1 THE MODULAR GROUP

By a <u>fundamental</u> <u>domain</u> U for a group Γ of isometries of H^2 we understand an open subset U of H^2 such that U and $\gamma(U)$ are disjoint for all $\gamma \neq \iota$ and such that each Γ-orbit meets $\bar{U}$. As a basic example, let us construct a fundamental domain for the modular group $PSl_2(\mathbb{Z})$.

$\zeta = exp\, 2\pi i/3$

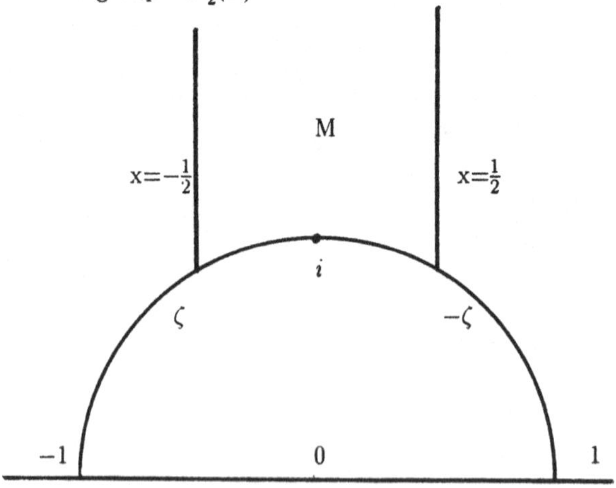

The modular figure 1.1 The three geodesics in the upper half-plane H^2

$$Re[z] = \tfrac{1}{2}, \quad Re[z] = -\tfrac{1}{2}, \quad |z| = 1$$

bound an open set M which is a fundamental domain for the group $PSl_2(\mathbb{Z})$.

Proof Let us fix a point $w = yi$, $y > 1$ and introduce the three transformations

$$\sigma(z) = -z^{-1}, \quad \tau(z) = z+1, \quad \omega(z) = z-1$$

We ask the reader to verify that the three geodesics bounding M are the perpendicular bisectors for $w,\omega(w)$, $w,\sigma(w)$ and $w,\tau(w)$. In order to find a point of $\bar{M}$ in the orbit through a given point $z \in H^2$ we shall minimise the hyperbolic distance $d(w,\mu(z))$, $\mu \in PSl_2(\mathbb{Z})$. Precisely if $\nu \in Sl_2(\mathbb{Z})$ is chosen such that

$$d(w,\nu(z)) \leq d(w,\mu(z)) \qquad ; \mu \in Sl_2(\mathbb{Z})$$

Then $\nu(z) \in \bar{M}$ as follows from the inequality

$$d(w,\nu(z)) \leq d(w,\omega^{-1}(z)) = d(\omega(w),\nu(z))$$

and similar inequalities with ω replaced by σ and τ.

Let us consider an orbit $\mathcal{O}$ of $Sl_2(\mathbb{Z})$ in H^2 and let us show that $Im[z]$ is a bounded function of $z \in \mathcal{O}$ which assumes its upper bound whenever $z \in \bar{M}$. In concrete terms, let us consider a $z \in \bar{M}$ and $\mu \in \Gamma$ represented by

$$\mu(z) = \frac{az+b}{cz+d} \qquad ; \ z \in \bar{M}, \ \mu = \begin{bmatrix} a & b \\ c & d \end{bmatrix} \in Sl_2(\mathbb{Z})$$

Let us note the inequality

$$|cz+d|^2 = c^2|z|^2 + 2Re[z]\,cd + d^2 \geq c^2 + d^2 - |cd| = (|c| - |d|)^2 + |cd| \geq 1$$

and draw the conclusion from the elementary formula

$$Im\left[\frac{az+b}{bz+d}\right] = Im[z]\,|cz+d|^{-2}\,det\begin{bmatrix} a & b \\ c & d \end{bmatrix}$$

When $|z| > 1$ and $c \neq 0$, then the inequality above is strict which gives us that $Im[\mu(z)] < Im[z]$. When $c = 0$, the transformation μ takes the form $z \mapsto z+b$ and $Im[z]$ is preserved. We conclude that the orbit $\mathcal{O}$ meets $\bar{M}$ in at least one point and M in at most one point as required. Taking into account that M is open, we find that an orbit which meets ∂M will not meet M. We ask the reader to verify that the figure $\tilde{M}$, which is the union of M and the part of ∂M which lies in the half-plane $Re[z] \geq 0$, meets each $Sl_2(\mathbb{Z})$-orbit in precisely one point. The identification of ∂M can be realised by the transformations σ,τ introduced above. □

1.2

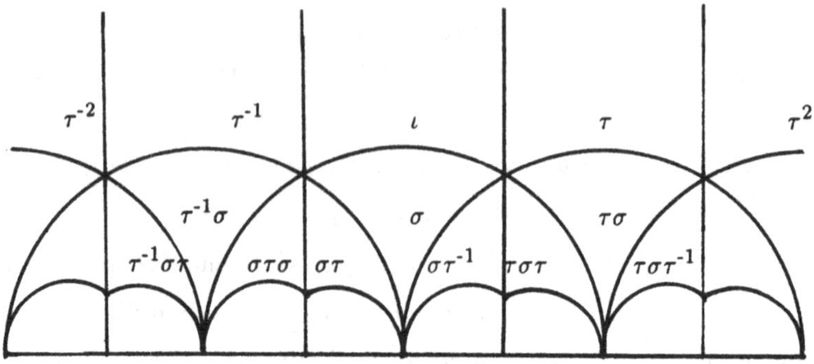

We shall see in the next section that the modular group is generated by σ and τ. The transformation $\rho = \tau\sigma$ is a rotation of order 3. Thus we have the relations

1.3 $$\sigma^2 = \iota \ , \quad \rho^3 = \iota \qquad ; \ \rho = \tau\sigma$$

It follows from Poincaré's theorem VII.3.7 that this set of relations is complete.

The hyperbolic triangle Δ with vertices i, ζ, ∞ is a fundamental domain for the <u>extended</u> <u>modular</u> <u>group</u> $PGl_2(\mathbb{Z})$ as follows from the fact that reflection β in the y–axis divides the fundamental domain for the modular group into two halves. Let us introduce reflections α,β,γ in the sides of Δ as given by the figure

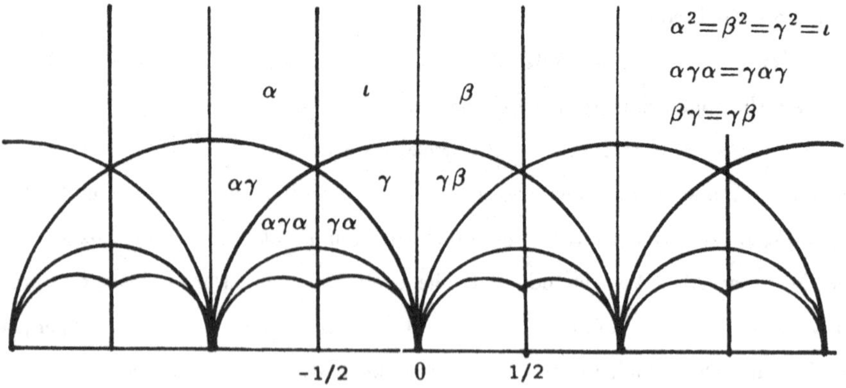

Let us construct a fundamental domain for some subgroups of $PSl_2(\mathbb{Z})$ using the general lemma

Lemma 1.4 Let Γ denote a discrete group of isometries of H^2 and Π a subgroup of Γ of finite index. If D is a fundamental domain for Γ and $S \subseteq \Gamma$ is a full set of representatives for $\Pi\backslash\Gamma$, then the interior U of

$$F = \bigcup_{\alpha \in S} \alpha\bar{D}$$

is a fundamental domain for Π.

Proof Since S is finite, we conclude that F is the closure of $\bigcup_{\alpha \in S} \alpha D$. Since this set is open, it must be contained in U (the interior of F), and it follows that $F = \bar{U}$. Let us show that the Π-orbit of a given point $z \in H^2$ meets F. To this end choose $\gamma \in \Gamma$ with $\gamma z \in \bar{D}$ and write $\gamma^{-1} = \pi^{-1}\alpha$ with $\alpha \in S$ and $\pi \in \Pi$ to get that $\alpha^{-1}\pi z \in D$ or $\pi z \in \alpha\bar{D}$.

Let us analyse a $\pi \in \Pi$ and a point $u \in F$ with $\pi u \in U$. Pick $\alpha \in S$ such that $u \in \alpha\bar{D}$ and let us observe that $\pi^{-1}F$ is a neighbourhood of u. It follows that $\pi^{-1}F$ meets αD or $F \cap \pi\alpha D \neq \emptyset$. Since F is the closure of $\bigcup_{\beta} \beta D$, we conclude that $\pi\alpha D$ meets βD for some $\beta \in S$. This means $\pi\alpha = \beta$ which implies $\pi = \iota$. □

The level 2 modular group Let us describe a fundamental domain for the group G(2) given by the exact sequence

1.5
$$0 \to G(2) \to PGl_2(\mathbb{Z}) \to Gl_2(\mathbb{F}_2) \to 0$$

The group $Gl_2(\mathbb{F}_2)$ is isomorphic to the permutation group S_3 through the action of $Gl_2(\mathbb{F}_2)$ on the three lines in $\mathbb{F}_2 \oplus \mathbb{F}_2$. This group is lifted back to $PGl_2(\mathbb{Z})$ by the dihedral group D_3 generated by α and γ whose matrices are

$$\begin{bmatrix} -1 & -1 \\ 0 & 1 \end{bmatrix}, \begin{bmatrix} 0 & 1 \\ 1 & 0 \end{bmatrix}$$

According to lemma 1.4 the union of the six translates of Δ by D_3 is a fundamental domain for G(2). It follows from the illustration above that this union is the hyperbolic triangle $0, -1, \infty$.

The *level 2* modular group $\Gamma(2) = G(2)^+$ has fundamental domain the quadrangle $1,0,-1,\infty$ as follows from 1.4 applied to the reflection β. We shall see in the next section that $\Gamma(2)$ is generated by

$$\mu(z) = z + 2 \ , \quad \nu(z) = \frac{z}{2z+1}$$

In fact we shall see as a consequence of Poincaré's theorem VII.3.7 that the level 2 modular group $\Gamma(2)$ is *free on these two generators*.

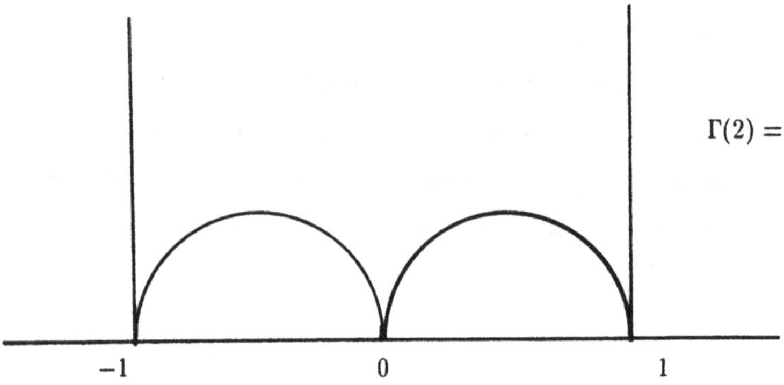

$\Gamma(2) = G(2)^+$

V.2 LOCALLY FINITE DOMAINS

Let Γ denote a discrete group of isometries of the hyperbolic plane H^2 and let us consider a fundamental domain $P \subseteq H^2$ which is locally finite in the sense that any point z of H^2 has a neighbourhood which meets $\sigma(P)$ for at most finitely many $\sigma \in \Gamma$. We shall relate the group theory of Γ and the topology of H^2/Γ to the geometry of $\bar{P}$.

Proposition 2.1 Let $P \subseteq H^2$ be a locally finite fundamental domain for the discrete group Γ. The canonical projection is a homeomorphism

$$p: \bar{P}/\Gamma \xrightarrow{\sim} H^2/\Gamma$$

Proof Since our map p is continuous, we must prove that p maps open subsets of $\bar{P}/\Gamma$ to open subsets of H^2/Γ. To this end it suffices to show that for any given (relative) open subset B of $\bar{P}$ the set

$$V = \cup_{\gamma \in \Gamma} \gamma(B)$$

is an open subset of H^2. It suffices to consider a point $b \in B$ and construct an open disc with centre b entirely contained in V. Let us use the fact that P is locally finite to see that the following subset S of Γ is finite

$$S = \{\sigma \in \Gamma \mid \sigma^{-1}(b) \in \bar{P}\}$$

Since P is locally finite we can choose $\epsilon>0$ so small that

$$D(b;\epsilon) \cap \sigma(\bar{P}) = \emptyset \qquad ; \sigma \notin S$$

The disc $D(b;\epsilon)$ is contained in the union of all Γ-translates of $\bar{P}$, thus

$$D(b;\epsilon) \subseteq \bigcup_{\sigma \in S} \sigma(\bar{P})$$

Let us take advantage of the fact that S is finite to decrease $\epsilon > 0$ further to get

$$D(\sigma^{-1}(b);\epsilon) \bigcap \bar{P} \subseteq B \qquad ; \sigma \in S$$

Let us combine these two pieces of information to get

$$D(b;\epsilon) = \bigcup_{\sigma \in S} \sigma(\bar{P}) \cap D(b;\epsilon) = \bigcup_{\sigma \in S} \sigma(\bar{P} \cap D(\sigma^{-1}(b);\epsilon)) \subseteq \bigcup_{\sigma \in S} \sigma(B)$$

This concludes the construction of the required disc. □

Corollary 2.2 Let $P \subseteq H^2$ be a locally finite fundamental domain for the discrete group Γ. Then H^2/Γ is compact if and only if P is bounded.

Proof A closer look at the proof of 2.1 reveals that the projection $P \to H^2/\Gamma$ maps (relative) open subsets of P onto open subsets of H^2/Γ. If H^2/Γ is compact then we can use Borel–Heine to find a finite set of open discs $D_1,\ldots,D_n$ with centres $z_1,\ldots,z_n$ in $\bar{P}$ such that each Γ-orbit in H^2 meets one of the sets $\bar{P} \cap D_1,\ldots,\bar{P} \cap D_n$. It follows that P is contained in the union of these discs. □

Proposition 2.3 Let $P \subseteq H^2$ be a locally finite fundamental domain for the discrete group Γ of isometries of H^2. The group Γ is generated by

$$\{\gamma \in \Gamma \mid \gamma(\bar{P}) \cap \bar{P} \neq \emptyset\}$$

Proof Let Π denote the subgroup of Γ generated by this subset. For $z \in H^2$ choose $\sigma \in \Gamma$ with $z \in \sigma(\bar{P})$ and observe that the class $[\sigma] \in \Gamma/\Pi$ is independent of σ. In other words we have defined a map

$$p: H^2 \to \Gamma/\Pi$$

This map is surjective since the Γ-translates of $\bar{P}$ cover H^2. The map π is locally constant: at a given point $z \in H^2$ we can use the fact that P is locally finite to find a finite subset S of Γ such that $z \in \sigma(\bar{P})$ for all $\sigma \in S$ and such that $N = \bigcup_{\sigma \in S} \sigma(\bar{P})$ is a neighbourhood of z in H^2; it follows that p takes the value $[\sigma]$, $\sigma \in S$. We can use the fact that H^2 is connected to conclude that p is constant. Taking into account that p is surjective, we conclude that $\Pi = \Gamma$. □

Let us conclude this section with some results on unbounded fundamental domains. The problem is to relate cusp points for Γ to "boundary" points of a locally finite fundamental domain P. The <u>horocyclic boundary</u> $\partial_h P$ is the set of points $R \in \partial H^2$ such that any horodisc with center R meets P. We ask the reader to compare this material with section IV.5 on cusps.

Proposition 2.4 Let P be a locally finite fundamental domain for Γ and $A \in \partial H^2$ for which there exists a point $L \in \bar{P}$ such that the geodesic segment $[L,A[$ is contained in $\bar{P}$. A transformation $\gamma \in \Gamma^+$ with $\gamma(A) = A$ is parabolic.

Proof The transformation γ is elliptic, hyperbolic or parabolic. A proper elliptic transformation has no fixed points on ∂H^2. Let us assume that γ is proper hyperbolic with translation axis h. Upon replacing γ by γ^{-1} we may assume that γ translates towards A. Let the horocycle with centre A through L intersect h in W and put $r = d(L,W)$. Let us contradict that P is locally finite by showing that

$$D(W;r) \bigcap \gamma^{-n}(\bar{P}) \neq \emptyset \qquad ; \ n \in \mathbb{N},\ n \geq 1$$

For $n \geq 1$, let $L_n \in [L,A[$ denote the point of intersection between $[L,A[$ and the horocycle with centre A through $\gamma^n(W)$. From the inequality

$$d(L_n, \gamma^n(W)) < d(L,W)$$

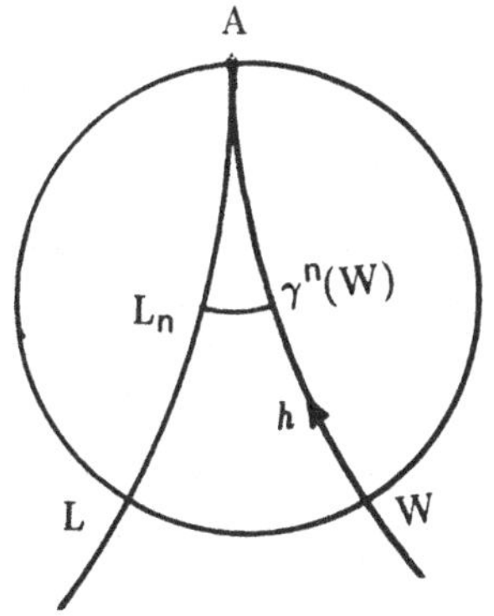

it follows that $\gamma^{-n}(L_n) \in D(W;r)$ as required. □

Example 2.5 Let Γ denote the group of isometries of the Poincaré half-plane generated by $\gamma(z) = 2z$, $z \in H^2$. The region P bounded by the two curves

$$x + i(2|x| + 1)\ ,\ \ x + i(2|x| + 2) \qquad ; x \in \mathbb{R}$$

is a locally finite fundamental domain. Observe that $\gamma(\infty) = \infty$ and that any horodisc with centre ∞ meets P. Thus $\infty \in \partial_h(P)$ but ∞ is not a cusp point for Γ.

Proposition 2.6 Let $A \in \partial H^2$ be a cusp for the discrete group Γ and D a horodisc with centre A. A locally finite fundamental domain P for Γ meets (at least one but) at most finitely many Γ-translates of D.

Proof Let us assume that the horodisc D is closed and use the remarks following IV.5.3 to pick an open horodisc U contained in D such that there exists a full Γ-orbit outside $\bar{U}$. Let $\gamma \in \Gamma$ be a parabolic element with fixed point A. Moreover, we pick a geodesic h emanating from A and let C be the compact subset of $D - U$ bounded by h and $\gamma(h)$. Observe that $D-U$ is union of translates of C by powers of γ. Let us enumerate the Γ-translates of P which meet C

$$\sigma_1(P),\ldots,\sigma_n(P) \qquad ; \sigma_1,\ldots,\sigma_n \in \Gamma$$

Let us now assume that $\sigma^{-1}(D)$ meets P or what amounts to the same thing that $\sigma(P)$ meets D. Assume for a moment that $\sigma(P) \subseteq U$; this implies $\sigma(\bar{P}) \subseteq \bar{U}$ which contradicts the existence of a full Γ-orbit outside $\bar{U}$. In conclusion $\sigma(P)$ meets $D - U$. It follows that we can find $s \in \mathbb{Z}$ such that $\gamma^s\sigma(P)$ meets C. This proves that $\gamma^s\sigma = \sigma_i$ for some $i = 1,\ldots,n$, and we conclude that $\sigma(D) = \sigma_i\gamma^{-s}(D) = \sigma_i(D)$. It follows that P meets at most finitely Γ-translates of D. □

Corollary 2.7 Let Γ denote a discrete group and P a locally finite fundamental domain for Γ. The Γ-orbit of a cusp meets the horocyclic boundary $\partial_h P$.

Proof Let us pick a horodisc K with its centre at a cusp A and let

$$K,\ \sigma_1(K),\ldots,\sigma_n(K) \qquad ; \sigma_1,\ldots,\sigma_n \in \Gamma$$

be a complete enumeration of the Γ-translates of K which meet P. For any horodisc L contained in K we find that at least one of the horodiscs

$$L,\ \sigma_1(L),\ldots,\sigma_n(L) \qquad ; \sigma_1,\ldots,\sigma_n \in \Gamma$$

will meet P since $\bar{P}$ meets any Γ-orbit. From this it follows that at least one of the points $A,\sigma_1(A),\ldots,\sigma_n(A)$ belongs to $\partial_h P$. □

V.3 CONVEX DOMAINS

In this section we shall study a group Γ of isometries of H^2 and a fundamental domain P, which is convex and locally finite. Recall that a subset R of H^2 is convex[1] if for any two points A,B ∈ R the geodesic arc [A,B] ⊆ R. The main objective is to relate the geometry of ∂P to Γ. It turns out that although no polygonal structure is supposed from the outset, the boundary comes out as a geodesic polygon. Let us start with a basic result on open convex subsets of H^2.

Hahn–Banach principle 3.1 Let U be an open convex subset H^2. Through any point A ∈ H^2 outside U passes a geodesic not meeting U.

Proof Let S^1 denote the unit circle in the tangent space $T_A(H^2)$. Let us define a map $p{:}U \to S^1$ as follows: for Q ∈ U draw the oriented geodesic from Q to A and let $p(Q) \in S^1$ denote its unit tangent vector. The map p is continuous and the image $V = p(U)$ is an open subset of S^1 which does not contain antipodal points. Let $\mathcal{A}{:}S^1 \to S^1$ denote the antipodal map, and observe that V and $\mathcal{A}(V)$ are non-empty open disjoint subsets of S^1. Since S^1 is connected we can find a point $s \in S^1$ not in $V \cup \mathcal{A}(V)$. It follows that neither s nor $\mathcal{A}(s)$ belongs to V. It follows that the geodesic through A with tangent vector s does not meet U. □

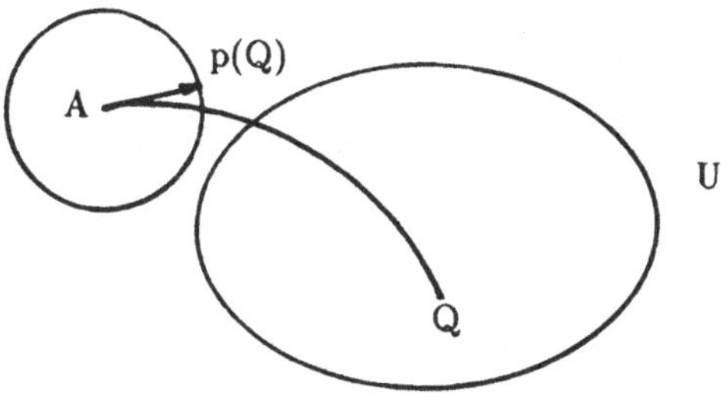

[1]In terms of Klein's disc model II.5, the set R is convex in the hyperbolic sense if and only if R is a convex subset of the ambient Euclidean plane.

Corollary 3.2 Let U be an open convex subset of H^2. For points $A \in U$ and $B \in \partial U$ the geodesic segment $[A,B[$ is contained in U.

Proof If $Q \in [A,B[$ is not in U we can draw a geodesic h through Q not meeting U. The closed half-plane H with $\partial H = h$ and $A \in H$ will contain U but not B. This makes B an exterior point for U. □

Corollary 3.3 Let U be a convex open subset of H^2. The closure $\bar{U}$ is a convex subset of H^2.

Proof Let us first show that we can write

$$\bar{U} = \bigcup_{H,\, H \supseteq U} H$$

where the union is taken over the set of all closed half-planes H in H^2 containing U. To this end consider a point $w \notin \bar{U}$ with the intention of constructing a closed half-plane $H \supseteq U$ with $w \notin H$. Pick a base point $u \in U$ and a point $z \in]u,w[$ outside U and use the Hahn–Banach principle to draw a geodesic h through z not meeting U. The closed half-plane H with $\partial H = h$ and $u \in H$ does the job. The representation of $\bar{U}$ as a union of closed half-spaces shows immediately that $\bar{U}$ is a convex subset of H^2. □

Corollary 3.4 Let U be a convex open subset of H^2. The interior of the closure $\bar{U}$ equals U.

Proof We must show that any neighbourhood of a given point $z \in \partial U$ meets the complement of $\bar{U}$. To see this, use the Hahn–Banach principle to draw a geodesic h through z not meeting U. Since U is connected, it is contained in a closed half-plane, say, H bounding h. It follows that H contains $\bar{U}$, but any neighbourhood of z meets the complement of H. □

Lemma 3.5 Let P be a locally finite convex fundamental domain for the discrete group γ. A set of the form

$$\bar{P} \cap \sigma(\bar{P}) \qquad ; \sigma \neq \iota \, , \sigma \in \Gamma$$

is either empty or a closed geodesic segment. A set of the form

$$\bar{P} \cap \sigma(\bar{P}) \cap \tau(\bar{P}) \qquad ; \ \iota \neq \sigma \neq \tau \neq \iota, \ \sigma,\tau \in \Gamma$$

is empty or reduced to a point.

Proof From $\sigma(P) \cap P = \emptyset$ it follows that $\partial P \cap \sigma(P) = \emptyset$ and $\sigma^{-1}(P) \cap \partial P = \emptyset$. Write $\bar{P} = P \cup \partial P$ and $\sigma(\bar{P}) = \sigma(P) \cup \sigma(\partial P)$ to get that

$$\bar{P} \cap \sigma(\bar{P}) = \partial P \cap \sigma(\partial P) \qquad ; \sigma \in \Gamma \, , \sigma \neq \iota$$

We conclude that $\bar{P} \cap \sigma(\bar{P})$ is a convex subset of ∂P. Let us show that $\bar{P} \cap \sigma(\bar{P})$ is contained in a geodesic line. If not, we can find three points of $\bar{P} \cap \sigma(\bar{P})$ not on a geodesic. An inner point z of the triangle based on these points furnishes an interior point of ∂P , contradicting 3.4.

Let us investigate an interior point z of $s = \bar{P} \cap \sigma(\bar{P})$. Pick points $x,y \in s$ such that $z \in \,]x,y[$ and consider a point $u \in P$. The interior Δ_u of the triangle with base [x,y] and vertex u is contained in P, 3.4. Similarly, the interior $\Delta_{\sigma u}$ of the triangle with base [x,y] and vertex $\sigma(u)$ is contained in $\sigma(P)$. It follows that we can find an open disc D with centre z such that $D \cap s$ is a diameter which divides D into two open half-discs D' and D" entirely contained in P and $\sigma(P)$ respectively. The half-discs D' and D" are disjoint from $\tau(P)$. From $D \subseteq \bar{D}' \cup \bar{D}''$ we conclude that $D \cap \tau(P) = \emptyset$. We can now conclude the proof by recalling that the set $\bar{P} \cap \sigma(\bar{P}) \cap \tau(\bar{P})$ is convex. □

Definition 3.6 Let P be a locally finite convex fundamental domain for the discrete group Γ. By a <u>side</u> of P we understand a closed geodesic segment, not reduced to a point, of the form $\bar{P} \cap \sigma(\bar{P})$ with $\sigma \in \Gamma - \{\iota\}$. By a <u>vertex</u> of P we understand a point which can be presented in the form $\bar{P} \cap \sigma(\bar{P}) \cap \tau(\bar{P})$, $\sigma,\tau \in \Gamma - \{\iota\}$, $\sigma \neq \tau$.

Proposition 3.7 The boundary ∂P of a locally finite convex fundamental domain is the union of its sides.

Proof Let $z \in \partial P$ and consider the following subset S of Γ

$$S = \{\sigma \in \Gamma - \{\iota\} \mid z \in \sigma(\bar{P})\}$$

Since P is locally finite we can find an open disc D with centre z such that

$$D \subseteq \bar{P} \cup \bigcup_{\sigma \in S} \sigma(\bar{P})$$

Observe that $S \neq \emptyset$: the alternative gives $D \subseteq \bar{P}$ which according to 3.4 gives $z \in P$, contradicting $z \in \partial P$. As a consequence we have that

3.8 $$\partial P \subseteq \bigcup_{\sigma \in \Gamma - \{\iota\}} \sigma(\bar{P})$$

Returning to the situation at hand, we cannot have $\bar{P} \cap \sigma(\bar{P}) = \{z\}$ for all $\sigma \in S$ since $D - \{z\}$ is connected. Thus we have found a side of P through $z \in \partial P$. □

A side s of P generates a geodesic h. The boundary points for s in h are called the the <u>end points</u> for the side s. The number of end points for s is 0,1 or 2.

Proposition 3.9 Let s be a side of the locally finite convex fundamental domain P. A point $z \in s$ is a vertex for P if and only if z is an end point for s.

Proof We have already shown during the proof of 3.5 that an interior point of s is not a vertex of P. Let us investigate a point z of $s = \bar{P} \cap \sigma(\bar{P})$, which is not a vertex of P. From the previous proof we find an open disc D with centre z such that $D \subseteq \bar{P} \cup \sigma(\bar{P})$. This gives us a partition

$$D = D \cap P \bigcup D \cap \sigma(P) \bigcup D \cap s$$

Since $D - D \cap s$ is disconnected and $D \cap s$ is a geodesic arc through the centre z of D, we conclude that $D \cap s$ is a diameter of D. □

Corollary 3.10 A vertex A of P is contained in exactly two sides of P.

Proof Consider a open disc D with centre A which meets only those sides of P which have vertex Az. Let these sides be enumerated cyclically $s_1,s_2,\dots,s_r$. Note also that $D\cap P$ is connected since it is a convex set. Let us change the enumeration above such that $D\cap P\subseteq\angle(s_1,s_2)$. Observe that $s_i\cap D\subseteq\bar{P}$ for $i=1,\dots,r$ and conclude that $r=2$. □

A side s of P can be put in the form $s=\bar{P}\cap\sigma(\bar{P})$ where $\sigma\in\Gamma-\{\iota\}$. According to 3.6 the transformation σ is unique and will be called the <u>side transformation</u> σ_s generated by s. Let us put $*s=\bar{P}\cap\sigma^{-1}(\bar{P})$ and observe that $\sigma_s(*s)=s$. The assignment $s\mapsto *s$ is an involution on the set of sides of P which is called the <u>side pairing</u> of P.

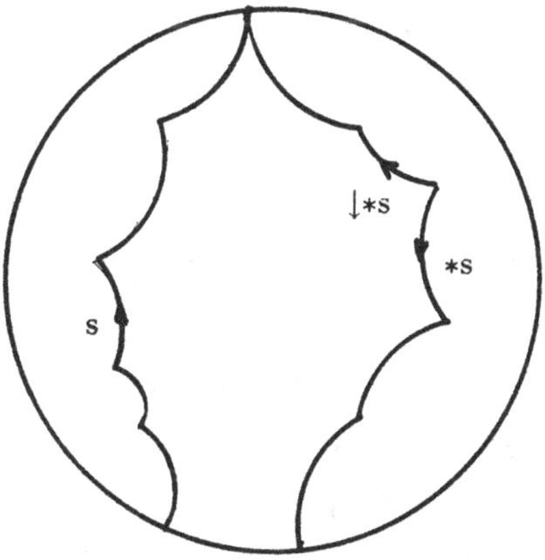

By an <u>edge</u> of P we understand an oriented side. The operator $*$ extends in a natural way to an operator on the set of $\mathcal{E}$ of edges of P. We shall in particular be interested in the subset $\mathcal{E}_+$ of edges whose initial vertex is finite. For an edge s with initial vertex A we let $\downarrow s$ denote the second edge with initial vertex A, compare 3.8. Let us combine these two operators to obtain a new operator

$$\Psi s=\downarrow *s \qquad ; \ s\in\mathcal{E}_+$$

The operator Ψ is a bijection since it is composed of two involutions.

Lemma 3.11 The operator $\Psi: \mathcal{S}_+ \to \mathcal{S}_+$ has finite orbits.

Proof Let us consider a sequence of iterates of Ψ on $s \in \mathcal{S}$

$$s_1, \ldots, s_r, \ldots \qquad ; \; s_i = \Psi^i s \, , \; s \in \mathbb{Z}$$

Observe that the side transformation $\sigma_i = \sigma_{s_i}$ maps the initial vertex A_i of s_i onto the initial vertex A_{i-1} of s_{i-1}. It follows that the sequence $A = A_1, \ldots, A_r, \ldots$ of initial points all belong to the finite set $P \cap \Gamma A$. □

Key lemma 3.12 Let s_1 be an edge of P with initial vertex A and let

$$s_1, \ldots, s_r, \ldots \qquad \sigma_1, \ldots, \sigma_r, \ldots$$

be the sequence of Ψ-iterates of s_1 and corresponding side transformations. There exists $n \geq 2$ with $\sigma_1 \ldots \sigma_n = \iota$ and such that the first of the two unions

$$\bigcup_{i=1}^{n} \sigma_1 \sigma_2 \ldots \sigma_i(\bar{P}) \; , \quad \bigcup_{i=1}^{n} \sigma_1 \sigma_2 \ldots \sigma_i(P)$$

is neighbourhood of A while the second union is a disjoint union.

Proof Let us make use of the orientation of H^2 which has P on the positive side of $\downarrow s_1$. Let us define the sign of an edge s of P by $sign(s) = +1$ when P is on the positive side of s and $sign(s) = -1$ otherwise. Observe that

$$sign(*s) = -\, det(\sigma_s)\, sign(s) \qquad ; \; s \in \mathcal{S}$$

Using $sign(\downarrow s) = -\, sign(s)$ we get for $s = s_{i-1}$ that

$$sign(\downarrow s_i) = det(\sigma_{i-1})\, sign(\downarrow s_{i-1}) \qquad ; \; i = 2, \ldots$$

Let us investigate the angle $\angle_{or}(\downarrow s_i, s_i)$ with vertex A_i. We have that

$$\angle_{or}(\downarrow s_i, s_i) \equiv sign(\downarrow s_i) \angle_{int} A_i \quad mod \; 2\pi \qquad ; \; i = 1, \ldots$$

Observe that $\sigma_1 \ldots \sigma_{i-1} \bar{P}$ has vertex $A = \sigma_1 \sigma_2 \ldots \sigma_{i-1}(A_i)$ and angle

$$\angle_{or}(\sigma_1 \ldots \sigma_{i-1} \downarrow s_i, \sigma_1 \ldots \sigma_{i-1} s_i) = det(\sigma_1 \ldots \sigma_{i-1}) \angle_{or}(\downarrow s_i, s_i)$$

From the above formulas and the normalisation $sign(\downarrow s_1) = 1$ we get that

$$\angle_{or}(\sigma_1 \ldots \sigma_{i-1} \downarrow s_i, \sigma_1 \ldots \sigma_{i-1} s_i) \equiv \angle_{int} A_i \quad mod \; 2\pi$$

Let us also observe that

$$\sigma_1 \ldots \sigma_{i-1} \downarrow s_i = \sigma_1 \ldots \sigma_{i-1} * s_{i-1} = \sigma_1 \ldots \sigma_{i-2} s_{i-1}$$

It is now clear that the integer n for which $\sum_{i=1}^{n} \angle_{int} A_i = 2\pi$ does the job. □

Let $s_1,\ldots,s_r$ be a full edge cycle, i.e. a full orbit for Ψ on $\mathcal{S}_+$. The corresponding sequence $A_1,\ldots,A_r$ of initial points is called a vertex cycle. The second vertex cycle which starts out with A_1 is $A_1,A_r,A_{r-1},\ldots,A_2$, as follows by observing that $\downarrow s_1,\downarrow s_r,\downarrow s_{r-1},\ldots,\downarrow s_2$ form a full edge cycle.

Theorem 3.13 An interior point z of $\bar{P}$ is Γ–equivalent to no other point of $\bar{P}$. An interior point z of a side s of P is Γ–equivalent to the point $\sigma_s^{-1}(z)$ of $\bar{P}$, but to no other point of $\bar{P}$. Two vertices of P are equivalent under Γ if and only if they belong to the same vertex cycle.

Proof Let us investigate an interior point z of a side $s=\bar{P}\cap\sigma(\bar{P})$. A point $w\in\bar{P}$ equivalent to z can be written in the form $w=\tau(z)$, $\tau\in\Gamma$. This implies $z\in\tau^{-1}(\bar{P})\cap\bar{P}\cap\sigma(\bar{P})$ and we conclude that $\tau=\iota$ or $\tau=\sigma^{-1}=\sigma_s^{-1}$. Let us now turn to the case where A is a vertex of P and let us consider $\gamma\in\Gamma$ such that $\gamma(A)\in\bar{P}$. This means that A belongs to the closure of $\gamma^{-1}(P)$ and we conclude from 3.12 that we can find $i=1,\ldots,n$ such that

$$\gamma^{-1}(P)\bigcap\sigma_1\ldots\sigma_i(\bar{P})\neq\emptyset$$

This gives $\gamma^{-1}=\sigma_1\sigma_2\ldots\sigma_i$. Evaluation of this formula at the vertex A_{i+1} gives $\gamma^{-1}(A_{i+1})=A$ or $A_i=\gamma(A)$. □

Corollary 3.14 Let P be a locally finite convex fundamental domain for Γ. The group Γ is generated by the side transformations σ_s, as s runs through the sides of P.

Proof According to 2.3 it suffices to show that the subgroup of Γ generated by the side transformations contains any $\gamma\in\Gamma$ for which there exists a $z\in\bar{P}$ with $\gamma(z)\in\bar{P}$. But this problem has already been dealt with during the proof of 3.13. □

Corollary 3.15 Let $s_1,...,s_r$ denote a full edge cycle for the operator Ψ, and let $\sigma_1,...,\sigma_r$ denote the corresponding side transformations. The cycle map

$$\sigma = \sigma_1 ... \sigma_r$$

is a rotation around the initial vertex A_1 of s_1. The order of σ is given by

$$2\pi = ord(\sigma)(\angle_{int} A_1 + ... + \angle_{int} A_r)$$

where $A_1,...,A_r$ are the initial vertices of the edge cycle.

Proof Let us return to the proof of 3.12. Evaluation of the formula $\sigma_1 .... \sigma_n = \iota$ on the edge s_{n+1} gives us $s_1 = s_{n+1}$. Taking into account that r is the length of the cycle, we find that $q = n/r$ is an integer. From the proof of 3.12 we get that

$$sign(s_{r+1}) = det(\sigma_1 ... \sigma_r)\ sign(s_1)$$

Since $s_{r+1} = s_1$, we get that $det\sigma = 1$ which makes σ a rotation. Observe that

$$2\pi = \sum_{i=1}^{n} \angle_{int} A_i = q \sum_{i=1}^{r} \angle_{int} A_i$$

and the result follows. □

Proposition 3.16 Let P be a locally finite convex fundamental domain for Γ and $A \in \partial_h P$ a cusp for Γ. The number of sides of P with end A is two.

Proof Let us first observe that for any point $Q \in P$ and any point $A \in \partial_h P$ the geodesic segment $[Q,A[$ is contained in P as follows from 3.1.

Let us assume that $A \in \partial_h P$ is a cusp and let us pick a horocycle D with center D. Recall from 2.6 that D meets at most finitely many translates of $\bar{P}$ and conclude that D meets at most finitely many sides of P. This allows us to shrink D such that all sides of P which meets D have ends at A. Let us observe that $D \cap P \neq \emptyset$ and $D \not\subseteq P$ and conclude from the connectedness of D that $D \cap \partial P \neq \emptyset$. It follows from 3.7 that $D \cap \partial P$ is traced on D by a number of geodesics $h_1,...,h_r$ through A. It follows from this that $D \cap P$ is fibred by traces of geodesics. Observe also $P \cap \gamma P = \emptyset$ for a generator γ of Γ_A. The remaining details are left of the reader. □

V.4 DIRICHLET DOMAINS

We shall present a method due to Dirichlet for construction of fundamental domains for a discrete group Γ of isometries of H^2. Let us recall from IV.1.9 that for a compact subset K of H^2 we have that $\sigma(K)$ meets K for finitely many σ only. This allows us to fix a point $w \in H^2$ with $\Gamma_w = \{\iota\}$.

For a non-trivial $\sigma \in \Gamma$ let $L_\sigma(w)$ denote the perpendicular bisector for w, $\sigma(w)$ and let $H_\sigma(w)$ denote the open half-plane of the complement of $L_\sigma(w)$ containing w.

4.1

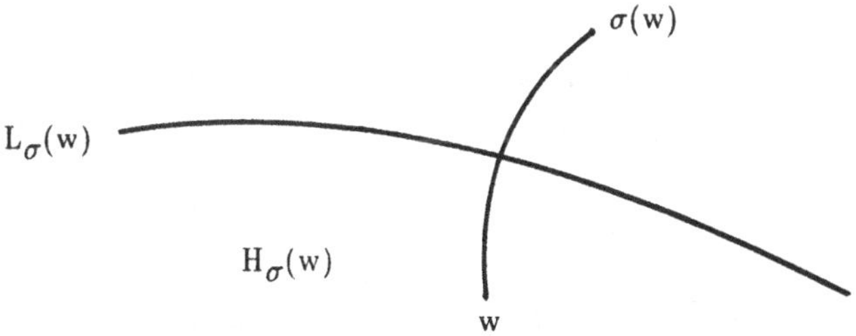

The Dirichlet domain with centre w is the set P(w) given by

4.2
$$P(w) = \bigcap_{\sigma \in \Gamma, \sigma \neq \iota} H_\sigma(w)$$

Proposition 4.3 The Dirichlet domain P(w) is a fundamental domain for Γ in H^2. Its closure $\bar{P}(w)$ is given by

$$\bar{P}(w) = \bigcap_{\sigma \in \Gamma, \sigma \neq \iota} \bar{H}_\sigma(w)$$

Proof Let us verify that the intersection $P(w) \cap D(w;r)$ is open for all $r > 0$. Observe that the set $S = \{\sigma \in \Gamma - \{\iota\} \mid \sigma(w) \in D(w;2r)\}$ is finite and that

$$P(w) \cap D(w;r) = \bigcap_{\sigma \in S} H_\sigma(w)$$

We find that P(w) is open, since $P(w) \cap D(w;r)$ is open for all r>0. Next, we ask the reader to verify the formula

$$\gamma(P(w)) = P(\gamma(w)) \qquad ; \gamma \in \Gamma$$

From this it follows immediately that P(w) and $\gamma(P(w))$ are disjoint for $\gamma \neq \iota$.

It is now time to verify the formula for $\bar{P}(w)$. Observe that the left hand side is included in the right hand side since the intersection of a family of closed sets is closed. To verify the opposite inclusion, consider a point $z \in H^2$ which belongs to $\bar{H}_\sigma(w)$ for all $\sigma \in \Gamma - \{\iota\}$. Notice that [w,z] belongs to $\bar{H}_\sigma(w)$ for all σ and conclude that [w,z[belongs to P(w); it follows that $z \in \bar{P}(w)$.

Let us show that any orbit $\mathcal{A}$ meets $\bar{P}(w)$. To this end pick a $z \in \mathcal{A}$ with the shortest possible distance from w. This gives

$$d(z,w) \leq d(\sigma^{-1}(z),w) \qquad ; \sigma \in \Gamma, \sigma \neq \iota$$

or $d(z,w) \leq d(z,\sigma(w))$ for all $\sigma \neq \iota$, which shows that $z \in \bar{H}_\sigma$ for all $\sigma \in \Gamma - \{\iota\}$. It follows that $z \in \bar{P}(w)$, since $\bar{P}(w)$ is the intersection of such sets. □

Proposition 4.4 Let Γ be a discrete group of isometries of H^2 and let w be a point of H^2 such that $\Gamma_w = \{\iota\}$. The Dirichlet domain P with centre w is a locally finite convex fundamental domain.

Proof Let us fix a number $r > 0$ and show that the open disc D(w;r) with centre w and radius $r > 0$ meets $\sigma(P)$ for at most finitely many $\sigma \in \Gamma$. If $\sigma(P)$ meets P, we can find $z \in P$ with $\sigma(z) \in D(w;r)$; this gives us

$$d(w,\sigma(w)) \leq d(w,\sigma(z)) + d(\sigma(z),\sigma(w)) < r + d(z,w)$$

On the other hand, from the fact that $z \in P$ it follows that

$$d(z,w) < d(z,\sigma^{-1}(w)) = d(\sigma(z),w) < r$$

But indeed, only finitely many $\sigma \in \Gamma$ satisfy $d(\sigma(w),w) \leq 2r$. □

The modular figure from 1.1 is in fact a Dirichlet domain for the modular group as follows from the following lemma.

Lemma 4.5 With the notation of 4.2 let S be a subset of $\Gamma-\{\iota\}$ such that

$$P_S(w) = \bigcap_{\delta \in S} H_\delta(w)$$

meets each Γ-orbit in at most one point. Then $P_S(w) = P(w)$.

Proof Let us consider a $\tau \in \Gamma$ and investigate two alternatives. If $L_\tau(w)$ does not meet $P_S(w)$ then $P_S(w) \subseteq H_\tau(w)$ as follows from the fact that $P_S(w)$ is connected and contains w. If $L_\tau(w) \cap P_S(w) \neq \emptyset$ then we can find a point z which is an interior point of $P_S(w) - P(w)$. Since $z \in \sigma(\bar{P}(w))$ for some $\sigma \in \Gamma$, $\sigma \neq \iota$, we can find a point of $\sigma(P(w))$ in $P_S(w) - P(w)$ contradicting the assumption that no orbit meets $P_S(w)$ twice. □

The reader will find much more information about Dirichlet domains in [Beardon]. In particular, the change of base point for the Dirichlet domain is discussed and many applications are given. See also [Macbeath, Hoare].

V EXERCISES

EXERCISE 1.1 The group $PGl_2(\mathbb{Z})$ contains the congruence group G(2) as a normal subgroup.

1° Show that $PGl_2(\mathbb{Z})$ has a unique subgroup N of index 2 containing G(2).

2° Show that the hyperbolic quadrangle i, $1+i$, $exp(\pi i/3)$, ∞ is a fundamental domain for N. Show that the sides of the quadrangle can be paired by a rotation about $exp(\frac{1}{3}\pi i)$ and a glide reflection fixing ∞, both belonging to N.

EXERCISE 1.2 Show that $PSl_2(\mathbb{Z})$ contains three subgroups of index 3 which contain $\Gamma(2)$. Find fundamental domains for these groups.

VI COVERINGS

In this chapter we shall introduce a number of geometric concepts which come up in the following problem, central for the proof of Poincaré's theorem in chapter VII: Let there be given a connected Hausdorff space X and a local homeomorphism $f:X \to H^2$; give conditions for this to be a global homeomorphism.

The first thing to do is to introduce a metric on X. This makes X into a hyperbolic surface with its own intrinsic geometry. The main result is the monodromy theorem, which asserts that $f:X \to H^2$ is a global homeomorphism if and only if X is a complete hyperbolic surface.

The geometric concepts we introduce create their own universe with its own life. We are led to consider the problem of space forms : to classify all geometries, which locally look like the hyperbolic plane. This will be done in terms of conjugacy classes of discrete subgroups of $PGl_2(\mathbb{R})$.

VI.1 HYPERBOLIC SURFACES

For the convenience of the reader we shall recall the definition of the arc length of a curve $\gamma:[a,b] \to X$ in a metric space X. To this end we consider a subdivision $a = a_0 \leq a_1 ... \leq a_n = b$ of our interval and form the sum

$$\sum_{i=1}^{n} d(\gamma(a_{i-1}),\gamma(a_i)) \qquad ; a = a_0 \leq a_1 ... \leq a_n = b$$

We say that the curve $\gamma:[a,b] \to X$ is rectifiable if γ is continuous and sums of the above form are bounded. The length of a rectifiable curve $\gamma:[a,b] \to X$ is defined to be the supremum over sums of this form taken over all subdivisions of [a,b]

1.1 $$l(\gamma) = sup \sum_{i=1}^{n} d(\gamma(a_{i-1}),\gamma(a_i))$$

It follows from the triangle inequality that $d(x,y) \leq l(\gamma)$. We shall mostly be concerned with metric spaces which have the following property.

Definition 1.2 We say that a metric space X has the shortest length property if, first of all, any two points x and y of X can be joined by a rectifiable curve γ. Moreover, it is required that

$$d(x,y) = inf_{\gamma} \; l(\gamma)$$

where the infimum is taken over all rectifiable curves in X joining x to y.

The hyperbolic plane H^2 has the shortest length property: the distance d(x,y) between $x,y \in H^2$ is realized as the arc length of the geodesic arc [x,y]. We shall be concerned with metric spaces X with the shortest length property which locally look like H^2.

Lemma 1.3 Let X be a metric space with the shortest length property and D an open disc in H^2 with centre a and radius r. For any isometry $\sigma: D \to X$ of D onto an open neighbourhood of $\sigma(a)$ in X we have $d(x,\sigma(a)) \geq r$ for all points $x \in X - \sigma(D)$.

Proof Let there be given a rectifiable curve γ in X which joins $\sigma(a)$ to the point x outside $\sigma(D)$. For a number $\rho \in]0,r[$ let K denote the compact disc in H^2 with radius ρ and centre a and let us prove that γ meets $\sigma(\partial K)$. Observe that $\sigma(K)$ is a compact subset of X and conclude that $V = X - \sigma(K)$ is an open subset since X is Hausdorff. This gives us a partition of X into the closed subset $\sigma(\partial K)$ and the two open subsets V and $U = \sigma(K - \partial K)$. Observe that γ meets V and U and conclude from the connectedness of the image of γ that γ meets $\sigma(\partial K)$ in y say. This gives us $l(\gamma) \geq d(\sigma(a),y) = \rho$. Take supremum over $\rho \in]0,r[$ to get that $l(\gamma) \geq r$. Finally, take infimum over all γ and conclude that $d(\sigma(a),x) \geq r$. □

Definition 1.4 By a hyperbolic surface we understand a metric space Y with the shortest length property, for which every point has an open neighbourhood isometric to an open disc in the hyperbolic plane H^2.

For a point y of a hyperbolic surface Y and a real number $r > 0$, $D(y;r)$ denotes the metric disc of points $v \in Y$ with $d(y,v) < r$. For $r > 0$ sufficiently small, $D(y;r)$ is isometric to a hyperbolic disc of radius r as follows from 1.3.

Local homeomorphisms. Let X and Y be topological spaces. A continuous map $f:X \to Y$ is called a local homeomorphism if each point $x \in X$ has an open neighbourhood which is mapped by f homeomorphically onto an open neighbourhood of $f(x)$. When X and Y are metric spaces, a map $f:X \to Y$ is called a local isometry if any point $x \in X$ has an open neighbourhood which is mapped isometrically onto an open neighbourhood of $f(x)$.

We shall consider a connected Hausdorff space X and a hyperbolic surface Y and demonstrate that a local homeomorphism $f:X \to Y$ induces a natural metric on X, which makes X into a hyperbolic surface and $f:X \to Y$ into a local isometry.

Let us first show that any two points $u,v \in X$ can be joined by a continuous curve γ such that $f\gamma$ is rectifiable in Y. To see this, fix the point u and separate the space X into the set U of points $v \in X$ which can be joined to u by such a curve and the set V of points $v \in X$ which can't. The assumption that X is connected ensures that $V = \emptyset$. Let us put

1.5 $$d(u,v) = inf_{\gamma}\ l(f\gamma) \qquad ; u,v \in X$$

where the infimum is taken over all continuous curves γ in X joining x to y such that $f\gamma$ is rectifiable in Y. We shall establish, during the proof of 1.6, that this defines a metric on X.

Proposition 1.6 Let X be a connected Hausdorff space and Y a hyperbolic surface and $f:X \to Y$ a local homeomorphism. The distance function given in 1.5 defines a metric on X, which turns X into a hyperbolic surface and $f:X \to Y$ into a local isometry.

Proof The distance $d(u,v)$ between $u,v \in X$, as defined in 1.5, is obviously symmetrical in u and v and satisfies the triangle inequality. We shall show below that distinct points have non-zero distance. Let us first record that

$$d(u,v) \geq d(f(u),f(v)) \qquad ; u,v \in X$$

Let us fix a point $x \in X$ and let W be an open neighbourhood of x which is mapped by f homeomorphically onto a hyperbolic disc $D(\sigma(x);r)$ in Y $(r > 0)$. For points u and v in W let λ be the curve in W such that $f\lambda$ is the geodesic arc from f(u) to f(v). This gives $d(u,v) \leq l(f\lambda) = d(f(u),f(v))$ and we conclude that

$$d(u,v) = d(f(u),f(v)) \qquad ; u,v \in W$$

Let us show that $d(x,z) > 0$ for any point $z \neq x$. If $z \in W$, this follows from the formula above. For $z \notin W$, let γ be any continuous curve in X with $f\gamma$ rectifiable which joins x and z. Pick $\rho \in]0,r[$ and let us show that $f\gamma$ meets the circle S in Y with centre $y = \sigma(x)$ and radius ρ. To see this, let us prove that γ meets the inverse image S' of S by $f:W \to Y$ (we shall give an outline only since a similar argument has been used in the proof of 1.3). Observe that the complement of S' in X is partitioned into two open sets: the inverse image D' of $D(y;\rho)$ by $f:W \to Y$ and the complement in X of the inverse image K' of $\bar{D}(y,\rho)$ by $f:W \to Y$. Finally use that the image of γ is connected to conclude that γ meets S'. It follows that $l(f\gamma) \geq \rho$. Let us make a variation of γ and conclude that $d(x,z) \geq \rho$.

We claim that a continuous curve $\gamma:[a,b] \to X$ is rectifiable in X if and only if $f\gamma$ is rectifiable in Y. To see this, use the Borel–Heine theorem to introduce a subdivision $a = a_0 \leq a_1 \ldots \leq a_n = b$ of [a,b] such that the restriction $\gamma:[a_{i-1},a_i] \to X$ takes values in a hyperbolic disc W as considered above. It follows that

$$l(\gamma) = \sum_i l(\gamma_i) = \sum_i l(f\gamma_i) = l(f\gamma)$$

This reveals that X has the shortest length property. □

VI.2 HOPF-RINOW THEOREM

In this section we shall be concerned with the geodesics on a hyperbolic surface X. The basic problem is to join two points x and y of X by a geodesic curve of arc length equal to the distance between x and y. Substantial results will only be obtained when X is complete in the sense that any Cauchy sequence on X is convergent.

To begin let us start with a geodesic curve $\gamma: J \to X$ defined on the interval J, compare II.1, and let us show that the following inequality is satisfied

2.1 $$d(\gamma(s),\gamma(t)) \leq |t-s| \qquad ; \quad s,t \in J$$

This follows from a straightforward subdivision argument based on the theorem of Borel–Heine and the triangle inequality.

Proposition 2.2. Let X be a hyperbolic surface. If two geodesic curves α,β: $\mathbb{R} \to X$ agree on a non-empty open interval $J \subseteq \mathbb{R}$ then they agree on all of $\mathbb{R}$.

Proof Let us assume that $0 \in J$ and let us introduce the following set

$$E = \{s \in]0,+\infty[\mid \alpha \text{ and } \beta \text{ agree on } [0,s]\}$$

From the very definition of E it follows that E is a non-empty interval with 0 as its left end point. Let us show that E is an open subset of $\mathbb{R}$. To this end we consider a point $s \in E$ and a small open disc D in X with centre $\alpha(s) = \beta(s)$. We conclude from our study (made during the proof of II.4.8) of geodesics in H^2 that α and β agree in a neighbourhood of $s \in \mathbb{R}$.

Let us show that E is unbounded. To this end we must rule out the alternative $E =]0,p[$ for some $p \in \mathbb{R}$: by the definition of E we find that α and β agree on $[0,p[$. By continuity of α and β, this implies that α and β agree on $[0,p]$, and we get $p \in E$, a contradiction. This argument shows that α and β agree on $[0,+\infty[$. The interval $]-\infty,0]$ is handled in a similar manner. □

Proposition 2.3 Let X be a complete hyperbolic surface. A geodesic curve γ:J$\to$X defined on an open interval J can be extended to a geodesic $\mathbb{R} \to$ X.

Proof Let us first remark that the result is true for $X = H^2$ as follows from the proof of II.4.8. In the general case let us assume that $0 \in J$ and try to extend γ to the right of J. To this end put

$$\mathfrak{J} = \{ x \in]0,+\infty[\mid \gamma \text{ extends to }]0,x[\}$$

The set $\mathfrak{J}$ is a subinterval of $]0,+\infty[$. Let us prove that $a \in \mathfrak{J}$ implies $a+\epsilon \in \mathfrak{J}$ for some $\epsilon>0$. From the inequality 2.1 we conclude that $\gamma:[0,a[\to X$ is uniformly continuous. From the completeness of X it follows[1] that γ can be extended to a continuous curve on [0,a]. In order to extend γ to the right of the point a consider a hyperbolic disc in X with centre $\gamma(a)$ and use the result for H^2. It follows that $\mathfrak{J} =]0,+\infty[$. Extension of γ to the left of J can be done in a similar way. □

Let us now consider two fixed points x and y of a hyperbolic surface X. The distance d = d(x,y) can almost (1.2) be realized as a the length of a curve connecting x to y, but not quite in general. Consider for example the punctured Poincaré disc D−{0} and the points x = −1/2 and y = 1/2.

On a complete surface, however, we can realize the distance as the length of a geodesic connecting x to y:

Hopf-Rinow theorem 2.4 Let X be a complete hyperbolic surface. For any two points x and y on X of distance d = d(x,y), there exists a geodesic curve $\sigma:\mathbb{R}\to X$ with $\sigma(0) = x$ and $\sigma(d) = y$.

---

[1] We are referring to the following general fact: let X be a complete metric space, any uniformly continuous curve $\gamma:[0,a[\to X$ can be extended to a continuous curve on [0,a]. A proof can be based on the observation that a uniformly continuous map γ preserves Cauchy sequences.

Proof Let us introduce an abbreviation which will facilitate the exposition and underline the strategy of the proof

$$\mathcal{HR}(x,z,y) : \left\{ \begin{matrix} \text{x and z satisfy the Hopf–Rinow theorem} \\ \text{and} \qquad d(x,z) + d(z,y) = d(x,y) \end{matrix} \right\}$$

Let us first show that for any two distinct points x and y of X, we can find $z \neq x$ such that $\mathcal{HR}(x,z,y)$. To this end choose an $r \in]0,d[$ such that the disc $D(x;\rho)$ is hyperbolic for some $\rho > r$. The circle $S(x;r)$ with centre x and radius r is compact as follows from 1.3. Let the continuous function

$$z \mapsto d(z,y) \quad ; \quad z \in S(x;r)$$

take its minimum at the point $z \in S(x;r)$. Consider a rectifiable curve $\gamma:[a,c] \to X$ with $\gamma(a) = x$ and $\gamma(c) = y$. Choose a point $b \in]a,c[$ with $d(x,\gamma(b)) = r$. We have

$$l(\gamma) = l(\gamma_{[a,b]}) + l(\gamma_{[b,c]}) \geq d(x,\gamma(b)) + d(\gamma(b),y) \geq r + d(z,y)$$

Take infimum over all rectifiable curves γ from x to y to get

$$d(x,y) \geq d(x,z) + d(z,y)$$

The opposite inequality is the triangle inequality and we have verified $\mathcal{HR}(x,z,y)$. Let us establish the basic relation

2.5 $$\mathcal{HR}(x,z,y) \ \& \ \mathcal{HR}(z,w,y) \Rightarrow \mathcal{HR}(x,w,y) \qquad ; x,y,z,w \in X$$

Simple manipulations of the two identities and the triangle inequality yield

$$d(x,w) + d(w,y) \leq d(x,z) + d(z,w) + d(w,y) = d(x,z) + d(z,y) = d(x,y)$$

Another application of the triangle inequality gives us

$$d(x,w) + d(w,y) = d(x,y)$$

Put this information back into the first inequality and deduce that in fact

$$d(x,z) + d(z,w) = d(x,w)$$

It follows from this and lemma 2.6 that x and w satisfy the Hopf–Rinow theorem.

To conclude the proof, let x and y be points with $d = d(x,y)$. Pick a point $v \neq x$ satisfying $\mathcal{HR}(x,v,y)$, put $e = d(x,v)$ and use 2.3 to construct a geodesic curve $\sigma:\mathbb{R} \to X$ with $\sigma(0) = x$ and $\sigma(e) = v$. We shall study the set

$$J = \{ s \in [0,d] \mid d(x,\sigma(s)) = s \ \& \ d(\sigma(s),y) = d - s \}$$

It is easily seen that the set J is a closed subinterval of [0,d] containing 0. Let us show that $s \in]0,d[\cap J$ implies $t \in J$ for some $t > s$. Observe that $z = \sigma(s)$ satisfies $\mathcal{HR}(x,z,y)$. Pick a point $w \neq z$ such that $\mathcal{HR}(z,w,y)$ and conclude from 2.8 that $\mathcal{HR}(z,w,y)$. It follows from 2.6 and 2.3 that w belongs to the image of σ. The parameter $t \in \mathbb{R}$ with $\sigma(t) = w$ satisfies our requirements. It follows from this that J=[0,d]. In particular $d(\sigma(d),y) = 0$, i.e. $\sigma(d) = y$. □

Lemma 2.6 Let $\gamma:[a,c]\to X$ be a curve in the metric space X and let there be given $b \in [a,c]$ such that

$$d(\gamma(a),\gamma(b)) + d(\gamma(b),\gamma(c)) = d(\gamma(a),\gamma(c))$$

If the restrictions $\gamma:[a,b]\to X$ and $\gamma:[b,c]\to X$ are distance preserving, then the curve $\gamma:[a,c]\to X$ is distance preserving.

Proof Let us first prove the following simple inequality

$$d(\gamma(s),\gamma(t)) \le t - s \qquad ; s,t \in [a,c],\ s \le t$$

It suffices to treat the case $s \in [a,b]$ and $t \in [b,c]$; the triangle inequality yields

$$d(\gamma(s),\gamma(t)) \le d(\gamma(s),\gamma(b)) + d(\gamma(b),\gamma(t)) = b - s + t - b = t - s$$

which completes the proof of the inequality. We need to prove the opposite inequality as well. To this end let us observe that

$$c - a = (c - b) + (b - a) = d(\gamma(a),\gamma(c)) \le d(\gamma(a),\gamma(s)) + d(\gamma(s),\gamma(t)) + d(\gamma(t),\gamma(c))$$

Let us combine this with the first inequality to get that

$$c - a \le s - a + d(\gamma(s),\gamma(t)) + c - t$$

which may be rewritten $t - s \le d(\gamma(s),\gamma(t))$. □

VI.3 UNIFORMIZATION

In this section we shall show that a complete hyperbolic surface X admits a local isometry of the form $f:H^2 \rightarrow X$, where H^2 is the hyperbolic plane. But first a result to the effect that plane figures cannot be isometric without being congruent. Such figures can't be deformed: they are rigid.

Rigidity 3.1 Let D be a subset of H^2 not contained in a hyperbolic line. Any isometry $\sigma:D \rightarrow H^2$ has a unique extension to an isometry of H^2.

Proof Let us pick three points A,B,C of D not on a geodesic. We can use lemma II.4.8 to pick an isometry $\tau:H^2 \rightarrow H^2$ which agrees with σ at the three points A,B,C. It follows from the proof of II.4.9 that the restriction of τ to D is actually equal to σ. □

Uniformization theorem 3.2 Let D be an open disc in the hyperbolic plane H^2. Any isometry $\phi:D \rightarrow X$ of D into a complete hyperbolic surface X has a unique extension to a local isometry $H^2 \rightarrow X$.

Proof Let O denote the centre of D and let us use 2.3 to extend the isometry $D \rightarrow X$ to a map $\phi:H^2 \rightarrow X$ whose restriction to each hyperbolic line through O is a geodesic. Let us fix a point A of H^2 and proceed to find an open disc with centre A which is mapped isometrically into X by ϕ. To this end use the Borel–Heine theorem to find an $r > 0$ such that for all $C \in [O,A]$ the metric disc $D(\phi(C);2r)$ is a hyperbolic disc. Let us first prove that

$$d(\phi(A),\phi(B)) = d(A,B) \qquad ; \quad B \in D(A;r)$$

To this end consider a subdivision of the segment $[O,A]$,

$$O = A_0,\ A_1,\ldots,A_{n-1},\ A_n = A \qquad ;\ d(A_{i-1},A_i) < r\ ,\ i = 1,\ldots,n$$

and let $B_0,B_1,\ldots,B_n$ be the perpendicular projections of the sequence of As onto the line through O and B. Let us prove[2] that for $i = 1,\dots,n$

$$\Delta A_{i-1}B_{i-1}A_i \approx \Delta\phi(A_{i-1})\phi(B_{i-1})\phi(A_i)$$
$$\Delta B_{i-1}A_iB_i \approx \Delta\phi(B_{i-1})\phi(A_i)\phi(B_i)$$

where $\approx$ means that the triangles are congruent in the sense that distances between pairs of corresponding vertices are equal.

This is done by induction on i. To accomplish the induction step, let us focus on the hyperbolic discs $D_i = D(A_i,2r)$ and $\mathfrak{D}_i = (\phi(A_i);2r)$. Observe at first that the restriction $\phi:[A_i,A_{i+1}]\to X$ has image in $\mathfrak{D}_i$ and conclude that

$$d(A_i,A_{i+1}) = d(\phi(A_i),\phi(A_{i+1}))$$

From the induction hypothesis we get that

$$d(\phi(A_i),\phi(B_i)) = d(A_i,B_i) \le d(A_n,B_n) \le r$$

Let us combine this with the inequality

$$d(\phi(B_i),\phi(T)) \le d(B_i,T) < r \qquad ; \; T \in [B_i,B_{i+1}]$$

to conclude that the restriction $\phi:[B_i,B_{i+1}]\to X$ has image in $\mathfrak{D}_i$. This gives

$$d(\phi(B_i),\phi(B_{i+1})) = d(B_i,B_{i+1})$$

Let us now focus on the following configuration in D_i

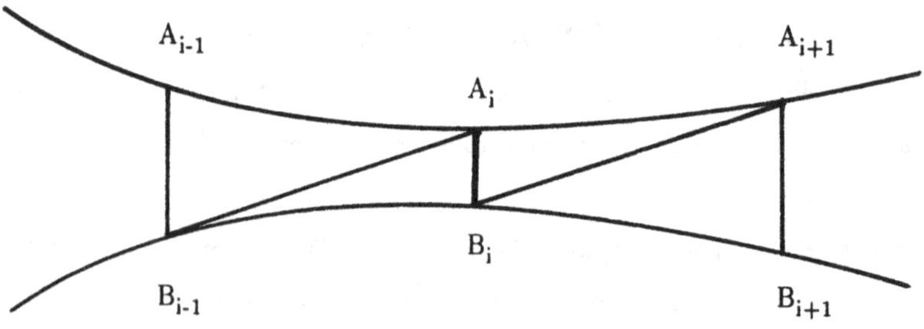

and the image configuration in $\mathfrak{D}_i$. From the induction hypothesis we conclude that

$$\angle A_{i-1}A_iB_{i-1} = \angle\phi(A_{i-1})\phi(A_i)\phi(B_{i-1})$$
$$\angle B_{i-1}A_iB_i = \angle\phi(B_{i-1})\phi(A_i)\phi(B_i)$$

with the conclusion that

---

[2]The details of the proof are elementary and can be administered in a number of ways. The reader is advised to work out his own version.

$$\angle A_{i-1}A_iB_i \quad = \quad \angle\phi(A_{i-1})\phi(A_i)\phi(B_i)$$

From this we draw the same conclusion for the complementary angle

$$\angle A_{i+1}A_iB_i \quad = \quad \angle\phi(A_{i+1})\phi(A_i)\phi(B_i)$$

We can now use elementary geometry to conclude that

$$\Delta A_{i+1}A_iB_i \quad = \quad \Delta\phi(A_{i+1})\phi(A_i)\phi(B_i)$$

In a similar way we can first deduce that

$$\angle B_{i-1}B_iA_i \quad = \quad \angle\phi(B_{i-1})\phi(B_i)\phi(A_i)$$

$$\angle A_iB_iA_{i+1} \quad = \quad \angle\phi(A_i)\phi(B_i)\phi(A_{i+1})$$

with the conclusion that

$$\angle A_{i+1}B_iB_{i+1} \quad = \quad \angle\phi(A_{i+1})\phi(B_i)\phi(B_{i+1})$$

Another reference to elementary geometry gives us

$$\Delta A_{i+1}B_iB_{i+1} \quad \approx \quad \Delta\phi(A_{i+1})\phi(B_i)\phi(B_{i+1})$$

which concludes the induction. Let us now prove that

$$\Delta A_nB_nB \quad \approx \quad \Delta\phi(A_n)\phi(B_n)\phi(B)$$

Observe that $\angle B_n = \angle\phi(B_n)$ is a right angle, and conclude as above that the corresponding sides have equal length. In conclusion $d(A,B) = d(\phi(A),\phi(B))$. Let us remark that at the same time we have proved that the following two angles are equal $\angle A_{n-1}AB = \angle\phi(A_{n-1})\phi(A)\phi(B)$.

It remains to prove that the distance between any two points B and C of D(A;r) is preserved. To see this, observe that the proof above shows that the angle $\angle OAB$ is preserved by ϕ. A similar remark applies to the point C and the result follows by hyperbolic trigonometry, III.5.1. □

The proof of the uniformization theorem is adapted from the Euclidean case given in [Nikulin, Shafarevich] II, §10. The reader is advised to consult this source for further comments.

VI.4 MONODROMY

In this section we shall be concerned with general properties of a local isometry $f:X \to Y$ between hyperbolic surfaces. It follows from rigidity 3.1 that a local isometry is an open map in the sense that any open subset U of X is mapped by f onto an open subset f(U) of Y. Let us establish an important inequality

4.1 $$\boxed{d(f(x),f(y)) \leq d(x,y)} \qquad ; x,y \in X$$

Proof Consider a point $x \in X$ and an $r > 0$ such that the metric disc $D(x;r)$ is a hyperbolic disc and such that the restriction $f:D(x;r) \to Y$ is an isometry. From the fact that f is open it follows that $D(x;r)$ is mapped homeomorphically onto an open neighbourhood of f(x) in Y. We conclude from lemma 1.3 that

$$f(D(x;r)) = D(f(x);r)$$

Let there be given points $x,y \in X$ and a continuous curve $\gamma:[a,b] \to X$ with $\gamma(a) = x$ and $\gamma(b) = y$. We ask the reader to use the Borel–Heine theorem to find an $r > 0$ such that for all $s \in [a,b]$ the point $\gamma(s)$ is the centre of a hyperbolic disc with radius r such that f maps $D(\gamma(s);r)$ isometrically onto $D(f\gamma(s);r)$. A simple subdivision argument shows that γ is rectifiable in X if and only if $f\gamma$ is rectifiable in Y. For a rectifiable curve γ in X we find that

$$l(f\gamma) = l(\gamma)$$

The formula 4.1 follows from variation of γ. □

Geodesic lifting property 4.2 Let $f: X \to Y$ be a local isometry between complete hyperbolic surfaces and consider points $x \in X$ and $y \in Y$ with $f(x) = y$. For any geodesic curve $\eta:\mathbb{R} \to Y$ with $\eta(0) = y$, there exists a unique geodesic curve $\xi: \mathbb{R} \to X$ with $f\xi = \eta$ and $\xi(0) = x$.

Proof Choose $r>0$ such that f induces an isometry $D(x;r) \xrightarrow{\sim} D(y;r)$. Next, choose an open interval J around 0 in $\mathbb{R}$ such that $\eta(J) \subseteq D(y;r)$. Let $\xi:J \to D(x;r)$

be given by $f\xi = \eta$. Let us use 2.3 to extend ξ to all of $\mathbb{R}$ and use 2.2 to conclude that the identity $f\xi = \eta$ is generally valid. □

Proposition 4.3 A local isometry $f:X \to Y$ between complete hyperbolic surfaces is surjective. Moreover, let there be given a point $y \in Y$ and a $\rho > 0$ such that $D(y;\rho)$ is hyperbolic, then for any $x \in X$ with $f(x) = y$ the disc $D(x;\rho)$ is mapped isometrically onto $D(y;\rho)$ by f.

Proof We have already observed that f(X) is open. Let us show that f(X) is closed. To this end consider a point w in the closure of f(X). Pick a hyperbolic disc D with centre w and a point $y \in D \cap f(X)$. Let us join the centre w of D with y along the radius of D and extend this to a geodesic $\eta:\mathbb{R} \to Y$. According to 4.2 we can find a geodesic $\xi:\mathbb{R} \to X$ with $\eta = f\xi$. From this it follows that all points of γ belong to f(X). Since f(X) is open and closed we can use the fact that Y is connected to conclude that $f(X) = Y$.

Let us now investigate the point x. For small values of $r > 0$, the map f induces an isometry $D(x;r) \tilde{\to} D(y;r)$. The inverse of this induces an isometry $s:D(y;r) \to X$ with $fs = \iota$ and $s(y) = x$. Let us use the uniformization theorem 3.2 to extend s to a local isometry $s:D(y;\rho) \to X$. The extension will still satisfy $s(y) = x$ and $fs = \iota$ as follows from 2.2 applied to the various radii in the disc $D(y;\rho)$. In fact s is an isometry: for $u,v \in D(y;\rho)$ we get from the inequality 4.1 applied twice

$$d(u,v) = d(f(s(u)),f(s(v)) \le d(s(u),s(v)) \le d(u,v)$$

which shows that $d(u,v) = d(s(u),s(v))$. We conclude from the proof of 4.1 that $s(D(y;\rho)) = D(x;\rho)$. Let us recall the relation $fs = \iota$ and it follows that $D(x;\rho)$ is mapped by f isometrically onto $D(y,\rho)$. □

We shall see that a local isometry $f:X \to Y$ between complete hyperbolic surfaces X and Y is a covering projection in the sense of homotopy theory. For more topological results and references see VI.8.

Corollary 4.4 Let $f:X \to Y$ denote a local isometry between complete hyperbolic surfaces and let $y \in Y$ and $\rho > 0$ be given such that the metric disc $D(y;\rho)$ is hyperbolic. For distinct points x and z of the fibre $f^{-1}(\{y\})$

$$D(x;\rho) \cap D(z;\rho) = \emptyset \qquad ; x,z \in f^{-1}\{y\} , x \neq z$$

Moreover, any point $w \in X$ with $f(w) \in D(y;\rho)$ is contained in the disc $D(x;\rho)$ for some $x \in f^{-1}\{y\}$.

Proof Let us put $D = D(y;\rho)$ and make a general remark on two continuous sections r,s of $f:X \to Y$ over $D \subseteq Y$. Let us show that

$$r(D) \cap s(D) \neq \emptyset \quad \Rightarrow \quad r = s$$

We may write a common point u of $r(D)$ and $s(D)$ in the form $u=r(v_1)=s(v_1)$, where $v_1,v_2 \in D$. This, however, gives $f(u) = v_1 = v_2$. Thus we have found a $v \in D$ with $r(v) = s(v)$. Observe that the connected set D is partitioned into the two sets

$$\{w \in D \mid r(w) \neq s(w) \} , \quad \{v \in D \mid r(v) = s(v)\}$$

The first set is open since r and s are continuous and X is Hausdorff. To see that the second set is open consider a point $v \in D$ with $r(v) = s(v)$; choose a neighbourhood U of $r(v) = s(v)$ such that the restriction of $f:X \to Y$ to U is injective and choose a neighbourhood V of v in D with $r(V) \subseteq U$ and $s(V) \subseteq U$. We have that

$$w = f(r(w)) = f(s(w)) \qquad ; \; w \in V$$

which gives $r(w) = s(w)$ for all $w \in V$. In conclusion we have partitioned the connected set D into two open subsets of which the second is non-empty. It follows that $r = s$.

To conclude the proof of the first statement, observe that at a point of the fibre $x \in f^{-1}\{y\}$ we can use the restriction of f to the disc $D(x;\rho)$ to get an isometry f_x: $D(x;\rho) \xrightarrow{\sim} D(y;\rho)$. The inverse $s_x = f_x^{-1}$ defines a section s_x of $f:X \to Y$ over D with $s_x(D) = D(x;\rho)$.

Consider a point $w \in X$ with $d(f(w),y) = r < \rho$. Let $\eta:\mathbb{R} \to Y$ denote a geodesic with $\eta(0) = f(w)$ and $\eta(r) = y$. Let us use 4.2 to find a geodesic $\xi:\mathbb{R} \to X$ with $f\xi = \eta$ and $\xi(0) = w$. This gives $f(\xi(r)) = \eta(r) = y$. In conclusion we have found a geodesic $\xi: \mathbb{R} \to X$ with $\xi(0) = w$ and $\xi(r) = x \in f^{-1}\{y\}$ where $r \in]0,\rho[$. It follows that $w \in D(x;\rho)$ as required. □

Proposition 4.5 Let $f:X \to Y$ be a local isometry of complete hyperbolic surfaces. The metric on Y can be recovered from that of X by the formula

$$d(y,w) = \mathit{inf} \; \{ \; d(x,v) \mid x,v \in X, \; f(x) = y, \; f(v) = w \; \}$$

Proof Use the Hopf–Rhiniw theorem 2.4 to pick a geodesic curve $\eta:[0,d] \to Y$ with $\eta(0) = y$ and $\eta(d) = w$, where $d = d(y,w)$. Let $\xi:[0,d] \to X$ be a lifting of η to X, compare 4.2. This gives points $x = \xi(0)$ and $v = \xi(d)$ with $f(x) = y$ and $f(v) = w$. It follows from the inequality 2.1 that $d(x,v) \leq d$. The required identity is now a consequence of the general inequality 4.1. □

A refinement of the argument above gives us

4.6 $$f(D(x;r)) = D(f(x);r) \qquad ; x \in X \, , \; r > 0$$

Proposition 4.7 Let Z be a complete hyperbolic surface. Any closed and bounded subset of Z is compact.

Proof By 3.2 we can find a local isometry $f:H^2 \to Z$. It follows from 4.3 that f is surjective and from 4.6 that all closed metric discs in Z are compact. □

Monodromy theorem 4.8 Let X be a complete hyperbolic surface. Any local isometry $\theta:X \to H^2$ is a bijection.

Proof Let us first show that any local isometry $\sigma:H^2 \to H^2$ is an isometry. To this end choose an open disc $D \subseteq H^2$ such that the restriction $\sigma':D \to H^2$ is an isometry. Next use 3.1 to extend σ' to a global isometry $\sigma'':H^2 \to H^2$. Observe that $\sigma = \sigma''$ as follows by restriction to hyperbolic lines through the centre of D.

In the general case, let us use the uniformization theorem 3.2 to construct a local isometry $\pi:H^2 \to X$. The composite $\theta\pi:H^2 \to H^2$ is a local isometry and consequently a bijection. We conclude from this that π is injective and from 4.3 that π is surjective. This shows that θ is a bijection as well. □

VI.5 ORBIT SPACES

Let us consider a hyperbolic surface X and a group G of isometries of X which acts discontinuously on X. By this we mean that any point $x \in X$ has an open neighbourhood U such that

5.1 $$\sigma(x) \notin U \quad \text{for all but finitely many } \sigma \in G$$

For the G-orbit $\mathcal{A}$ through x we conclude the existence of $r > 0$ such that

$$D(x;r) \cap \mathcal{A} = \{x\}$$

For two distinct points p,q of $\mathcal{A}$ we can choose $\sigma \in G$ with $\sigma(p) = x$ and conclude that $d(\sigma(p),\sigma(q)) > r$. Since σ is an isometry we can conclude that

5.2 $$d(p,q) \geq r \qquad ; p,q \in \mathcal{A}, \ p \neq q$$

Proposition 5.3 A discontinuous group G of isometries of the hyperbolic surface X acts with closed orbits on X.

Proof Let $\mathcal{A}$ be a G-orbit on X and z a point of X outside $\mathcal{A}$. With the notation of 5.2 the disc $D(z;r/2)$ intersects $\mathcal{A}$ in at most one point. From this it is easy to find a disc with centre z not meeting $\mathcal{A}$. □

We intend to put a metric on the space X/G of orbits. To get started let us introduce the distance between two G-orbits $\mathcal{A}$ and $\mathcal{B}$ by the formula

5.4 $$d(\mathcal{A},\mathcal{B}) = \inf\{\, d(x,y) \mid x \in \mathcal{A},\ y \in \mathcal{B} \,\}$$

Let us fix points $x \in \mathcal{A}$ and $y \in \mathcal{B}$. We can rewrite the definition above as

$$d(\mathcal{A},\mathcal{B}) = \inf\{\, d(\sigma(x),\tau(y)) \mid \sigma,\tau \in G \,\}$$

Since we can write $d(\sigma(x),\tau(y)) = d(x,\sigma^{-1}\tau(y))$, we conclude that

5.5 $$d(\mathcal{A},\mathcal{B}) = \inf\{\, d(x,y) \mid y \in \mathcal{B} \,\} \qquad ; x \in \mathcal{A}$$

Proposition 5.6 The distance function defines a structure of metric space on the orbit space X/G. This metric has the shortest length property.

Proof The distance between two orbits is symmetrical by 5.4. In order to prove the triangle inequality for orbits $\mathcal{A},\mathcal{B},\mathcal{C}$ let us fix a point $y \in \mathcal{B}$. This gives

$$d(\mathcal{A},\mathcal{C}) \leq d(x,z) \leq d(x,y) + d(y,z) \qquad ; \; x \in \mathcal{A}, \; z \in \mathcal{C}$$

Fix $x \in \mathcal{A}$ and take infimum over all $z \in \mathcal{C}$ and conclude from 5.5 that

$$d(\mathcal{A},\mathcal{C}) \leq d(x,y) + d(\mathcal{B},\mathcal{C}) \qquad ; \; x \in \mathcal{A}$$

Take infimum over all $x \in \mathcal{A}$ and apply 5.5 to get

$$d(\mathcal{A},\mathcal{C}) \leq d(\mathcal{A},\mathcal{B}) + d(\mathcal{B},\mathcal{C})$$

If $\mathcal{A} \neq \mathcal{B}$ we conclude from 5.5 and 5.3 that $d(\mathcal{A},\mathcal{B}) > 0$. Let $f:X \to X/G$ denote the projection. Observe that f decreases distances to conclude that a rectifiable curve γ in X joining x to v maps to a rectifiable curve $f\gamma$ in X/G with

$$l(\gamma) \geq l(f\gamma) \geq d(f(x),f(v))$$

For points $y,w \in Y$ choose $x,v \in X$ with $f(x) = y$, $f(v) = w$ and $d(x,v) = d(y,w)$. For $\epsilon > 0$ given, choose a rectifiable curve γ in X joining x to v which satisfies $l(\gamma) \leq d(x,z) + \epsilon$. From the inequality above we get that

$$l(f\gamma) \leq d(y,w) + \epsilon$$

which shows that X/G has the shortest length property. □

It follows from 5.1 that the canonical projection $f:X \to X/G$ satisfies

5.7 $$f(D(x;r)) = D(f(x);r) \qquad ; x \in X \,,\; r>0$$

Let us make a closer analysis of this in terms of the <u>stabiliser</u> of x given by

$$G_x = \{\, \sigma \in G \mid \sigma(x) = x \,\}$$

Let us notice that G_x acts on any disc $D(x;r)$ with centre x and radius $r > 0$. Thus we can form the orbit space $D(x;r)/G$.

Proposition 5.8 Let G be a group of isometries of X acting discontinuously on X. For any point x of X there exists $r > 0$ such that f induces an isometry of $D(x;r)/G_x$ onto $D(y;r)$.

Proof Let us use 5.1 to choose $r > 0$ such that

$$d(x,\sigma(x)) \geq 4r \qquad ; \quad \sigma \in G - G_x$$

Using the triangle inequality we get for $y \in D(x;r)$ and $\sigma \in G - G_x$ that

$$4r \leq d(x,\sigma(x)) \leq d(x,\sigma(y)) + d(\sigma(x),\sigma(y)) \leq d(x,\sigma(y)) + r$$

which allows us to conclude that $d(x,\sigma(y)) \geq 3r$ for $y \in D(x;r)$ and $\sigma \in G - G_x$. From this it follows that for $z,y \in D(x;r)$ we have that

$$inf\{\ d(z,\sigma(y)) \mid \sigma \in G\ \} = inf\{\ d(z,\sigma(y)) \mid \sigma \in G_x\ \}$$

which is the required result. □

Proposition 5.9 Let G be a discontinuous group of isometries of the hyperbolic surface X. The orbit space X/G is complete if and only if X is complete.

Proof When X is complete, we get from 5.7 that any closed metric disc in X is compact. It follows from 5.4 that any closed metric disc in X/G is compact. To prove that a Cauchy sequence $\{z_n\}$ in X/G is convergent we can refer to the elementary fact that a Cauchy sequence is bounded, i.e. contained in some closed metric disc in X/G. Let us finally recall the fact that a Cauchy sequence in a compact metric space is convergent.

Suppose conversely that X/G is complete and let us investigate a Cauchy sequence $\{x_n\}$ in X. The sequence $\{p(x_n)\}$ is a Cauchy sequence in X/G and converges towards a point $y \in X/G$, say. Let $\mathcal{Y}$ be the corresponding G-orbit in X and choose r>0 such that the discs $D(x;3r)_{x \in \mathcal{Y}}$ are disjoint. Let us choose an $n \in \mathbb{N}$ such that $d(p(x_n),y) < r$ and such that $d(x_n,x_p) < r$ for all $p \geq n$. Let the point $x \in \mathcal{Y}$ be given by the condition $d(x_n,x) < r$. It follows that

$$x_p \in D(x;2r) \qquad ; \quad p \geq n$$

Upon shrinking r we can obtain that the closure of $D(x;2r)$ is compact, compare 1.3. It follows that the sequence $\{x_n\}$ is convergent in X. □

VI.6 CLASSIFICATION

In this section we bring our theory of complete hyperbolic surfaces to culmination by classifying all such surfaces in terms of torsion free[3] discrete subgroups of $PGl_2(\mathbb{R})$. Let us observe that a torsion free discrete subgroup Γ of $PGl_2(\mathbb{R})$ acts discontinuously on H^2 with trivial stabilisers. It follows from 5.8 that the orbit space H^2/Γ is a complete hyperbolic surface.

It is useful to observe that the canonical projection $\pi: H^2 \to H^2/\Gamma$ has the following mapping property: a map $f: H^2/\Gamma \to Y$ into a hyperbolic surface Y is a local isometry if and only if $f\pi: H^2 \to Y$ is a local isometry.

Theorem 6.1 Any complete hyperbolic surface X is isometric to a surface of the form H^2/Γ where Γ is a torsion free discrete subgroup of $PGl_2(\mathbb{R})$. Two such subgroups Γ and Σ define isometric surfaces H^2/Γ and H^2/Σ if and only if Γ and Σ are conjugated subgroups of $PGl_2(\mathbb{R})$.

Proof Given a hyperbolic surface X. Use the uniformization theorem 3.2 to pick a local isometry $\pi: H^2 \to X$. Let us introduce the subgroup Γ of $PGl_2(\mathbb{R})$ by

$$\Gamma = \{ \sigma \in PGL_2(\mathbb{R}) \mid \pi\sigma = \pi \}$$

Let us investigate the action of Γ in a neighbourhood of a point $z \in H^2$. To this end pick an open disc D with centre z such that the restriction of $\pi: D \to X$ is injective. Let us show that

$$\sigma(z) \notin D \quad \text{for all } \sigma \in \Gamma - \{\iota\}$$

Assume for a moment that $\sigma \in \Gamma$ satisfies $\sigma(z) \in D$. From $\pi(\sigma(z)) = \pi(z)$ we conclude that $\sigma(z) = z$, which implies $\sigma(D) = D$. This gives $\pi(\sigma(x)) = \pi(x)$ for all $x \in D$ and therefore $\sigma(x) = x$ for all $x \in D$; thus 3.1 gives us $\sigma = \iota$. This shows that Γ acts discontinuously on H^2 without fixed points. We can now conclude that Γ is a discrete subgroup of $PGl_2(\mathbb{R})$ which is torsion free: elements of finite order of $PGl_2(\mathbb{R})$ are rotations and reflections, but both types have fixed points.

---

[3] A group Γ is torsion free if all non-trivial elements have infinite order.

Consider points z,w $\in H^2$ with $x = \pi(z) = \pi(w)$. Let us show that we can find a transformation $\gamma \in \Gamma$ with $\gamma(z) = w$. The restrictions $\pi_z: D(z;r) \to D(x;r)$ and $\pi_w: D(w;r) \to D(x;r)$ are isometries for small values of $r > 0$. Introduce the isometry

$$\theta = \pi_w^{-1} \pi_z : D(z;r) \xrightarrow{\sim} D(w;r)$$

and use rigidity 3.1 to extend θ to a global isometry σ of H^2 into itself. The identity $\pi\sigma = \pi$ holds in D(z;r), but it extends to all of H^2 by rigidity 3.1.

We can now use $\pi: H^2 \to X$ to produce a bijection $\theta: H^2/\Gamma \xrightarrow{\sim} X$. But 4.5 allows us to conclude that θ is an isometry.

Let us now study local isometries from H^2/Γ to H^2/Σ. Let us first show that a $\sigma \in PGl_2(\mathbb{R})$ with $\sigma\Gamma\sigma^{-1} \subseteq \Sigma$ induces a map

6.2 $$H^2/\Gamma \to H^2/\Sigma \;;\quad h \mapsto \sigma(h) \qquad ;\ \sigma\Gamma\sigma^{-1} \subseteq \Sigma$$

In other words, we must verify that " $x \equiv y \ mod\ \Gamma \Rightarrow \sigma(x) \equiv \sigma(y)\ mod\ \Sigma$ ". To this end we write $y = \gamma(x)$ with $\gamma \in \Gamma$ to get $\sigma(y) = \sigma\gamma(x)) = (\sigma\gamma\sigma^{-1})\sigma(x)$ and observe that $\sigma\gamma\sigma^{-1} \in \Sigma$ to get $\sigma(x) \equiv \sigma(y)\ mod\ \Sigma$ as required. It follows from the remarks from beginning of this section that the map 6.2 is a local isometry.

Let us finally verify that the transformation 6.2 is an isometry if and only if $\sigma\Gamma\sigma^{-1} = \Sigma$. To see this make the factorization of 6.2

$$H^2/\Gamma \to H^2/\sigma\Gamma\sigma^{-1} \to H^2/\Sigma$$

where the first map is the isomorphism induced by σ and the second map is simple projection. This second map is bijective if and only if $\sigma\Gamma\sigma^{-1} = \Sigma$. □

Corollary 6.3 Let $\Gamma \subseteq PGl_2(\mathbb{R})$ denote a discrete torsion free subgroup. The automorphism group of H^2/Γ is isomorphic to $N\Gamma/\Gamma$ where the <u>normaliser</u> $N\Gamma$ of Γ is given by

$$N\Gamma = \{\ \sigma \in PGl_2(\mathbb{R}) \mid \sigma\Gamma\sigma^{-1} = \Gamma\ \}$$

VI.7 INSTANT JET SERVICE

Let us consider a group Γ of isometries of a hyperbolic surface X and a closed subset Z of X with the property that the family $(\sigma Z)_{\sigma \in \Gamma}$ forms a locally finite covering of X. This assumption implies that Γ acts discontinuously on X.

The action of Γ induces an equivalence relation $\mathcal{G}$ on Z and the projection $f:X \to X/\Gamma$ induces a homeomorphism $Z/\mathcal{G} \xrightarrow{\sim} X/\Gamma$, as follows from the proof of V.2.1. We are interested in transporting the metric from X/Γ to $Z/\mathcal{G}$.

Let us illustrate this by the analogous example of a cylinder realized as the orbit space of the Euclidean plane X under the group Γ generated by a single parallel translation τ; the set Z is a plane strip perpendicular to the translation direction and of width equal to the length of the translation. The corresponding examples in hyperbolic geometry are treated in 7.7. These examples are basic for the proof given in VII.3 of the Poincaré theorem in the case of unbounded polygons.

Definition 7.1 Let $\mathcal{G}$ be an equivalence relation on the metric space Z. The <u>distance</u> between two equivalence classes $\mathcal{Z}$ and $\mathcal{W}$ is defined by

$$d(\mathcal{Z},\mathcal{W}) = \mathit{inf} \sum_{i=1}^{n} d(z_{i-1},w_i)$$

where the infimum is taken over finite sequences in Z of the form

$$z_0,\dots,z_{n-1},w_1,\dots,w_n \qquad ; z_0 \in \mathcal{Z},\ w_n \in \mathcal{W}$$

where $z_i \equiv w_i \ \mathit{mod}\ \mathcal{G}$ for all $i = 1,\dots,n-1$.

This "instant jet service" distance on $Z/\mathcal{G}$ is symmetrical and satisfies the triangle inequality. Two distinct equivalence classes may have distance 0:

Example 7.2 Consider the closed interval $Z=[0,1]$ with the identifications $t \equiv 1-t$, $t \in]0,1[$. The "instant jet service" distance between $\{0\}$ and $\{1\}$ is 0.

Lemma 7.3 Let Z be a metric space equipped with an equivalence relation $\mathcal{G}$, and let $f:Z\to W$ be a distance decreasing map into a second metric space. If f is constant on the equivalence classes modulo $\mathcal{G}$, then f induces a distance decreasing map $\tilde{f}:Z/\mathcal{G}\to W$.

Proof Let us first observe that a sum of the type which occurs in 7.1 gives rise to the following inequality, where we have put $z_n = w_n$

$$\sum_{i=1}^{n} d(z_{i-1},w_i) \geq \sum_{i=1}^{n} d(f(z_{i-1}),f(w_i)) \geq \sum_{i=1}^{n} d(f(z_{i-1}),f(z_i)) \geq d(f(z_0),f(w_n))$$

Take infimum over all sequences of the type considered and conclude that the map $\tilde{f}:X/\mathcal{G}\to W$ is distance decreasing. □

Proposition 7.4 Let Γ be a group of isometries of a hyperbolic surface X and Z a closed subset of X whose Γ-translates form a locally finite covering of X. The projection $f:X\to X/\Gamma$ induces an isometry $f:Z/\mathcal{G} \xrightarrow{\sim} X/\Gamma$ where $\mathcal{G}$ is the equivalence relation on Z induced by the action of Γ on X and the metric on $Z/\mathcal{G}$ is instant jet service.

Proof A simple application of lemma 7.3 shows that $f:Z/\mathcal{G} \xrightarrow{\sim} X/\Gamma$ decreases distances. Let us prove that the inverse map f^{-1} decreases distances. To this end we consider points $z,w \in Z$ with equivalence classes $\mathcal{Z},\mathcal{W} \in Z/\mathcal{G}$ and show that

$$d(z,\sigma(w)) \geq d(\mathcal{Z},\mathcal{W}) \qquad ; \sigma \in \Gamma$$

Let us fix $\sigma \in \Gamma$ and consider a rectifiable curve $\gamma:[a,b]\to X$ with $\gamma(a) = z$ and $\gamma(b) = \sigma(w)$. We shall show below that there exists a subdivision of $[a,b]$ $a = t_0 \leq t_1 \ldots \leq t_n = b$ and elements $\iota = \sigma_1,\ldots,\sigma_n = \sigma$ of Γ such that

7.5
$$\gamma(t_i) \in \sigma_i(Z) \cap \sigma_{i+1}(Z) \qquad ; i = 1,..,n-1$$

Let us use 7.5 to pick a sequence of points $z = z_0,..,z_{n-1},w_1,..,w_n = w$ of Z with

$$\gamma(t_i) = \sigma_i(w_i) = \sigma_{i+1}(z_i) \qquad ; \ i = 1,..,n-1$$

This gives us the estimate

$$l(\gamma) \geq \sum_{i=1}^{n} d(\gamma(t_{i-1}),\gamma(t_i)) = \sum_{i=1}^{n} d(\sigma_i(z_{i-1}),\sigma_i(w_i)) = \sum_{i=1}^{n} d(z_{i-1},w_i) \geq d(\mathcal{Z},\mathcal{W})$$

Finally, make a variation of γ and deduce that $d(x,\sigma(w)) \geq d(\mathcal{Z},\mathcal{W})$ as required.

Let us start the construction of a subdivision of the interval [a,b] subject to 7.5 by observing that $\gamma([a,b])$ meets only finitely many Γ-translates of Z as follows from Borel–Heine. When $\gamma(b) \in Z$ we can use $n = 2$ and $t_1 = b$. Let us assume that $\sigma(w) \notin Z$ and let $t_1 \in [a,b]$ be the last time γ leaves Z:

$$t_1 = \mathit{sup}\{t \in [a,b] \mid \gamma(t) \in Z\}$$

Let us observe that any union of Γ-translates of Z is a closed set as follows from the fact that the covering of X with Γ-translates of Z is locally finite. We can now conclude that

$$\gamma(t_1) \in \bigcup_{\sigma \neq \iota} \sigma(Z)$$

since the alternative would make $\gamma(t_1)$ an interior point of Z, contradicting its definition. Choose $\sigma_2 \neq \sigma_1 = \iota$ such that $\gamma(t_1) \in \sigma_2(Z)$. When $\gamma(b) \in \sigma_2(Z)$ we can use $n = 3$ with $t_2 = b$. If $\gamma(b) \notin \sigma_2(Z)$, let $t_2 \in [a,b]$ be the last time γ leaves $\sigma_2(Z)$ and observe that

$$\gamma(t_2) \in \bigcup_{\sigma \neq \sigma_1,\sigma_2} \sigma(Z)$$

since the alternative makes $\gamma(t_2)$ an interior point of $\sigma_1(Z) \cup \sigma_2(Z)$ contradicting the definition of t_2 and $\gamma(b) \notin \sigma_2(Z)$, that is $t_2 < b$. This procedure terminates since γ meets only finitely many Γ-translates of Z. □

We are going to stydy some examples which are usefull for the understanding of Poincarés theorem in VII.3 in case of unbounded polygones.

Example 7.6 Let h and k be the geodesics in the Poincaré half-plane H^2 given by $Re[z] = 1$ and $Re[z] = a$ where $a > 1$. Observe that $z \mapsto az$ transforms h into k. Let $\mathfrak{R}$ denote the equivalence relation in the strip Z in H^2 bounded by h and k based on the identification $z \sim az$, $z \in h$. We are going to show that $Z/\mathfrak{R}$ is <u>not</u> complete. (For a different approach see exercise 5.2).

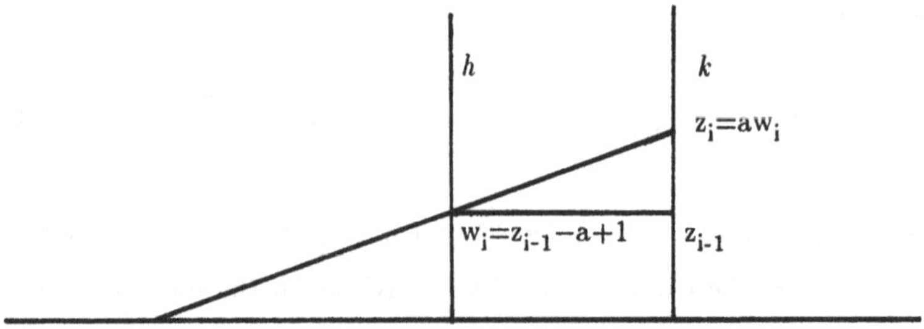

Consider the sequence $\{z_i\}_{i \in \mathbb{N}}$ in Z given by the recursion formulas above starting from z_0=a+ib, b > 0. It will be useful to observe that $Im[z_i] = a^i b$, $i \in \mathbb{N}$. Let us recall from II.8.2

$$d(z,w) \leq 2\, sinh\, \tfrac{1}{2}d(z,\ w) = |w - z|\,(Im[w]Im[z])^{-1/2}$$

Let us use this to estimate the distance in $Z/\mathfrak{R}$

$$d([z]_i,[z_{i-1}]) = d([w_i],[z_{i-1}]) \leq d(w_i,z_{i-1}) \leq (a-1)b^{-1}a^{1-i} \qquad ; i \geq 1$$

Recall that $a > 1$ to see that the infinite series $\sum d([z_{i-1}],[z_i])$ is convergent and conclude that $[z_i]_{i \in \mathbb{N}}$ is a Cauchy sequence in $Z/\mathfrak{R}$. In order to show that our sequence is not convergent, consider the function $f:Z \to \mathbb{R}$ given by $f(z) = y/x$. The function f induces a continuous function on $Z/\mathfrak{R}$ (see exercise 5.2). If our sequence was convergent, then f would be bounded on the sequence, which is not the case.

Proposition 7.7 Let Z denote the closed subset of H^2 bounded by two non-intersecting geodesics *h* and *k* with ends A,B and C,D in the relative positions shown below. Let σ be an even isometry with $\sigma(A) = D$ and $\sigma(B) = C$. The equivalence relation given by the identification $z \sim \sigma(z)$, $z \in h$, is denoted $\mathfrak{R}$. If the ends of *h* and *k* are distinct, then $Z/\mathfrak{R}$ is complete. If *h* and *k* have a common end $A = D$, then $Z/\mathfrak{R}$ is complete if and only if σ is a horolation.

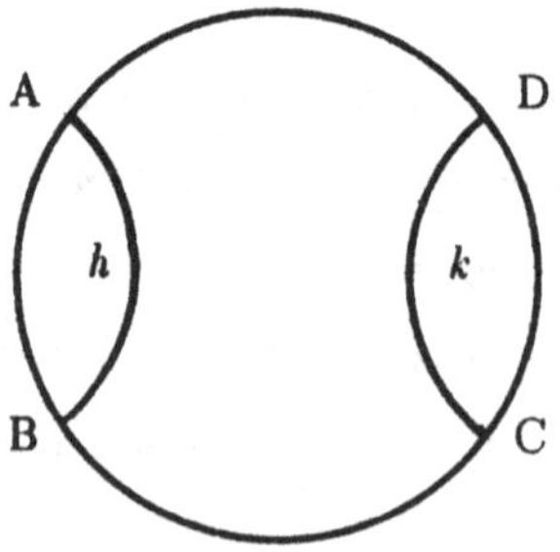

Proof Suppose that the ends of h and k are distinct. Consider a sequence $(w_n)_{n \in \mathbb{N}}$ of points of Z representing a Cauchy sequence in $Z/\mathfrak{R}$. We shall prove that the sequence is bounded in H^2. Once this is done, we conclude that $\{w_n\}_{n \in \mathbb{N}}$ is contained in a compact subset of Z. It follows that our Cauchy sequence is contained in a compact subset of $Z/\mathfrak{R}$, which makes it convergent.

Let us introduce the geodesic s through A and D and consider the function $f_s{:}H^2 \to \mathbb{R}$ which is zero on the half-plane containing B and C, while its value at a point z on the other half-plane is the hyperbolic distance from z to s. We ask the reader to show that $f_s{:}X \to \mathbb{R}$ is distance decreasing and we conclude from lemma 7.3 that the sequence $f_s(z_n)_{n \in \mathbb{N}}$ is a Cauchy sequence. This implies that we can find a constant c such that

$$f_s(z_n) \leq c \qquad ; n \in \mathbb{N}$$

The same conclusion is available for the geodesic through B and C.

In order to show that our initial sequence is bounded away from the points A and D, pick a horocycle $\mathcal{A}$ with centre A and define the polar function $p_{\mathcal{A}}{:}H^2 \to \mathbb{R}$ which is zero outside the horodisc bounded by $\mathcal{A}$ while the value at a point z of the horodisc is the distance from z to $\mathcal{A}$ measured along the geodesic through z and A. Let us pick a second horocycle $\mathfrak{D}$ with centre D disjoint from $\mathcal{A}$ and such that $\sigma(\mathcal{A} \cap h) = \mathfrak{D} \cap k$. The function $p = \frac{1}{2}(p_{\mathcal{A}} + p_{\mathfrak{D}})$ is distance decreasing and constant on the equivalence classes modulo $\mathfrak{R}$. It follows from lemma 7.3 applied to $p{:}Z \to \mathbb{R}$ that p is bounded on our sequence. The same conclusion is available for the points B and C. Thus we have proved that our sequence is bounded in H^2.

Let us now assume that $A = D$ and that σ is a horolation with centre A. Pick a horocycle $\mathcal{A}$ with centre A and observe that the function $p_{\mathcal{A}}$ is constant on the equivalence classes of $\mathfrak{R}$. As above we conclude that $Z/\mathfrak{R}$ is complete.

When $A = D$ and σ is a translation we can use example 7.6 to conclude that $Z/\mathfrak{R}$ is not complete. □

VI.8 HOMOTOPY

In this section we shall study homotopy classes of paths on a complete hyperbolic surface X. The key fact of topological nature is 4.4 according to which a local isometry $f:X \to Y$ between complete hyperbolic surfaces is a covering projection in the sense of 8.1 below. For proofs of 8.2 and 8.3 see [Massey].

Definition 8.1 A continuous map $f:X \to Y$ is called a covering projection if every $y \in Y$ has an open neighbourhood D such that for each $x \in f^{-1}\{y\}$ there exists a continuous section $s_x: D \to X$ with $s_x(y) = x$ inducing a homeomorphism

$$f^{-1}\{y\} \times D \xrightarrow{\sim} f^{-1}(D) \qquad ; \quad (x,v) \mapsto s_x(v)$$

Unique lifting lemma 8.2 Let $f:X \to Y$ denote a covering projection and $r,s:W \to X$ continuous maps with $fs = fr$. If W is connected and X is Hausdorff, then $s(w) = r(w)$ for some $w \in W$ implies that $s = r$.

Homotopy lifting lemma 8.3 Let $f:X \to Y$ be a covering projection. For any topological space T, any solid diagram of continuous maps

$$\begin{array}{ccc} T \times [0] & \xrightarrow{\phi} & X \\ \downarrow & & \downarrow f \\ T \times [0,1] & \xrightarrow{\Psi} & Y \end{array}$$

can be completed with a dashed arrow to a commutative diagram in the category of continuous maps.

Theorem 8.4 Let x and y be distinct points of the complete hyperbolic surface X. Any homotopy class of continuous curves joining x to y can in a unique way be represented by a geodesic curve joining x to y.

Proof Let us use the uniformization theorem 3.2 to find a local isometry $f:H^2 \to X$ and let us use 4.3 to pick a base point $z \in H^2$ with $f(z) = x$. For a curve $\gamma:[a,b] \to X$ with $\gamma(a) = x$ and $\gamma(b) = y$ there is a unique curve $\beta:[a,b] \to H^2$ with $\beta(a) = z$ and $f\beta = \gamma$. It follows from 8.2 and 8.3 that the homotopy type of β is determined by the point $\beta(b) \in f^{-1}\{y\}$. The image in Y of the geodesic line segment $[\beta(a),\beta(b)]$ represents the homotopy class of β. The uniqueness in the statement follows from the geodesic lifting property 4.2. □

Definition 8.5 By a closed geodesic we understand a geodesic curve $\gamma:\mathbb{R} \to X$ such that there exists $a > 0$ with $\gamma(x+a) = \gamma(x)$ for all $x \in \mathbb{R}$.

Let us recall that two continuous curves $\alpha,\beta:S^1 \to X$ are freely homotopic if there exists a continuous map $\Psi:S^1 \times [0,1] \to X$ such that

$$\Psi(s,0) = \alpha(s)\ ,\quad \Psi(s,1) = \beta(s) \qquad ; s \in S^1$$

Theorem 8.6 Let X be a compact hyperbolic surface. A closed curve in X not homotopic to zero is freely homotopic to a unique closed geodesic in X.

Proof With the notation from the proof of 8.1, let us consider a curve $g:[0,1] \to X$ with $g(0) = g(1) = x$. Let $h:[0,1] \to H^2$ be a lifting of g with $h(0) = z$ and let $\gamma \in \Gamma$ be determined by $h(1) = \gamma z$. Since g is not homotopic to zero we have $\gamma \neq \iota$. We conclude from IV.5.6 that γ is a translation with translation axis say k. We intend to show that the restriction of the projection $f:H^2 \to X$ to k defines a closed geodesic $k \to X$ which is freely homotopic to g. A free homotopy is formed by the restriction of $f:H^2 \to X$ to the hyperbolic quadrangle given below

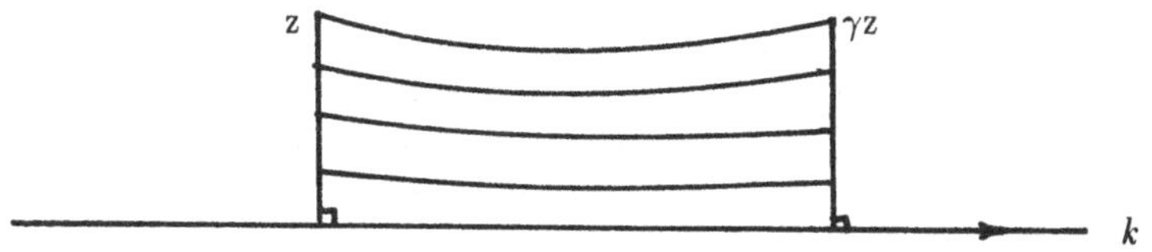

The horizontal coordinate lines in the quadrangle are formed by γ-hypercycles. □

VI EXERCISES

EXERCISE 1.1 Let $\gamma:[a,c]\rightarrow X$ be a continuous curve in a metric space X.

1° Show that for any point b between a and c the restrictions $\gamma_1:[a,b]\rightarrow X$ and $\gamma_2:[b,c]\rightarrow X$ are rectifiable if and only if $\gamma:[a,c]\rightarrow X$ is rectifiable.

2° With the notation above assume that γ is rectifiable and show that

$$l(\gamma) = l(\gamma_1) + l(\gamma_2)$$

3° Show that a rectifiable curve $\gamma:[a,c]\rightarrow X$ which satisfies

$$l(\gamma) = d(\gamma(a),\gamma(b))$$

is a geodesic curve. Hint: Show that γ is distance preserving.

EXERCISE 1.2 Let us fix a point $u+iv$ of the Poincaré half-plane and consider the real function $\psi(x,y)$ in two variables given by

$$\psi(x,y) = d(x+iy,u+iv)^2$$

1° Show that the Hessian of this function at the point $u+iv$ is

$$d^2\psi(x,y) = \tfrac{1}{v}(dx^2 + dy^2)$$

2° Show that a continuously differentiable curve $\gamma:[a,b]\rightarrow H^2$ is rectifiable with curve length

$$l(\gamma) = \int_a^b \sqrt{(x'(t)^2 + y'(t)^2)/y(t)}\; dt$$

where $\gamma(t) = (x(t),y(t))$, $t \in [a,b]$.

EXERCISE 1.3 Show that a circle in H^2 with radius r has arc length 2π *sinh* r, compare VI.1. Show for example first that the length k_n of the edge of an inscribed n-gon is given by the formula

$$sinh(k_n) = sinh(r)\ 2\,sin(\pi/n)$$

and use this to calculate the length of the circle as $\lim_{n\rightarrow\infty} nk_n$

EXERCISE 1.4 Consider a fixed real number $y > 0$ and let $\mathcal{Y}$ denote the Euclidean line in the upper half-plane through iy and parallel to the x-axis.

1° Show that $\mathcal{Y}$ is a horocycle, and that the arc length is given by

$$length_{\mathcal{Y}}\,(iy, x+iy) = \frac{x}{y} \qquad ; \quad x \geq 0$$

2° Verify the formula

$$length_{\mathcal{Y}}\,(iy, x+iy) = 2\,sinh\,\tfrac{1}{2}d(iy, x+iy) \qquad ; \quad x \geq 0$$

EXERCISE 2.1 Let X be a metric space and $\gamma:[a,b]\to X$ a geodesic curve. Show that the curve γ is rectifiable with arc length $l(\gamma) = b - a$.

EXERCISE 5.1 Let X be a hyperbolic surface and G a group of isometries of X which acts discontinuously on X and let $f:X\to X/G$ denote the canonical projection. Show that a subset U of X/G is open (in the sense of the metric on X/G) if and only if $f^{-1}(U)$ is an open subset of X.

EXERCISE 5.2 Let α be the isometry of the Poincaré half-plane given by $z\mapsto az$, where $a > 1$ is a fixed real number.

1° Show that the group Γ generated by α acts discontinuously on $X = \{z \in H^2 \mid Re[z] > 0\}$.

2° Let $f:X\to\mathbb{R}$ be the function given by $f(x+iy) = y/x$, $x+iy \in X$. Show that f induces a continuous function on the surface X/Γ.

3° Show that the surface X/Γ is not complete. Hint: Use 5.9.

EXERCISE 7.1 With the notations and assumptions of proposition 7.7. Show that the set $X = \bigcup_{n\in\mathbb{Z}} \sigma^n Z$ is open and connected. Prove that $Z/\mathcal{R}$ is complete if and only if $X = H^2$. Hint: Apply propositions 5.9 and 7.4.

EXERCISE 8.1 Let Γ be a discrete group of isometries of H^2 and $\gamma\in\Gamma$ a horolation. Explain how to construct from γ a closed curve in H^2/Γ unique up to free homotopy. Hint: Model the proof of 8.6.

VII POINCARÉ'S THEOREM

We have arrived at the central theme of the book, Poincaré's theorem, which allows us to construct a discrete group from a given hyperbolic polygon with side pairing and to write down generators and relations for the group.

The proof of the theorem is a beautiful application of the monodromy theorem VI.4.8. In order to display the geometric ideas of the proof, we have first treated the case where the polygon is compact. The general case requires a more technical argument based on completeness and is given in section 3. Let me mention that Poincaré's original proof [Poincaré] is incomplete in the non-compact case, compare [de Rham].

VII.1 COMPACT POLYGONS

Let us consider an ordinary polygon Δ in the hyperbolic plane H^2 with a side pairing. By this we understand an involution $s \mapsto *s$ on the set of edges $\mathcal{E}$ of Δ such that s and $*s$ have the same hyperbolic length and such that

1.1 $$*(s^{-1}) = (*s)^{-1} \qquad ; s \in \mathcal{E}$$

Recall that an edge s is an oriented side and that s^{-1} denotes the result of changing the orientation. For $s \in \mathcal{E}$ we let σ_s denote the isometry of H^2 which maps $*s$ to s in such a way, that locally, the half-plane bounded by $*s$ containing Δ is mapped to the half-plane bounded by s but opposite Δ. The isometry σ_s is called the side transformation determined by s. Let us at once observe that we have the

Reflection relations 1.2

$$\sigma_{s^{-1}} = \sigma_s \quad , \quad \sigma_{*s} = \sigma_s^{-1} \qquad ; s \in \mathcal{E}$$

Given an edge s of Δ, we let $\downarrow$s denote the edge with the same initial vertex as s but different final vertex. This defines for us a second involution on $\mathcal{E}$, $s \mapsto \downarrow s$. The combination of these two operators defines the <u>edge operator</u> $\Psi:\mathcal{E} \to \mathcal{E}$ given by

1.3 $$\Psi s = \downarrow * s \qquad ; s \in \mathcal{E}$$

The edge operator Ψ on $\mathcal{E}$ is bijective and $\mathcal{E}$ is finite so we conclude that Ψ has finite order. The cycles for Ψ on $\mathcal{E}$ are called <u>edge cycles</u>. The sequence of initial points of an edge cycle is called a <u>vertex cycle</u> (a vertex cycle may contain repetitions).

Lemma 1.4 Let $s_1,...,s_r$ be an edge cycle with vertex cycle $P_1,...,P_r$ and side transformations $\sigma_1,...,\sigma_r$ ($\sigma_i = \sigma_{s_i}$, $i = 1,..,r$). The <u>cycle map</u> $\sigma = \sigma_1...\sigma_r$ is a rotation around P_1 of angle (measured in the direction determined by Δ and s_1) congruent *mod* 2π to the sum of the interior angles

$$\angle_{int}P_1 + \angle_{int}P_2 + ... \angle_{int}P_r$$

Proof Let us choose the orientation of H^2 which places Δ on the positive side of s_1 near P_1. With the notation of III.2.4 we have

$$\angle_{or}(s_i,\downarrow s_i) \equiv sign(s_i) \angle_{int}P_i \quad mod\ 2\pi \qquad ; i = 1,..,r$$

where $sign(s_i) = \epsilon_i$ equals $+1$ (resp.-1) in case Δ lies on the positive(resp. negative) side of s_i. From the very construction of the side transformations it follows that

$$sign(s_m) = sign(s_{m+1})\ det(\sigma_m) \qquad ; \ m = 1,...,r$$

Let us multiply these equations together for $m = 1,...,r$ to get that

$$sign(s_1) = det(\sigma_1...\sigma_r)\ sign(s_{r+1})$$

Since $s_{r+1} = s_1$ we conclude that the cycle map is even. Let us proceed to calculate the rotation angle of the cycle map. Let us prove a general formula

$$\sum_{i=m...r} \angle_{int} P_i \equiv \epsilon_m \angle_{or} (s_m, \sigma_m...\sigma_r(s_1)) \qquad ; m = 1,...,r$$

To this end we recall that

$$s_i = \sigma_i(*s_i)\ ,\ \ s_{i+1} = \downarrow * s_i\ ,\ \ \downarrow s_i = \sigma_i(s_{i+1})$$

This will be done by decreasing induction on m. The case m = 1 runs as follows

$$\angle_{int}P_r = \epsilon_r\angle_{or}(s_r,\downarrow s_r) = \epsilon_r\angle_{or}(s_r,\sigma_r(s_1))$$

The induction step m+1 follows from the formulas we have just derived

$$\sum_{i=m,\dots r}\angle_{int} P_i \equiv \epsilon_m\angle_{or}(s_m,\downarrow s_m) + \epsilon_{m+1}\angle_{or}(s_{m+1}, \sigma_{m+1}\dots\sigma_r(s_1)) \equiv$$
$$\epsilon_m\angle_{or}(s_m,\downarrow s_m) + \epsilon_{m+1}\, det(\sigma_m)\angle_{or}(\downarrow s_m, \sigma_m\dots\sigma_r(s_1)) \equiv$$
$$\epsilon_m\angle_{or}(s_m,\downarrow s_m) + \epsilon_m\angle_{or}(\downarrow s_m,\sigma_m\dots\sigma_r(s_1)) \equiv \epsilon_m\angle_{or}(s_m, \sigma_m\dots\sigma_r(s_1))$$

This concludes the proof of our formula. Finally, put m = 1 in our formula and recall that $\epsilon_1 = 1$ to get the announced rotation angle. □

Examples

Consider an 8-gon with side pairing as indicated in the drawing.

$\alpha = \sigma_a, \beta = \sigma_b, \gamma = \sigma_c, \delta = \sigma_d.$

Edge cycle: $a{*}b^{-1}{*}a^{-1}b\ c\ {*}d^{-1}\ {*}c^{-1}\ d$

Vertex cycle: $P_1P_4P_3P_2P_5P_8P_7P_6$

Cycle map: $\alpha\beta^{-1}\alpha^{-1}\beta\gamma\delta^{-1}\gamma^{-1}\delta$

The identification space is a sphere with 2 handles (oriented surface of genus 2)

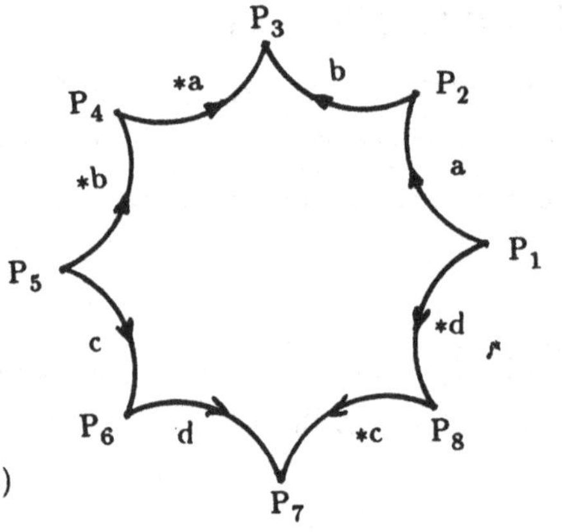

Our next example is an isoceles triangle with the side pairing indicated on the drawing, the base edge c is paired with itself *c=c. The side transformation $\alpha=\sigma_a$ is a rotation with centre C, while $\gamma=\sigma_c$ is a reflection in the geodesic through c.

Edge cycles	a^{-1},	${*}a^{-1}$,	$a\ c\ {*}a\ c^{-1}$
Vertex cycles	C ,	C ,	B A A B
Cycle maps	α ,	α^{-1} ,	$\alpha\gamma\alpha\gamma$

The identification space is a disc.

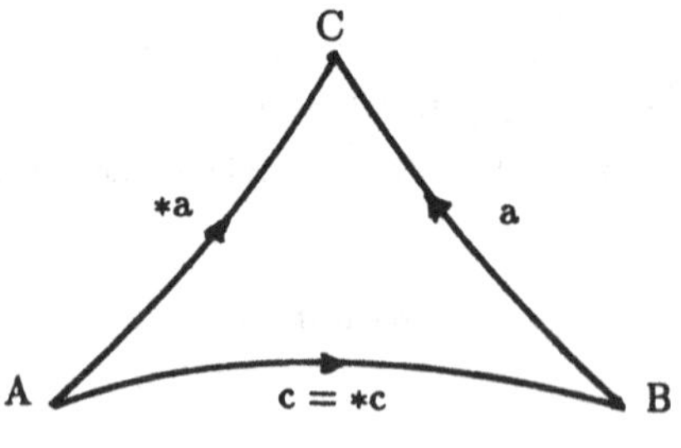

Let us consider a different side pairing on the same triangle given below: again $*c = c$. The transformation $\gamma = \sigma_c$ is a reflection as before, while the transformation $\alpha = \sigma_a$ is a glide reflection.

Edge cycle: $a\ a^{-1}\ c\ *a^{-1}\ *a\ c^{-1}$

Cycle map: $\alpha^2\gamma\alpha^{-2}\gamma$

Vertex cycle: B C A A C B

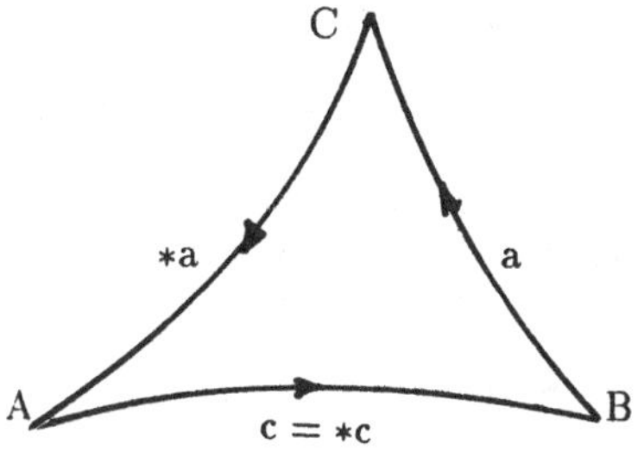

The identification space is a Möbius band.

Remark 1.5

Let us consider an edge cycle $s_1,s_2,\ldots,s_r$. From $s_{i+1} = \downarrow *s_i$, $i = 1,\ldots,r$, we conclude that $\downarrow s_{i+1} = *s_i$. This gives us

$$\Psi(\downarrow s_{i+1}) = \downarrow s_i \qquad ; \ i = 1,\ldots$$

It follows that $\downarrow s_1, \downarrow s_r, \downarrow s_{r-1},\ldots,\downarrow s_2$ form an edge cycle. Notice that the corresponding vertex cycle is $P_1,P_r,P_{r-1},\ldots,P_1$, which is nothing but the reverse cycle of the original vertex cycle. It follows that a vertex cycle makes up a full equivalence class of Δ under the equivalence relation generated by the side pairing. Let us observe that

$$\downarrow s_1, \downarrow s_r, \downarrow s_{r-1},\ldots,\downarrow s_2 = *s_r, *s_{r-1},\ldots,*s_1$$

and conclude that the cycle map is $\sigma_r^{-1}\sigma_{r-2}^{-1}\ldots\sigma_1^{-1}$, which is the inverse of the cycle map of the original cycle.

Cycle condition 1.6

We say that an edge cycle $s_1,\ldots,s_r$ satisfies the cycle condition if the angle sum along the corresponding vertex cycle $P_1,\ldots,P_r$ has the form

$$\sum_{i=1\ldots r} \angle_{\text{int}}\, P_i = \frac{2\pi}{n_c} \qquad ; \ n_c = 1,2,3\ldots$$

The cycle map $\sigma = \sigma_1\ldots\sigma_r$ of an edge cycle which satisfies the cycle condition is a rotation about P_1 of order n_c.[1]

[1]Cycles c with $n_c = 1$ are sometimes called <u>accidental</u> cycles

Poincaré's Theorem 1.7 Let Δ be a compact convex polygon Δ with a side pairing which satisfies the cycle conditions 1.6. Then, the group G generated by the side transformations σ_s, $s \in \mathcal{S}$, is a discrete group with Δ as a fundamental domain. A complete set of relations for these generators is the reflection relations 1.2 and the cycle relations

$$\sigma_c^{n_c} = \iota$$

one for each edge cycle c.

Proof Let Γ denote the universal group on the generators σ_s , $s \in \mathcal{S}$, subject to the reflection relations 1.2 and the cycle relations. The canonical map $\Gamma \to G$ is denoted $\gamma \mapsto \tilde{\gamma}$. On the basis of these data we shall construct a topological space X and a local homeomorphism $f: X \to H^2$, compare the introduction to chapter VI. The set underlying X is the quotient space for the equivalence relation on $\Gamma \times \Delta$ generated by relations of the form

1.8 $$(\gamma\sigma_s, h) \sim (\gamma, \sigma_s(h)) \qquad ; \ \gamma \in \Gamma, s \in \mathcal{S}, h \in *s$$

The equivalence class of the point (γ,k) is denoted $\gamma.k \in X$. Let us equip $\Gamma \times \Delta$ with the product topology: an open subset has the form $U = \bigcup_{\sigma \in \Gamma} \{\sigma\} \times U_\sigma$ where U_σ is an open subset of Δ, $\sigma \in \Gamma$. The topology on X is the quotient topology: open subsets of X correspond to open, $\sim$ stable subsets of $\Gamma \times \Delta$.

The evaluation map $(\gamma,h) \mapsto \tilde{\gamma}(h)$, $\Gamma \times \Delta \to H^2$ is continuous and compatible with our equivalence relation, so it will induce a continuous map $f: X \to H^2$

$$f(\gamma.h) = \tilde{\gamma}(h) \qquad ; \ \gamma \in \Gamma, \ h \in \Delta$$

The natural action of Γ on $\Gamma \times \Delta$ induces an action of Γ on X through the formula

$$\alpha(\beta.s) = (\alpha\beta).s \qquad ; \alpha,\beta \in \Gamma, s \in \Delta$$

Let us make the observation that $f: X \to H^2$ is Γ-equivariant in the sense that

$$f(\tau x) = \tilde{\tau} f(x) \qquad ; \tau \in \Gamma, x \in X$$

Let us prove that X is <u>connected</u>. To this end let us note that $s \mapsto \iota.s$ induces a continuous section of $f: X \to H^2$ over Δ and conclude that $\iota.\Delta$ is a connected subset of X. For $x \in X$ given, let us construct a connected subset of X containing x and meeting $\iota.\Delta$. It is clear from the construction that X is the union of all Γ-translates of $\iota.\Delta$. So let us pick $\tau \in \Gamma$ with $x \in \tau.\Delta$ and let us write $\tau = \sigma_1...\sigma_n$ where the σ_is are side transformations with respect to the edges $s_1,...,s_n$ of Δ. For $i = 1,...,n$ we get from 1.8 that $\iota.\Delta$ meets $\sigma_i.\Delta$, which implies that $\sigma_1...\sigma_{i-1}.\Delta$ meets $\sigma_1...\sigma_i.\Delta$. The chain

$$\iota.\Delta \cup \sigma_1.\Delta \cup \sigma_1\sigma_2.\Delta \cup ... \sigma_1\sigma_2...\sigma_{n-1}.\Delta \cup \tau.\Delta$$

is a connected set of the required sort.

We proceed to construct some special neighbourhoods of a given point $x \in X$. For convenience we introduce the following terminology concerning the equivalence class of $\Gamma \times \Delta$ represented by x. But first let us agree that for $z \in \Delta$ and $\epsilon > 0$ we shall put $\Delta(z;\epsilon) = \{x \in \Delta |\ d(x,z) < \epsilon\}$.

Local tesselation property 1.9 We say that an equivalence class $\mathfrak{C}$ of $\Gamma \times \Delta$ has the local tesselation property if the following three conditions are satisfied. 1° $\mathfrak{C}$ is a finite subset of $\Gamma \times \Delta$

$$\mathfrak{C} : (\eta_1,k_1),...,(\eta_m,k_m) \qquad ; \ \eta_i \in \Gamma,\ k_i \in \Delta$$

2° For $\epsilon>0$ sufficiently small, the set

$$\{\eta_1\} \times \Delta(k_1;\epsilon) \bigcup ... \bigcup \{\eta_m\} \times \Delta(k_m;\epsilon)$$

is a $\sim$ stable subset of $\Gamma \times \Delta$ and 3° the corresponding open subset of X

$$\eta_1.\Delta(k_1;\epsilon) \bigcup ... \bigcup \eta_m.\Delta(k_m;\epsilon)$$

is mapped by f bijectively onto the open disc $D(z;\epsilon)$, $z = \tilde{\eta}_1(k_1) = ... = \tilde{\eta}_m(k_m)$.

We are going to show that all equivalence classes of $\Gamma \times \Delta$ have the local tesselation property. Observe that if the equivalence class $\mathfrak{C}$ has the local tesselation property, then the same is true for the class $\gamma\mathfrak{C}$ for all $\gamma \in \Gamma$. Thus it suffices to verify that for each $z \in \Delta$, the equivalence class generated by (ι,z) has the local tesselation property. Let us investigate a number of special cases.

The full equivalence class of a point of the form (ι,z), $z \in Int\,\Delta$, consists of that point alone. Any subset of $\{\iota\} \times \Delta$ is $\sim$ stable in $\Gamma \times \Delta$ and the open subset $\iota.Int\Delta$ of X is mapped by $f:X \to H^2$ onto $Int\,\Delta$.

The full equivalence class in $\Gamma \times \Delta$ generated by a point of the form (ι,a), where a is an interior point of an edge s, consists of two points

$$(\iota,a)\ ,\ (\sigma_s,b) \qquad ; \ b = \sigma_s^{-1}(a)$$

For $\epsilon > 0$ sufficiently small, the following subset of $\Gamma \times \Delta$

$$\{\iota\} \times \Delta(a;\epsilon) \ \cup \ \{\sigma_S\} \times \Delta(b;\epsilon)$$

is open and stable. The image in X of this set

$$\iota.\Delta(a;\epsilon) \ \cup \ \sigma_S.\Delta(b;\epsilon)$$

is mapped by $f{:}X \to H^2$ onto $D(a;\epsilon)$.

Let us now consider a vertex P_1 of Δ. Pick an edge s_1 with initial point P_1 and iterate the edge operator Ψ to get a sequence of edges $s_1,...,s_r,...,s_d$, where r is the length of the edge cycle and $d = nr$ where n is defines in 1.6 such that

$$n \sum_{i=1,...,r} \angle_{int} P_i = 2\pi$$

We let $P_1,...,P_d$ denote the corresponding sequence of initial vertices and $\sigma_1,...,\sigma_d$ the sequence of side transformations. Let me stress that both sequences have period r. Observe that P_{i+1} is the initial vertex of s_{i+1} and $\downarrow s_{i+1} = *s_i$, this gives us $\sigma_i(P_{i+1}) = P_i$ for $i = 1,...,r$. We conclude that the full equivalence of class of the point (ι,P_1) in $\Gamma \times \Delta$ is

$$(\iota,P_1) \ , \ (\sigma_1,P_2) \ , \ (\sigma_1\sigma_2,P_3) \ ,..., \ (\sigma_1\sigma_2...\sigma_{d-1},P_d) \qquad ; d = rn$$

Let us prove that this equivalence class of $\Gamma \times \Delta$ satisfies the local tesselation property 1.9. The stability condition follows basically by observing that the boundary of $\Delta(P_i;\epsilon)$ is located on the edges s_i and $\downarrow s_i$ and that

$$\sigma_{i-1}(\downarrow s_i) = \sigma_{i-1}(*s_{i-1}) = s_{i-1} \quad , \quad s_i = \sigma_i(*s_i) = \sigma_i(\downarrow s_{i+1})$$

from which we deduce that

$$\sigma_{i-1}.\downarrow s_i = \iota.s_{i-1} \quad , \quad s_i = \sigma_i.\downarrow s_{i+1}$$

It is now time to calculate the image of $\sigma_1\sigma_2...\sigma_{i-1}.\Delta(P_i;\epsilon)$ in H^2, that is

$$\sigma_1\sigma_2...\sigma_{i-1}\Delta(P_i;\epsilon)$$

The sector $\Delta(P_i;\epsilon)$ is bounded by $\angle(s_i,\downarrow s_i)$. The image is bounded by the circle with centre P_1 and radius ϵ and the two edges

$$\sigma_1...\sigma_{i-1}(s_i) \quad , \quad \sigma_1...\sigma_{i-1}(\downarrow s_i) = \sigma_1...\sigma_{i-2}(s_{i-1})$$

Let us use the notation from the proof of lemma 1.4 to calculate the oriented angle between these two edges

$$\angle_{or}(\sigma_1...\sigma_{i-1}(s_i),\sigma_1...\sigma_{i-1}(\downarrow s_i)) = det(\sigma_1)...det(\sigma_{i-1})\angle_{or}(s_i,\downarrow s_i) \equiv$$
$$det(\sigma_1)...det(\sigma_{i-1})sign(s_i)\angle_{int}P_i = sign(s_1) \ \angle_{int} \ P_i$$

Observe that if these angles are added from $i = 1$ to $i = d = nr$ we get $sign(s_1)2\pi$. From this it follows that our class satisfies the local tesselation property 1.9.

Let us show that the open subsets of X exhibited in 1.9 form a basis for the topology on X: with the notation above, an open subset U of X which contains the point x representing the equivalence class $\mathfrak{C}$ can be thought of as a $\sim$ stable open subset V of $\Gamma\times\Delta$ containing $(\eta_1,k_1),\dots,(\eta_m,k_m)$. This implies that V contains $\{\eta_i\}\times\Delta(k_i;\epsilon)$ for $i=1,\dots,m$ and $\epsilon>0$ sufficiently small as required. As a consequence of the local tesselation property, we find that $f:X\to H^2$ maps open subsets of X to open subsets of H^2. Indeed f is a local homeomorphism.

It is time to show that X is a Hausdorff space: let $\mathfrak{D}\neq\mathfrak{C}$ be equivalence classes where $\mathfrak{C}$ is given by 1.9 and $\mathfrak{D}$ is given by

$$(\xi_1,h_1),\dots,(\xi_n,h_n) \qquad ;\ \xi_j\in\Gamma\,,\ h_j\in\Delta$$

Let us consider a fixed pair i,j and show that we can find $\epsilon_{ij}>0$, such that

$$\{\eta_i\}\times\Delta(k_i;\epsilon)\ \bigcap\ \{\xi_j\}\times\Delta(h_j;\epsilon)=\emptyset \qquad ;\ \epsilon\in\,]0,\epsilon_{ij}]$$

If $\eta_i\neq\xi_j$ any $\epsilon_{ij}>0$ will do. If $\eta_i=\xi_j$ then $k_i\neq h_j$ and it suffices to make sure that $\Delta(k_i;\epsilon)$ and $\Delta(h_j;\epsilon)$ are disjoint.

As another consequence of the local tesselation property, we conclude that the Γ-translates of $\iota.\Delta$ form a locally finite covering of X: let us consider the point $x\in X$ given by the equivalence class $\mathfrak{C}$ from 1.9. If we let U denote an open neighbourhood of x in X of the form displayed in 1.9, we have that

$$\sigma(\iota.\Delta)\ \bigcap\ U=\emptyset \qquad ;\ \sigma\neq\eta_1,\dots,\eta_m$$

In particular, we have $\sigma x\in U$ for finitely many $\sigma\in\Gamma$ only; that is the action of Γ on X is discontinuous.

Let us prove that X/Γ is complete. The continuous map $\Delta\to X$, $h\mapsto\iota.h$, induces a surjective map $\Delta\to X/\Gamma$ which implies that X/Γ is compact and therefore complete. We conclude from VI.5.9 that X is complete.

We can now use the monodromy theorem VI.4.8 to show that $f:X\to H^2$ is bijective. Surjectivity of f means that

$$H^2=\bigcup_{g\in G}\ g(\Delta)$$

From the very definition of the equivalence relation $\sim$ follows that

$$\sigma.\Delta\ \bigcap\ \iota.Int\Delta\ =\emptyset \qquad ;\ \sigma\in\Gamma,\ \sigma\neq\iota$$

The injective map $f:X\to H^2$ transforms disjoint sets into disjoint sets, thus we get

$$\tilde{\sigma}(\Delta)\ \bigcap\ Int\Delta\ =\emptyset \qquad ;\ \sigma\in\Gamma-\{\iota\}$$

This shows that the kernel of $\Gamma\to G$ is trivial. It follows that the canonical map from the universal group Γ to the group G is a bijection and that Δ is a fundamental domain for the action of G on H^2. □

Parallelogram groups 1.10

Let us consider a parallelogram ABCD in H^2. By this we mean a convex quadrangle □ABCD where the opposite sides have equal length. The elementary geometry of the parallelogram tells us that a half-turn with respect to the midpoint of AC transforms ΔABC into ΔCDA. Otherwise expressed, the diagonals have a common midpoint M and a half-turn with respect to M transforms the parallelogram into itself. The angle sum of the quadrangle is less that 2π, III.9.7. Let us assume that this has the form

$$\angle A + \angle B + \angle C + \angle D = \frac{2\pi}{n} \quad ; \ n = 2,3,...$$

We can apply Poincaré's theorem to the side pairing $*AD = BC$ and $*AB = DC$. The corresponding side transformations are denoted α and β.

Edge cycle : a b $*a^{-1}*b^{-1}$

Cycle map : $\alpha\beta\alpha^{-1}\beta^{-1}$

Vertex cycle : ABCD

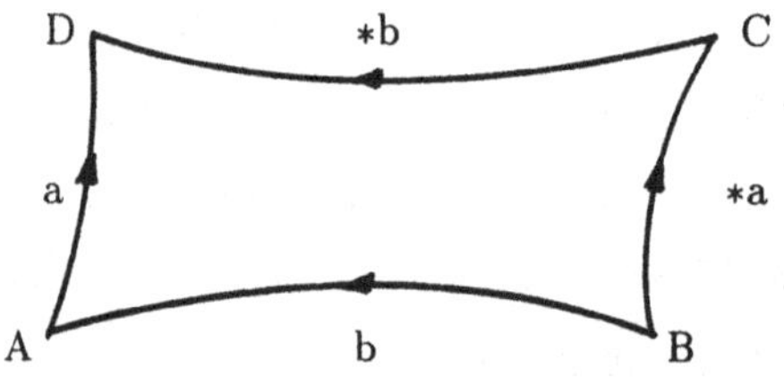

The cycle transformation $\alpha\beta\alpha^{-1}\beta^{-1}$ is a rotation around A of order n. We find that the parallelogram group G is the universal group on α and β subject to

$$(\alpha\beta\alpha^{-1}\beta^{-1})^n = \iota$$

VII.2 TRIANGLE GROUPS

Let us consider ΔABC in the hyperbolic plane H^2. Reflection in the geodesic BC is denoted α, reflection in CA is denoted β, while reflection in AB is denoted γ. We shall assume that the angles have measures as follows

2.1
$$\angle A = \frac{\pi}{p}, \quad \angle B = \frac{\pi}{q}, \quad \angle C = \frac{\pi}{r} \qquad ; \quad p,q,r \in \mathbb{N}$$

Since the angle sum of ΔABC is $< \pi$ we must of course assume that

2.2
$$\frac{1}{p} + \frac{1}{q} + \frac{1}{r} < 1 \qquad ; \quad p,q,r \in \mathbb{N}$$

Conversely, for any three integers p,q,r satisfying this condition, we can find ΔABC with angles of these measures; this triangle is unique up to congruence, compare III.5. The subgroup of $Isom(H^2)$ generated by the reflections α,β,γ is denoted G(p,q,r).

The isometry $\gamma\beta$ is a rotation around A of angle $2\pi/p$ which makes $\gamma\beta$ a rotation of order p, similar remarks apply to the vertices B and C. We can now write down three reflection relations and three cycle relations for our three generators α,β,γ.

2.3
$$\alpha^2 = \iota, \ \beta^2 = \iota, \ \gamma^2 = \iota, \ (\gamma\beta)^p = \iota, \ (\alpha\gamma)^q = \iota, \ (\alpha\beta)^r = \iota$$

Let us apply Poincaré's theorem to Δ ABC with the trivial pairing $*a = a$, $*b = b$, $*c = c$. The edge cycles are

$$c\,b^{-1}, \quad a\,c^{-1}, \quad a^{-1}\,b$$

This realises the relations 2.3 as reflection relations and cycle relations.

A picture of the tesselation of the Poincaré disc by the fundamental domain for the triangle group D(2,3,7) is given in the introduction.

Let us work out generators and relations for the even part $G^+(p,q,r)$ of the triangle group. It is quite clear that $G^+(p,q,r)$ is generated by

$$\rho_{2A} = \beta\gamma \,, \quad \rho_{2B} = \gamma\alpha \quad , \quad \rho_{2C} = \alpha\beta$$

From 2.3 we derive the following relations for the ρs

2.4
$$\rho_{2A}\, \rho_{2B}\, \rho_{2C} = \iota \,, \ \rho_{2A}{}^{p} = \iota \,, \ \rho_{2B}{}^{q} = \iota \,, \ \rho_{2C}{}^{r} = \iota$$

Let us show that the list 2.4 is a complete set of relations for $G^+(p,q,r)$. To this end we apply Poincaré's theorem to the quadrangle ABCD where the point D is reflection of B in the geodesic through A and C.

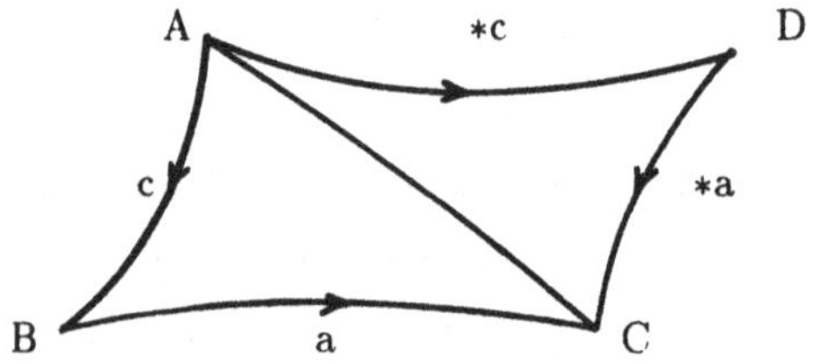

Edge cycles: $a{*c}^{-1}$, $c^{-1}{*a}$, a^{-1}, ${*a}^{-1}$, c, *c. The side transformations are $\sigma_a = \rho_{2C}$ and $\sigma_{*c} = \rho_{2A}$ as follows from

$$\rho_{2C}(D) = \alpha\beta(D) = \alpha(B) = B \,, \quad \rho_{2A}(B) = \beta\gamma(B) = \beta(B) = D$$

Thus Poincaré's theorem gives us the following complete set of relations

$$(\rho_{2C}\rho_{2A})^{q} = \rho_{2C}{}^{r} = \rho_{2A}{}^{p} = \iota$$

which is the same as 2.4.

We shall continue to give some applications of Poincaré's theorem to compact polygons. The following notation is useful and will be exploited in VII.6.

Definition 2.5 By a <u>boundary cycle</u> of a polygon Δ we understand a cyclic enumeration of its edges $a_1,\dots,a_n$ ($a_{n+1} = a_1$) where the final vertex of a_i equals the initial vertex of a_{i+1} for $i = 1,\dots,n$. Once the hyperbolic plane is oriented, the positive boundary cycle is defined by requiring that locally Δ is situated on the positive side of the edges of the boundary.

Hyperbolic pants Consider a rectangular hexagon $\mathcal{R}$ in H^2 with boundary cycle $a_1,...,a_6$ and side pairing $*a_i = a_i$, i=1,...,6. Poincaré's theorem tells us that reflections $\rho_1,...,\rho_6$ in the sides of $\mathcal{R}$ generate a discrete group Π with fundamental domain $\mathcal{R}$. A complete set of twelve relations is given below

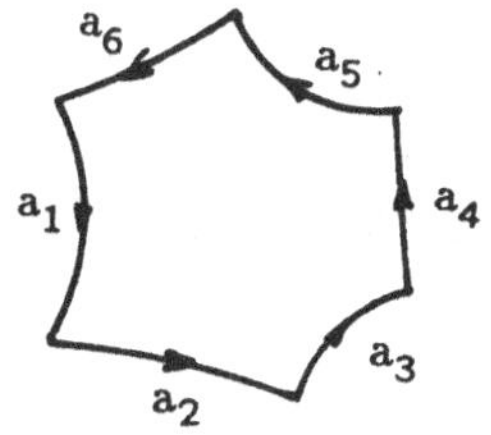

$$\rho_i^2 = \iota \quad , \quad \rho_i\rho_{i+1} = \rho_{i+1}\rho_i$$
$$; \ i = 1,...,6 \ (\rho_7 = \rho_1)$$

Consider a second copy of the same hexagon and sew the two hexagons together along the seems a_2,a_4,a_6. The result is a pair of hyperbolic pants. In order to realise this as an orbit space, consider the rectangular octagon $\Delta = \mathcal{R} \cup \rho_4(\mathcal{R})$ with the side pairing given below $(*a = a, *c = c, *e = e, *f = f)$.

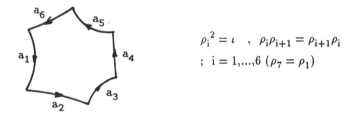

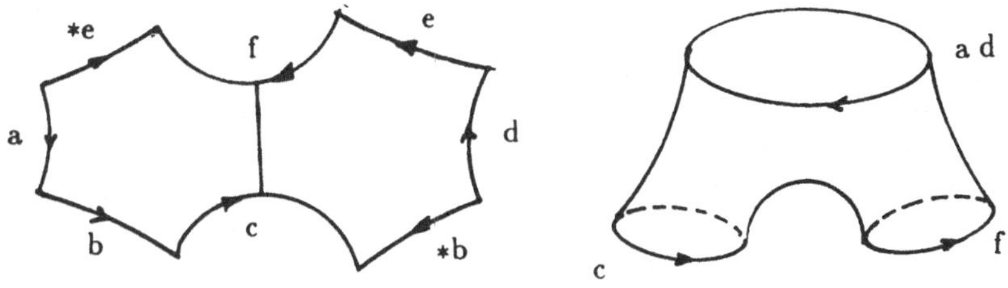

The boundary cycle is $a\ b\ c\ *b^{-1}\ d\ e\ f\ *e^{-1}$. The two side transformations are $\sigma_b = \rho_2\rho_4$ and $\sigma_e = \rho_4\rho_6$. We leave it to the reader to apply Poincaré's theorem and conclude that the group Γ generated by

$$\rho_1, \rho_3, \rho_5, \rho_{64}, \rho_{42}, \rho_{26} \qquad ; \ \rho_{ij} = \rho_i\rho_j$$

is a discrete subgroup of index 2 in Π with fundamental domain Δ. The cycle and reflection relations may be rewritten in the following form

2.6
$$\rho_1^2 = \iota, \ \rho_3^2 = \iota, \ \rho_5^2 = \iota, \ \rho_{64}\rho_{42}\rho_{26} = \iota$$
$$\rho_1\rho_{26} = \rho_{26}\rho_1 \ , \ \rho_3\rho_{42} = \rho_{42}\rho_3 \ , \ \rho_5\rho_{64} = \rho_{64}\rho_5$$

Torus with a hole Let us start with a right angled hexagon with boundary cycle $a_1a_2a_3a_4a_5a_6$ as before. This time we require a_2 and a_6 to have the same length; formula III.8.1 implies that a_3 and a_5 have the same length as well. Reflect this in the side a_1 to get an octagon with the side pairing as indicated

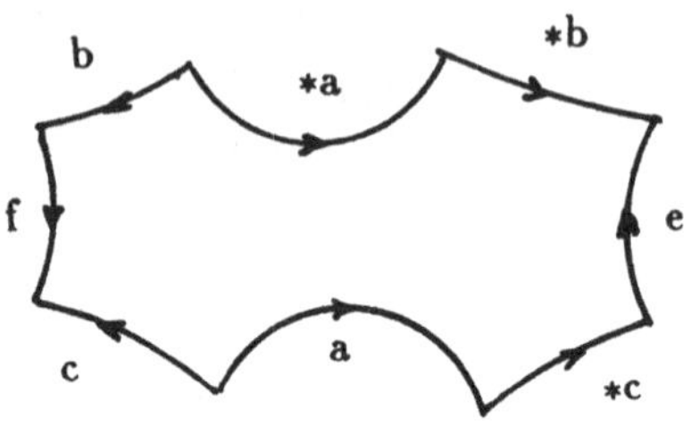

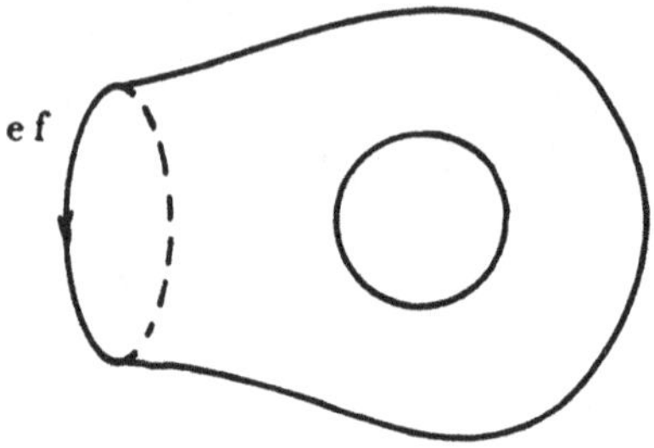

Edge cycles : $a\ b\ {*a}^{-1}{*c}$, $e\ {*c}^{-1}\ f^{-1}\ c^{-1}$

The group is generated by the side transformations $\alpha\ \beta\ \gamma\ \eta\ \phi$ subject to

2.7 $$\alpha\beta\alpha^{-1} = \gamma\ ,\ \eta^2 = \iota\ ,\ \phi^2 = \iota\ ,\ \gamma\eta^{-1}\gamma^{-1} = \phi$$

It follows that the group is generated by α, β ,η subject to the sole relation $\eta^2 = \iota$.

Example 2.8 Let us fix a pentagon Δ with angle sum π/n and consider the side pairing with boundary cycle

$$a\ {*b}^{-1}{*a}^{-1}\ b\ c$$

in particular the edge c is reflected in itself. The edge cycle is

$$a\ b\ {*a}^{-1}{*}\ b^{-1}c\ b^{-1}a^{-1}{*b}\ {*a}\ c^{-1}$$

With the convention $\alpha = \sigma_a$... we find a complete set of relations

$$(\alpha\beta\alpha^{-1}\beta^{-1}\gamma\ \beta\alpha\beta^{-1}\alpha^{-1}\gamma)^n = \iota,\ \ \gamma^2 = \iota$$

It is convenient to introduce $\epsilon = [\alpha\beta]^{-1}$ and $\epsilon^{-1}\gamma\epsilon = \delta$ to get

$$[\alpha\beta]\epsilon = \iota\ ,\ \delta = \epsilon^{-1}\gamma\epsilon\ ,\ (\delta\gamma)^n = \iota\ ,\ \delta^2 = \gamma^2 = \iota$$

The orbit space is a torus with a hole and a marked point on the boundary.

VII.3 PARABOLIC CYCLE CONDITION

Let us consider a polygon Δ in H^2 with some of its vertices on ∂H^2. To be sure the sides of the polygon are geodesic arcs in H^2 or arcs of ∂H^2. The edges supported by arcs on the boundary are called free edges. The set of edges including the free ones is denoted $\mathcal{S}$. Let there be given a side pairing that is an involution $s \mapsto *s$ on $\mathcal{S}$ with

3.1 $$*(s^{-1}) = (*s)^{-1} \qquad ; s \in \mathcal{S}$$

The free edges are supposed to be fixed points under $*$ while the non-free edges are subject to the following conditions: if the initial, resp. final vertex of the edge s is finite we require the initial, resp. final vertex of $*s$ to be finite as well ; if both ends of s are finite, we require s and $*s$ to have the same length.

We shall assume that for each edge $s \in \mathcal{S}$ there is given an isometry σ_s of H^2 subject to the conditions: for a non-free edge s the map σ_s maps the edge $*s$ onto s in such a way that locally the half-plane determined by $*s$ and Δ is mapped to the half-plane bounded by s but opposite Δ. For a free edge s we require $\sigma_s = \iota$.

Observe that unless both ends of the non-free edge $s \in \mathcal{S}$ are infinite then the isometry σ_s is uniquely determined by these requirements. In general the precise determination of σ_s is part of the structure.

The isometry σ_s is called the side transformation determined by s. It is required that these data satisfy the

Reflection relations 3.2

$$\sigma_{s^{-1}} = \sigma_s \ , \quad \sigma_{*s} = \sigma_s^{-1} \qquad ; s \in \mathcal{S}$$

For an edge s of Δ we let $\downarrow s$ denote edge with the same initial vertex as s but different final vertex. This defines for us a second involution on $\mathcal{S}$, $s \mapsto \downarrow s$. The composite of these two operators defines the edge operator $\Psi: \mathcal{E} \to \mathcal{E}$ given by

3.3 $$\Psi s = \downarrow * s \qquad ; s \in \mathcal{E}$$

Let us introduce the set $\mathcal{E}_\infty$ of edges with initial point on ∂H^2. Observe that $\mathcal{E}_\infty$ is stable under $*$ and $\downarrow$ and conclude that $\mathcal{E}_\infty$ is stable under the edge operator Ψ. The cycles $s_1,...,s_r$ of Ψ on $\mathcal{E}_\infty$ are of two sorts: free cycles which contain free edges and parabolic cycles which do not contain free edges. The cycle map $\sigma = \sigma_1...\sigma_r$ of a parabolic cycle is even, as follows from the proof of 1.4. Let us remark that σ is a parabolic transformation if the side transformations generate a discrete group with fundamental domain Δ, compare V.2.4. The cycle map of a free cycle is a reflection or ι as follows from the general

Lemma 3.4 Let $s_1,...,s_r$ be an edge cycle with side transformations $\sigma_1,...,\sigma_r$ and cycle map $\sigma = \sigma_1\sigma_2...\sigma_r$. If none of the edges of the cycle are fixed by $*$ then the vertex cycle $P_1,...,P_r$ has no repetitions. Conversely, if the vertex cycle $P_1,...,P_r$ contains a repetition, $P_1 = P_n$ say, then the edge cycle $s_1,...,s_r$ contains exactly two edges fixed by $*$, and the "half cycle maps"

$$\sigma_1...\sigma_{n-1} \ , \ \sigma_n...\sigma_r$$

are conjugates of the side transformations relative to the two edges fixed by $*$.

Proof From $P_1 = P_n$ and $s_1 \neq s_n$ we get $s_1 = \downarrow s_n$, and we conclude that

$$s_1 = \downarrow\downarrow * s_{n-1} = * s_{n-1}, \ s_2 = \downarrow * s_1 = \downarrow s_{n-1} = \downarrow\downarrow * s_{n-2} = * s_{n-2}$$

A simple induction on i=1,...,n-1 gives us

$$\boxed{s_i = * s_{n-i}} \qquad ; \ i = 1,...,n-1$$

Let us assume for a second that n is odd, say $n = 2m+1$. Put $i = m$ in the formula above to get that $s_m = * s_{m+1}$ or $* s_m = s_{m+1} = \downarrow * s_m$, which is absurd. In conclusion the number n must be even, $n = 2m$. From the boxed formula with $n = 2m$ and $i = m$ we get that $s_m = * s_m$. From the same source we get that

$$s_1,...,s_m,...,s_{2m-1} = s_1,...,s_{m-1},s_m,* s_{m-1},...,* s_1$$

and we conclude that

$$\sigma_1...\sigma_{n-1} = (\sigma_1...\sigma_{m-1})\sigma_m(\sigma_1...\sigma_{m-1})^{-1}$$

which shows that $\sigma_1...\sigma_{n-1}$ is a conjugate of σ_m.

Let us proceed to apply this result to the opposite edge cycle $*s_r*s_{r-1}...*s_1$. From $s_1 = s_n$ we get $\downarrow s_1 = \downarrow s_n$ or $*s_r = *s_{n-1}$. Thus we can conclude that the sequence $s_n,...,s_r$ consists of an odd number of edges and that the middle edge s_f is fixed, $*s_f = s_f$. Moreover, we find that $\sigma_n...\sigma_r$ is a conjugate of σ_f.

Conversely, let us suppose that our cycle $s_1,...,s_r$ contains and edge c with $*c = c$. Let us consider the full infinite sequence of edges defined by

$$c_i = \Psi^i c \qquad ; i \in \mathbb{Z}$$

Let us make a simple observation that $*c = c = \downarrow *c_{-1}$ which gives $c_1 = *c_{-1}$. By continuation of this process we obtain the formula

$$*c_{-m} = c_m \qquad ; m \in \mathbb{Z}$$

or $c_{1-m} = \downarrow c_m$. Thus we can rewrite our sequence as

$$\downarrow c_4, \downarrow c_3, \downarrow c_2, \downarrow c_1, c_1, c_2, c_3 ...$$

It follows that the sequence $P_1,...,P_r$ of initial points of $s_1,...,s_r$ contains a repetition, $P_1 = P_n$ say. With the notation above we must have $c = s_m$ or $c = s_f$. □

Example 3.5

Let us consider a pentagon with all vertices on ∂H^2

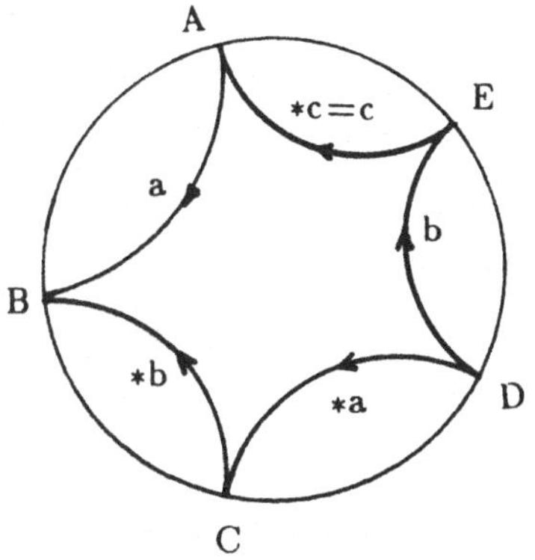

Boundary cycle :

$a *b^{-1}*a^{-1} b c$

Edge cycle :

$a b *a^{-1}*b^{-1}c b^{-1}a^{-1}*b *a c^{-1}$

Cycle transformation :

$\alpha\beta\alpha^{-1}\beta^{-1}\gamma \beta\alpha\beta^{-1}\alpha^{-1}\gamma$

The vertex cycle is ADCBEEBCDA. The "half-cycles" relative to a point P are reflections, providing a decomposition of the cycle map into a product of two reflections in geodesics through P. It follows that the cycle map is parabolic.

Example 3.6 Let us consider a quadrangle with a free edge cycle.

Edge cycle : a c *a b

Vertex cycle: ACCA

Cycle map: reflection β in b

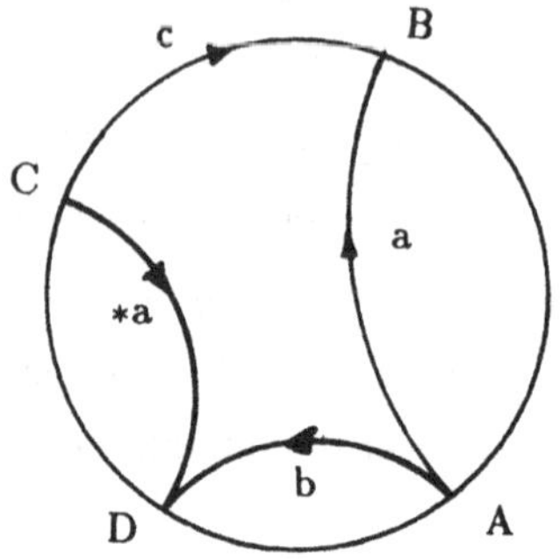

Theorem 3.7 The conclusion of Poincaré's theorem 1.7 remains valid for polygons with vertices at infinity, provided that each parabolic cycle $s_1,\dots,s_r$ generates a parabolic cycle map $\sigma_1\sigma_2\dots\sigma_{r-1}\sigma_r$.

Proof Let us return to the proof of 1.7 and observe that it suffices to prove that X/Γ is complete. According to VI.7.4 this space is isometric to the space $\iota.\Delta/\mathcal{G}$ where $\mathcal{G}$ is the equivalence relation on $\iota.\Delta$ induced by the action of Γ on X. Projection $\iota.\Delta\to\Delta$ decreases distances[2] and induces a distance decreasing bijection

$$\iota.\Delta/\mathcal{G} \to \Delta/\mathcal{R}$$

where $\mathcal{R}$ denotes the equivalence relation on Δ generated by the side pairing. On the other hand, the presence of the continuous section, $z\mapsto\iota.z$, $\Delta\to\iota.\Delta$ shows that the bijection above has a continuous inverse. Thus it suffices to prove that $\Delta/\mathcal{R}$ is complete.

Let us consider a sequence $(w_n)_{n\in\mathbb{N}}$ of points of Δ representing a Cauchy sequence in $\Delta/\mathcal{R}$. We intend to show that this sequence is bounded in H^2. Once this is done, we conclude that our Cauchy sequence is contained in a compact subset of $\Delta/\mathcal{R}$, which makes it convergent in $\Delta/\mathcal{R}$.

Consider first a free edge AB of Δ. Let us pick a hypercycle $\mathcal{C}$ with ends A and B, which does not meet $\partial\Delta$. The function $f_{\mathcal{C}}:H^2\to\mathbb{R}$ introduced in the proof of VI.7.7 is distance decreasing and induces a distance decreasing function on

[2] When Δ is convex, the projection $\iota.\Delta\to\Delta$ is an isometry, a priori.

$\Delta/\mathfrak{R}$, VI.7.3. It follows that $f_{\mathfrak{C}}$ is bounded on the sequence $(w_n)_{n \in \mathbb{N}}$. As a consequence we can find a hypercycle $\mathfrak{C}$ as above such that none of the points of our sequence belongs to the region bounded by $\mathfrak{C}$ and the free edge AB.

Let us consider a parabolic cycle $s_1,...,s_r$. Pick a horocycle $\mathcal{A}$ at the initial point P_1 of s_1 and let $\mathcal{A}_1,...,\mathcal{A}_r$ be defined recursively by

$$\mathcal{A}_i = \sigma_i \mathcal{A}_{i+1} \ , \ \mathcal{A}_r = \sigma_r \mathcal{A}_1 \qquad ; i = 1,...,n$$

Notice that $\mathcal{A} = \mathcal{A}_1$ since $\sigma = \sigma_1 .... \sigma_r$ is either a horolation or ι. If repetitions occur in the sequence $P_1,...,P_r$, the same repetitions occur in the sequence $\mathcal{A}_1,...,\mathcal{A}_r$ as follows from 3.4 : "half-cycle maps" are reflections.

With the notation from the proof of VI.7.7 let p_i denote the radial function defined by $\mathcal{A}_i$. Recall that p_i is zero outside the horodisc bounded by $\mathcal{A}_i$ and that p_i is distance decreasing. Let us form the distance decreasing function $p:\Delta \to \mathbb{R}$

$$p = \tfrac{1}{r}\,(p_1 + ... + p_r)$$

and observe that this function is constant on the equivalence classes modulo $\mathfrak{R}$ when the horocycle $\mathcal{A}$ is sufficiently small. We conclude from VI.7.3 that p defines a distance decreasing function on $\Delta/\mathfrak{R}$. It follows that the sequence $(p(w_n))_{n \in \mathbb{N}}$ is convergent and we conclude that p is bounded on $(w_n)_{n \in \mathbb{N}}$. In conclusion we can choose the horocycle $\mathcal{A}$ such that the sequence $(w_n)_{n \in \mathbb{N}}$ does not meet any of the horodiscs bounded by $\mathcal{A}_1,...,\mathcal{A}_r$.

Let us consider a free cycle $s_1,...,s_r$. It follows from 3.4 that the cycle contains one or two free edges. In the case of two free edges we get from 3.4 that all "half-cycle maps" are ι. If the number of free edges is 1 we get from 3.4 that the "half-cycle maps" are ι and reflection in a geodesic. It follows that the cycle map is a reflection in a geodesic with end P_1. Thus a free cycle can be treated just like a parabolic cycle with parabolic cycle map. Our journey has ended ! □

VII.4 CONJUGACY CLASSES

In this section we shall show how to use a fundamental domain to determine the conjugacy classes of parabolic and elliptic elements in a discrete group Γ. More precisely we are given a polygon Δ with a side pairing satisfying the hypothesis of Poincaré's theorem 3.7 and we look for conjugacy classes in the group Γ generated by the side transformations of Δ.

Let us start by the observation that an elliptic element is conjugated to an an element which stabilises a point of $\partial\Delta$ while a parabolic element is conjugated to an element which stabilises a point of $\partial_h\Delta$, V.2.7. This makes it relevant to determine the stabiliser Γ_P of point of an edge s of Δ.

Proposition 4.1 Let P be an interior point of the edge s. The stabiliser Γ_P is trivial except in the following two cases

$*s = s$: Γ_P is generated by the reflection σ_s.

$*s = s^{-1}$ and P is the midpoint of s : Γ_P is generated by σ_s.

Proof Let us assume that $P \in H^2$ is an inner point of the edge s. Observe that $\Delta \cup \sigma_s\Delta$ is a neighbourhood of P and conclude that a $\gamma \in \Gamma_P$ must equal ι or σ_s. When $\gamma = \sigma_s$, we conclude that P is a fixed point for σ_s and we get that $*s = s$ or $*s = s^{-1}$. In the first case σ_s is a reflection in s, in the second case σ_s is a half-turn in P.

Let us assume that $P \in \partial_h\Delta$. Observe first that an even $\gamma \in \Gamma_P$ is parabolic as follows from V.2.4. It follows that an odd $\gamma \in \Gamma_P$ must be a reflection in a geodesic through P since γ^2 can't be proper hyperbolic. It also follows that horocycles with centre P are stable under Γ. When P is an interior point of a free side s of Δ we conclude that Δ contains a full horocycle $\mathcal{A}$ with centre P. For $\gamma \in \Gamma_P$ we conclude that $Int\Delta$ meets $\gamma\, Int\Delta$, thus $\gamma = \iota$. □

Theorem 4.2 Let P be a vertex of the polygon Δ and let $s_1,.,.,s_r$ be a full edge cycle with vertex cycle $P = P_1,...,P_r$. If the vertex cycle is without repetitions then Γ_P is generated by the cycle map $\sigma_1...\sigma_r$. If the vertex cycle has a repetition, $P_1 = P_n$ say, then Γ_P is generated by the two "half-cycle maps"

$$\sigma_1...\sigma_{n-1} \quad , \quad \sigma_n...\sigma_r$$

Proof Let us first consider the case where P is a point of H^2. Let us investigate the following union of translates of Γ

$$\mathfrak{J} = \Delta \cup \sigma_1\Delta \cup \sigma_1\sigma_2\Delta \cup \ ... \cup \sigma_1\sigma_2...\sigma_{r-1}\Delta$$

From the formula $\sigma_i(P_{i+1}) = P_i$ it follows that each of these translates passes through the point P_1; in fact $P_1 = \sigma_1...\sigma_i(P_{i+1})$. Observe that

$$\sigma_1...\sigma_{i-1}\Delta \cap \sigma_1...\sigma_i(\Delta) \supset \sigma_1...\sigma_{i-1}(s_i) \qquad ; i = 1,...,r$$

as follows from $s_i = \sigma_i(*s_i)$. Let us consider a small circle $\mathcal{A}$ with centre P_1 and observe that $\mathfrak{L} = \mathfrak{J} \cap \mathcal{A}$ is a connected with $\mathfrak{L} \cap \sigma\mathfrak{L} \neq \emptyset$. More precise information is available: the arc $\mathfrak{L}$ subtends an angle $2\pi/n$ where $n = ord(\sigma)$. It follows that translates of $\mathfrak{J}$ of the form $\sigma^s\mathfrak{J}$, $s \in \mathbb{Z}$, cover an open neighbourhood of P. For $\gamma \in \Gamma_P$ we conclude that $\gamma\, Int\Delta$ meets one of these translates ; in fact γ must have the form

$$\gamma = \sigma^s \ \text{ or } \ \gamma = \sigma^s\sigma_1...\sigma_{n-1} \qquad ; n = 2,...,r,\ s \in \mathbb{Z}$$

In the second case, we must have $P_1 = \sigma_1...\sigma_{n-1}(P_1)$, i.e. $P_1 = P_n$. It follows that $\sigma_1...\sigma_{n-1}$ is a "half-cycle map".

Let us consider a parabolic cycle $s_1,...,s_r$. The main point is to show that the cycle map $\sigma_1...\sigma_n$ is non-trivial. Pick a horocycle $\mathcal{A}$ at the initial point P_1 of s_1 and let $\mathcal{A}_1,...,\mathcal{A}_r$ be defined recursively by

$$\mathcal{A}_i = \sigma_i\mathcal{A}_{i+1} \ , \ \mathcal{A}_r = \sigma_r\mathcal{A} \qquad ; i = 1,...,n$$

Notice that $\mathcal{A} = \mathcal{A}_1$ since $\sigma = \sigma_1....\sigma_r$ is either a horolation or ι. If repetitions occur in the sequence $P_1,...,P_r$, the same repetitions occur in the sequence $\mathcal{A}_1,...,\mathcal{A}_r$ as follows from 3.4: "half cycle maps" are reflections. We can now measure the the distance from s_1 to $\sigma(s_1)$ along the horocycle $\mathcal{A}$ and show that this equals the sum $\sum_{i=1,...,r} \theta_i$ where θ_i is the length of the horocyclic arc cut by $(s_i, \downarrow s_i)$ in $\mathcal{A}_i$. The detailed argument is like the proof of 1.4

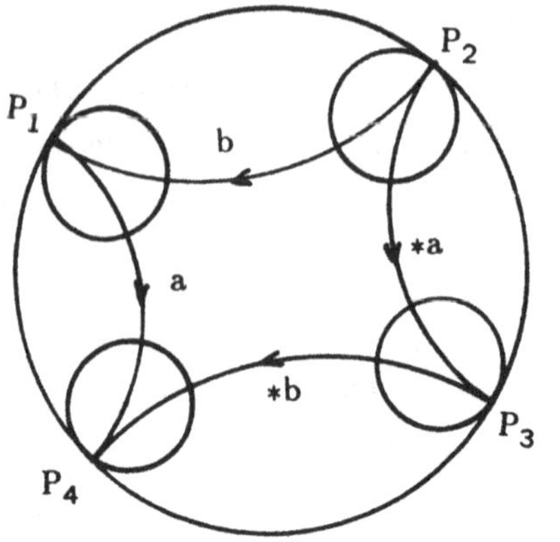

We can introduce the horocycle arc $\mathcal{L} = \mathcal{A} \cap \mathcal{J}$ and observe that $\mathcal{L} \cap \sigma\mathcal{L} \neq \emptyset$ implies that $\mathcal{A}$ is covered by translate $\sigma^s\mathcal{L}$, $s \in \mathbb{Z}$. We can proceed as above taking the second half of the proof of 4.2 into account.

Let us now turn to a free cycle. Let us treat the case where $*s_r = \downarrow s_1$ is a free edge. The full cycle can be written, compare the proof of 3.4

$$s_1 \ldots s_{n-1} s_n * s_{n-1} \ldots * s_1 s_r \qquad ; \; r = 2n$$

with $*s_n = s_n$. In the case where the edge s_n is non-free we have found that s_n is a reflection and it follows that $\sigma = \sigma_1 \ldots \sigma_{r-1}$ is a reflection. Observe that the horocyclic arc $\mathcal{L} = \mathcal{A} \cap \mathcal{J}$ contains $\mathcal{A} \cap \Delta$ and $\mathcal{A} \cap \sigma\Delta$. Since $\mathcal{L}$ is connected we find that $\mathcal{L} = \mathcal{A}$ and we can proceed as before. In the case where the edge s_n is free, let us observe that $Int\,\Delta$ and $\sigma_1 \ldots \sigma_{n-1} Int\,\Delta$ are disjoint. It follows that Δ and $\sigma_1 \ldots \sigma_{n-1}\Delta$ cover the two ends of $\mathcal{A}$ and we can conclude that $\mathcal{L} = \mathcal{A}$. We can finally conclude the proof as above. □

VII.5 SUBGROUPS OF THE MODULAR GROUP

In this section we shall apply our theory to some important subgroups of the modular group introduced in V.1. Let us take our point of departure at the fact that the hyperbolic triangle Δ with vertices i, ζ, ∞ is a fundamental domain for the extended modular group $G = PGl_2(\mathbb{Z})$. For unexplained notation see V.1.

Straightforward application of Poincaré's theorem to Δ gives us the generators α,β,γ for G and the complete set of relations

$$\alpha^2 = \beta^2 = \gamma^2 = \iota$$
$$\alpha\gamma\alpha = \gamma\alpha\gamma$$
$$\beta\gamma = \gamma\beta$$

5.1

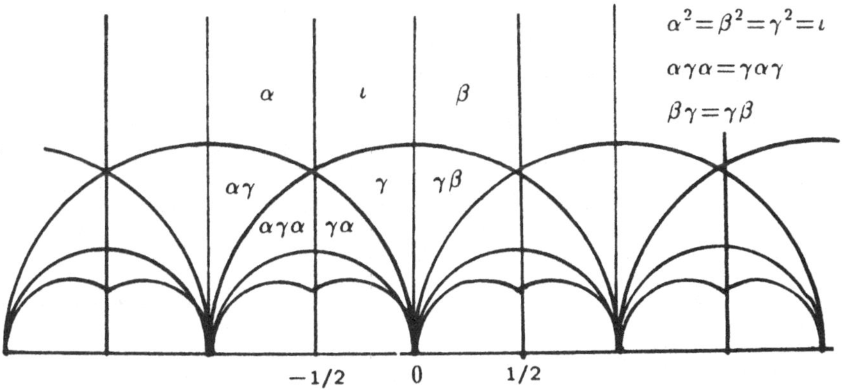

Commutator group

From the relations for G it follows by pure group theory that its commutator subgroup $[G,G]$ has index 4 in G with its factor group generated by β and γ (observe that $\alpha \equiv \gamma \ mod \ [G,G]$). It follows from V.1.4 applied to $\{\iota,\beta,\gamma,\beta\gamma\}$ that we can use the hyperbolic quadrangle $-\zeta$, 0, ζ, ∞ as the fundamental domain for $[G,G]$. Let us apply Poincaré's theorem to get:

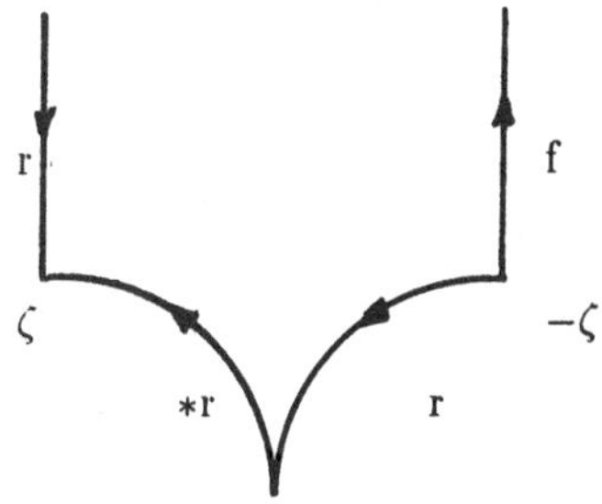

$\sigma_f = \rho = \tau\sigma = \beta\alpha\gamma\beta$

$\sigma_r = \alpha\gamma = \alpha\beta\beta\gamma = \sigma\rho^{-1}\sigma$

Boundary cycle : $r{*}r^{-1}{*}f^{-1}f$

Elliptic edge cycles : r^{-1} , f

Parabolic edge cycle : $r\ {*}f^{-1}$, ${*}r\ f^{-1}$

It follows that $[G,G]$ is generated by the third order rotations ρ and $\sigma\rho\sigma$ and that there is no relation between these two generators. For further reference let us note that the parabolic cycle transformations are

$$(\sigma\rho^{-1}\sigma)\,\rho^{-1} = \tau^{-2} \quad , \quad (\sigma\rho^{-1}\sigma)^{-1}\rho = \sigma\tau^{2}\sigma$$

It follows from our investigations that G has three subgroups of index 2.

The modular group Γ is generated by σ and ρ with the complete set of relations $\rho^3 = \iota$ and $\sigma^2 = \iota$ it follows by applying Poincaré's theorem to the modular figure from V.1. From these relations it follows that $[\Gamma,\Gamma]$ has index 6 in Γ with a factor group generated by $\sigma\rho = \tau$. In particular, we conclude that Γ has a unique subgroup Γ^2 of index 2, in fact $\Gamma^2 = [G,G]$.

From our investigations above it follows that Γ has a unique normal subgroup Γ^3 of index 3. The group Γ^3 contains $\sigma = \sigma^3$ and its conjugates $\tau\sigma\tau^{-1}$ and $\tau^{-1}\sigma\tau$. Let us show that Γ^3 is generated by these three elements of order two subject to no further conditions. From $\rho = \tau\sigma$ it follows that Γ/Γ^3 is cyclic of order 3 generated by τ. We can use V.1.4 to conclude that $\tau^{-1}\Delta \cup \Delta \cup \tau\Delta$ represents a fundamental domain.

Side transformations :

$$\tau^3,\ \tau^{-1}\sigma\tau,\ \sigma,\ \tau\sigma\tau^{-1}$$

Cycle relation :

$$(\tau^{-1}\sigma\tau)\ \sigma\ (\tau\sigma\tau^{-1})\tau^3 = \iota$$

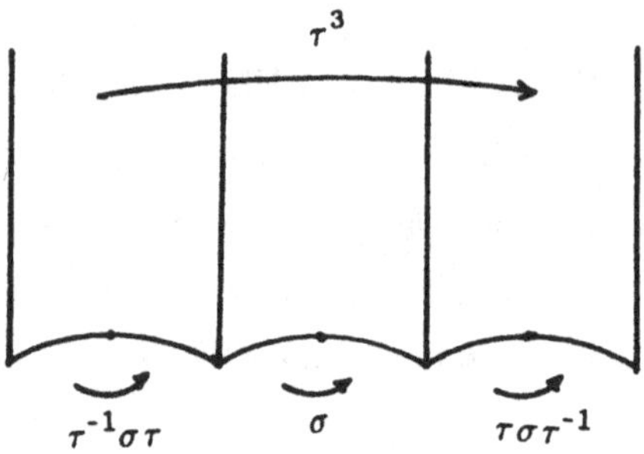

The cycle relation, which may be read $\tau^{-2}\rho^3\tau^2 = \iota$, eliminates the generator τ^3. The remaining cycle conditions simply express the fact that $\tau^{-1}\sigma\tau$, σ, $\tau\sigma\tau^{-1}$ all have order 2. We may express this by saying that Γ^3 is the free product of three cyclic groups of order 2.

Proposition 5.2 The commutator group $[\Gamma,\Gamma]$ has index 6 in Γ and the group $\Gamma/[\Gamma,\Gamma]$ is cyclic. The group $[\Gamma,\Gamma]$ is a free group on the two generators

$$\sigma\tau^{-1}\sigma\tau \ , \quad \sigma\tau\sigma\tau^{-1}$$

Proof The group $[\Gamma,\Gamma]$ has index 2 in Γ^3 with a factor group generated by σ. It follows from the discussion above that $[\Gamma,\Gamma]$ has the fundamental domain

$$\tau^{-1}\Delta \cup \Delta \cup \tau\Delta \cup \sigma\tau^{-1}\Delta \cup \sigma\Delta \cup \sigma\tau\Delta$$

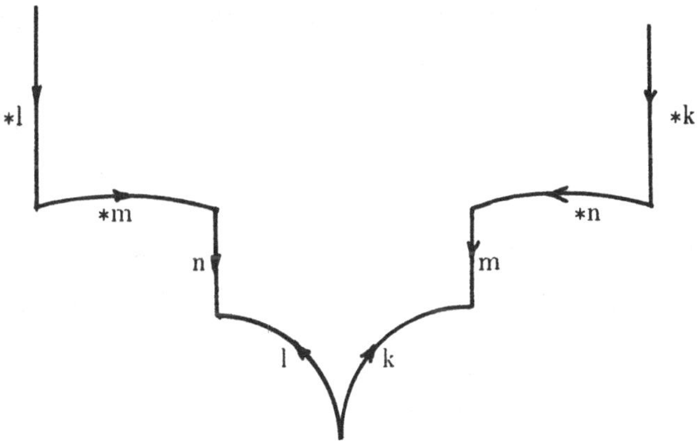

Two of the side pairings can at once be written down

$$\mu = \sigma\tau^{-1}\sigma\tau \ , \quad \nu = \sigma\tau\sigma\tau^{-1}$$

taking into account that $\mu(\tau^{-1}(\zeta)) = -\zeta$ and $\nu(\tau(-\zeta)) = \zeta$. Observe that

$$\mu(\infty) = 1,\ \mu(-1) = 0,\ \nu(\infty) = -1,\ \nu(1) = 0$$

and conclude that the remaining side transformations are $\lambda = \nu\mu$ and $\kappa = \mu\nu$. The counter clockwise boundary cycle, starting at ∞, can now be written as

$$*l\ *m\ n\ l^{-1}k\ m^{-1}{*n}^{-1}\ {*k}^{-1}$$

The edge cycles are $n^{-1}\ m\ {*l}^{-1}$ and $m^{-1}\ n\ {*k}^{-1}$. The cycle relations are

$$\nu\mu\lambda^{-1} = \iota \ , \quad \mu\nu\kappa^{-1} = \iota$$

For later reference let us notice that the parabolic cycle $*k$ l gives us the parabolic

$$\kappa^{-1}\lambda = \nu^{-1}\mu^{-1}\nu\mu = \tau^6$$

It follows that the stabiliser $[\Gamma,\Gamma]_\infty$ has index 6 in Γ_∞. □

Let us conclude this section with a general formula for the genus of the compact Riemann surface, IV.5.8, associated with a normal subgroup Π of finite index in the modular group Γ. The important invariants are the indices $[\Gamma_x:\Pi_x]$ of the stabiliser Π_x of the point x, where $x = i, \zeta, \infty$ ($\zeta = exp(2\pi i/3)$).

Genus formula 5.3 Let Π denote a normal subgroup of finite index of the modular group Γ. The horocyclic compactification X of H^2/Π is a Riemann surface with Euler characteristic

$$\chi(X) = 2[\Gamma:\Pi] - [\Gamma:\Pi] \sum_{x=i,\zeta,\infty} (1 - [\Gamma_x:\Pi_x]^{-1})$$

Proof Let us consider the triangulation $\mathcal{T}$ of the space X induced by the triangulation of H^2 formed by the hyperbolic triangle Δ with vertices i, ζ, ∞ and its translates by the extended modular group. The Euler characteristic is given by

$$\chi(X) = F - E + V$$

where the number of vertices in $\mathcal{T}$ is V, the number of unoriented edges is E and the number of unoriented faces is F. We have $F = 2[\Gamma:\Pi]$ and $E = \frac{1}{2}e$ where e is the number of edges. For a vertex v of $\mathcal{T}$ let e_v denote the number of edges with an initial point at v to get

$$\chi(X) = 2[\Gamma:\Pi] - \textstyle\sum_v (\frac{1}{2}e_v - 1)$$

The number e_v can also be calculated as the number of triangles from $\mathcal{T}$ with a vertex at v. From the fact that Π is a normal subgroup follows that Γ acts on X and $\mathcal{T}$. It also follows that we can separate the vertices of $\mathcal{T}$ into the three orbits of $v = [x]$, $x = i, \zeta, \infty$. This identifies the orbit of [x] and the set $\Gamma/\Pi.\Gamma_x$ and gives us the formula $e_v = 2[\Gamma_x:\Pi_x]$. The standard exact sequence of groups

$$0 \to \Gamma_x/\Pi_x \to \Gamma/\Pi \to \Gamma/\Pi.\Gamma_x \to 0$$

gives us the numerical formula

$$[\Gamma:\Pi] = [\Gamma:\Pi.\Gamma_x][\Gamma_x:\Pi_x]$$

Simple elimination between the three formulas completes the proof. □

The genus g of the compact Riemann surface X, given by the formula $\chi = 2 - 2g$, can be computed from the branch scheme

$$([\Gamma_i,\Pi_i],[\Gamma_\zeta,\Pi_\zeta],[\Gamma_\infty,\Pi_\infty])$$

In particular we find that the group $[\Gamma,\Gamma]$ has a branch scheme (2,3,6) and the genus formula gives $g = 1$. Observe also that the branch scheme (2,3,6) gives genus 1 independent of the index $[\Gamma:\Pi]$. The group Γ^2 has a branch scheme (2,1,2) while the group Γ^3 has a branch scheme (1,3,3).

Let us conclude this section with the introduction of a series of important normal subgroups of the modular group Γ. Let us consider a positive integer n and introduce $\Gamma(n)$, modular group of level n, by the exact sequence

5.4 $$0 \to \Gamma(n) \to PSl_2(\mathbb{Z}) \to PSl_2(\mathbb{Z}/n) \to 0$$

We leave it to the reader to prove that the reduction map $Sl_2(\mathbb{Z}) \to Sl_2(\mathbb{Z}/n)$ is surjective. This is a non-trivial exercise in elementary arithmetic. It is also left to the reader to show that the branch scheme of $\Gamma(n)$ is (2,3,n).

For more information on this topic refer the reader to the books of [Magnus] and [Newman] and the references given there. In particular [Newman] p.143: apart from the three exceptions Γ, Γ^2 and Γ^3, every normal subgroup of Γ is free.

VII.6 COMBINATORIAL TOPOLOGY

In this section we shall consider a discrete group of isometries Γ of H^2 for which H^2/Γ is compact. From the Dirichlet construction V.4 it follows that Γ has a compact fundamental polygon Δ, compare V.2.2. We shall introduce a simple cut and paste technique to modify Δ in such a way that the topological type of H^2/Γ can be read off the boundary cycle of Δ.

In order to justify the cut and paste technique we must generalise the concept of the fundamental polygon to allow for the sides of the polygon to be composed of finitely many geodesic arcs. This generalisation does not destroy the validity of Poincaré's theorem as follows from its proof.

The edges of the polygon are paired by the symbol $a \mapsto *a$ and we insist on the condition $*a \neq a^{-1}$. Let us enumerate one half of the edges (one edge from each side of the polygon)

$$e_1,\dots,e_n,*e_1,\dots,*e_n,f_1,\dots,f_m$$

its understood that $f_i = *f_i$, $i = 1,\dots,m$. The f_is are referred to as the unpaired letters. We can write the boundary cycle as a word of these $2n+m$ symbols and their inverses.

Let us assume that the boundary cycle of Δ has the form $ApQB{*p^{-1}}W$ (see the illustration on the next page), where some of the words A,B,Q,W may be empty. Let us introduce a curve p in the interior of Δ with the same initial point as p and the same final point as Q. Let me stress that a <u>curve</u> is oriented, without double points, and composed of finitely many hyperbolic arcs. Let the side transformation $\sigma^{-1} = \sigma_p^{-1}$ map the polygon pQp^{-1} into $*pQ*p^{-1}$ and let us construct a new fundamental polygon

$$\Delta' = (\Delta - pQp^{-1}) \cup *pQ*p^{-1}$$

The boundary cycle of Δ' is $ApBQ{*p^{-1}}W$. This justifies the first of the four rules given below. The remaining rules are left to the reader to justify.

6.1

A p QB $*p^{-1}$ W $\mapsto$ A p BQ $*p^{-1}$ W

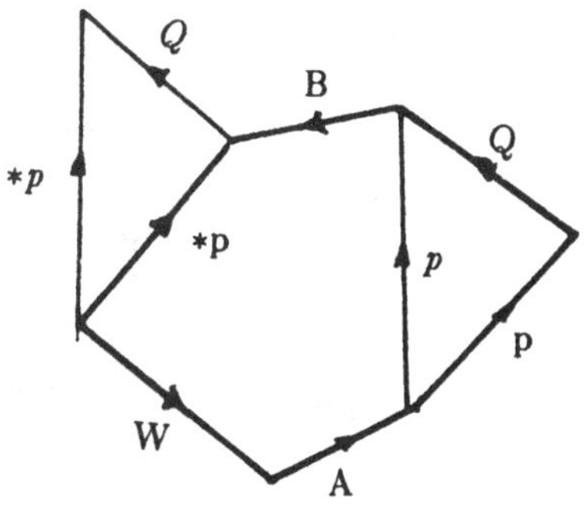

6.2

A p QB *p W $\mapsto$ A p B *p Q^{-1}W

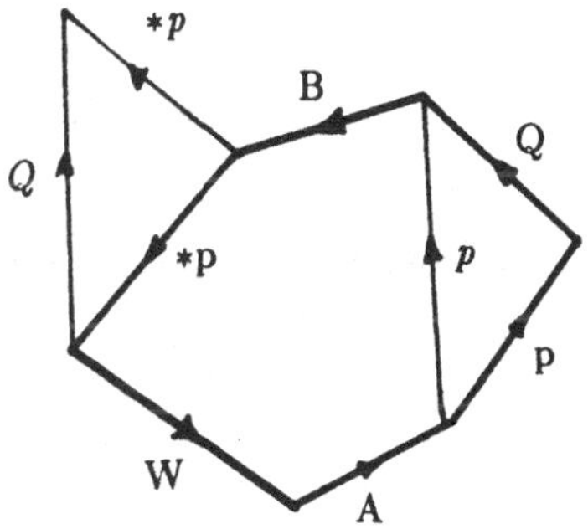

6.3

A Q p B $*p^{-1}$ W $\mapsto$ A p B $*p^{-1}$ QW

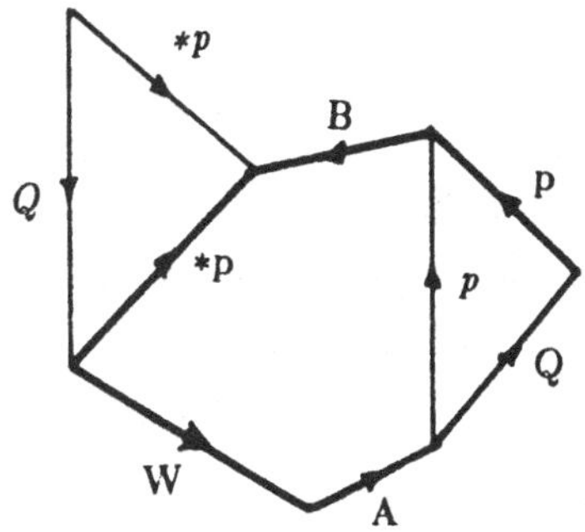

6.4

A Q p B *p W $\mapsto$ A p BQ^{-1} *p W

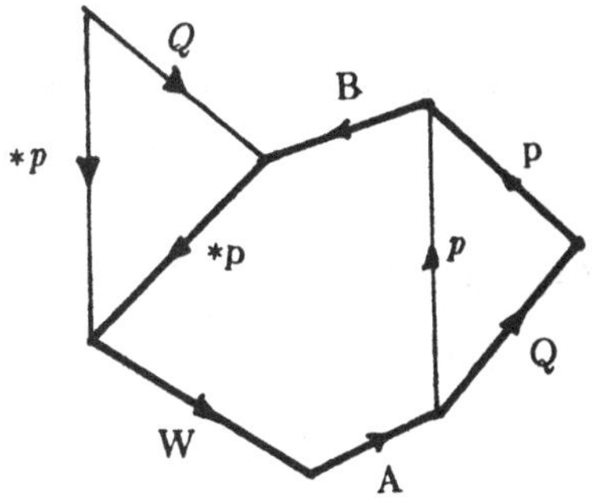

Theorem 6.5 The fundamental polygon Δ for Γ can be chosen in such a way that the boundary cycle for Δ has the form

$$\prod_k (d_k * d_k^{-1}) \prod_{j=1}^{g} (a_j b_j * a_j^{-1} * b_j^{-1}) \prod_{i=1}^{s} (c_i * c_i) \left(\prod_{h=1}^{r} e_h U_h * e_h^{-1}\right) V$$

where V and $U_1, \ldots, U_r$ are words of the unpaired letters.

Proof Let us depart from the boundary cycle of any fundamental polygon and apply the rules 6.1 through 6.4. Let us first observe that if the boundary cycle contains p and *p, then we can use 6.2 with $B = \emptyset$ to bring these symbols together in such a way that the boundary cycle contains the syllable p*p. This syllable can be moved freely around as follows from 6.4 and 6.2 with $B = \emptyset$

$$Qp*p \quad \mapsto \quad pQ^{-1}*p \quad \mapsto \quad p*pQ$$

Let us now assume that the boundary cycle doesn't contain pairs of the form p,*p. Observe that a syllable of the form $a*a^{-1}$ can be brought up front in virtue of 6.3 with $B = \emptyset$

$$Qa*a^{-1} \quad \mapsto \quad a*a^{-1}Q$$

Let us suppose that the boundary cycle contains the letters $a, *a^{-1}, b, *b^{-1}$. The heart of the matter is the relative position of these four symbols. To get the general idea we ask the reader to work out the two alternative reduction steps

$$a\,X\,b\,Y *a^{-1}\,Z *b^{-1} \mapsto a\,X\,b\,ZY *a^{-1}*b^{-1} \mapsto a\,ZYX\,b *a^{-1}*b^{-1} \mapsto a\,b *a^{-1}*b^{-1}ZYX$$

$$a\,X\,b\,Y *b^{-1}Z *a^{-1} \mapsto a\,ZX\,b\,Y *b^{-1}*a^{-1} \mapsto a\,ZX *a^{-1}\,b\,Y *b^{-1}$$

We trust the reader to complete the proof. □

Remark 6.6 The standard form of the boundary cycle as given in 1.5 is not unique. The following reductions can easily be deduced from our four rules

$$a\,b *a^{-1} *b^{-1}\,c *c \mapsto a\,b *a^{-1}\,c *b *c \mapsto a\,b\,c *b *a *c \mapsto$$

$$a\,c^{-1}\,b *b *a *c \mapsto a *c^{-1}\,c^{-1}\,b *b *a \mapsto *c^{-1}c^{-1}\,b *b\,a *a$$

The number $2g+s$ is unique, being the genus of the surface H^2/Γ. Let us also mention that the surface is orientable if $s = 0$ and unorientable if $s > 0$. The number of non-empty words on the list $U_1,\ldots,U_r,V$ equals the number of boundary components of H^2/Γ. These facts follow from Reidemeister's theory of surfaces: a boundary cycle for the surface H^2/Γ is obtained by deleting the symbol $*$ in the boundary cycle 6.5 for Δ, see for example [Zieschang, Vogt, Coldewey] p.114.

Let us conclude this section with some standard notations, see [Griffiths].

Sphere	$a *a^{-1}\ b *b^{-1}$
Torus	$a *b^{-1} *a^{-1}\ b$
Möbius band	$a *a\ c$
Projective plane	$a *a\ b *b^{-1}$
Klein bottle	$a *a\ b* b$, $a\ b *a^{-1} *b$

The first presentation of the Klein bottle is the result of gluing two Möbius bands together along their boundaries. The second presentation alludes to a cylinder whose boundary circles have been given different orientations and glued together accordingly.

VII.7 FRICKE-KLEIN SIGNATURES

In this section we consider a Fuchsian group Γ of H^2 with compact orbit space. We try to gather enough geometric information about Γ to be able to reconstruct Γ as an abstract group. The signature itself is a collection of geometric data sufficient to provide a reconstruction of the group by generators and relations. As a general principle, two discrete groups should have the same signature if and only if they are isomorphic as abstract groups.

The basic part of the signature is the genus g of the compact Riemann surface H^2/Γ given through the Euler characteristic

7.1
$$\chi(H^2/\Gamma) = 2 - 2g$$

Moreover, we let $n_1,\ldots,n_r$ denote the order of the conjugacy classes of elliptic elements of Γ. The fact that the number of conjugacy classes is finite follows from the presence of a compact Dirichlet polygon. The signature of Γ is

7.2
$$sign\ \Gamma = (\ g\,;n_1,\ldots,n_r)$$

Thus the signature is collection of integers with $g \geq 0$ and $n_i \geq 2$, $i = 1,\ldots,r$. In order to investigate which data can occur as the signature we consider a Dirichlet polygon Δ for Γ. We shall show that

7.3
$$Area(\Delta) = 2\pi[-\chi(H^2/\Gamma) + \sum_{i=1}^{r}(1 - \tfrac{1}{n_i})]$$

Proof Let the polygon Δ have v vertices. By the definition of area, III.9.9
$$Area(\Delta) = \pi(v-2) - \textstyle\sum_P \angle_{int} P$$
where the sum is over all vertices P of Δ. Let us recall from VII.3.5 that the vertex cycles of a Fuchsian group are without repetitions. This allows us to partition the sum above into vertex cycles and use V.3.15 to rewrite the formula

$$Area(\Delta) = \pi(v-2) - \sum_{i=1}^{r} \frac{2\pi}{n_i}$$

The polygon Δ modulo identification of sides defines a complex on H^2/Γ with

$$\chi(H^2/\Gamma) = V - E + F$$

where $V = r$, $E = \frac{1}{2}v$ and $F = 1$. Elimination of v and r gives us

$$Area(\Delta) = 2\pi(E-1) - V + 2\pi \sum_{i=1}^{r}(1-\frac{1}{n_i})$$

which is the required result, once its recalled that $F = 1$. □

As an interesting application of the area formula 7.3, we find a lower bound for the area of a Dirichlet polygon Δ for a Fuchsian group

7.4 $$Area(\Delta) \geq \frac{\pi}{21}$$

Proof According to 7.3, it amounts to the same to prove the estimate

$$2g - 2 + \sum_{i=1}^{r}(1-\frac{1}{n_i}) \geq \frac{1}{42}$$

Let us first remark that this is clear for $g \geq 1$. For $g = 1$ we must have $r \geq 1$ and the result follows since $(1-\frac{1}{n_1}) \geq \frac{1}{2}$. When $g = 0$ we have $r \geq 3$ since the area is positive. When $r \geq 5$, the estimate follows by observing that each term in the sum is at least $\frac{1}{2}$. When $r = 4$, the minimum is taken for signature (0;2,2,2,3), but

$$0 - 2 + \frac{1}{2} + \frac{1}{2} + \frac{1}{2} + 1 - \frac{1}{3} = \frac{1}{6}$$

It remains to treat the case $g = 0$ and $r = 3$. We must prove the inequality

$$f(k,m,n) = 1 - (\frac{1}{k} + \frac{1}{m} + \frac{1}{n}) \geq \frac{1}{42} \qquad ; 2 \leq k \leq m \leq n$$

for all strictly positive values of f(k,m,n). Observe that $f(3,3,4) = \frac{1}{12}$ and conclude that we only have to worry about the case $k = 2$. Observe that

$$f(2,2,n) < 0 \ , \ \ f(2,4,4) = 0 \ , \ \ f(2,4,5) = \frac{1}{20}$$

and conclude that we only have to worry about $k = 2$ and $m = 3$. We have

$$f(2,3,n) = \frac{1}{6} - \frac{1}{n}$$

In the end, the result follows from the evaluation $f(2,3,7) = \frac{1}{42}$. □

Theorem 7.5 The signature $(g\ ;\ n_1,\dots,n_r)$ of a Fuchsian group Γ satisfies

$$2g-2+\sum_{i=1}^{r}(1-\tfrac{1}{n_i}) > 0$$

Conversely, any sequence of integers $(g;n_1,\dots,n_r)$ with $g \geq 0$ and $n_i \geq 2$, $i=1,\dots,r$, which satisfies this inequality can be realised as the signature of a Fuchsian group.

Proof The first part of the theorem follows from 7.3. Conversely, for any data as above, we shall construct a suitable polygon and apply Poincaré's theorem. We fix a point O of H^2 and consider a circle with centre O and radius r, say. Let us introduce the angle

$$\theta = \frac{2\pi}{8g+2r}$$

and a sequence of vertices $A_1,\dots,A_g,B_1,\dots,B_r$ ($B_1 = A_{g+1}$, $B_{r+1} = A_1$) on the circle subject to the conditions

$$\angle A_iOA_{i+1} = 8\theta \ , \ \angle B_jOB_{j+1} = 2\theta \quad ; i=1,\dots,g,\ j=1,\dots,r$$

For the remaining specifications we refer to the drawing

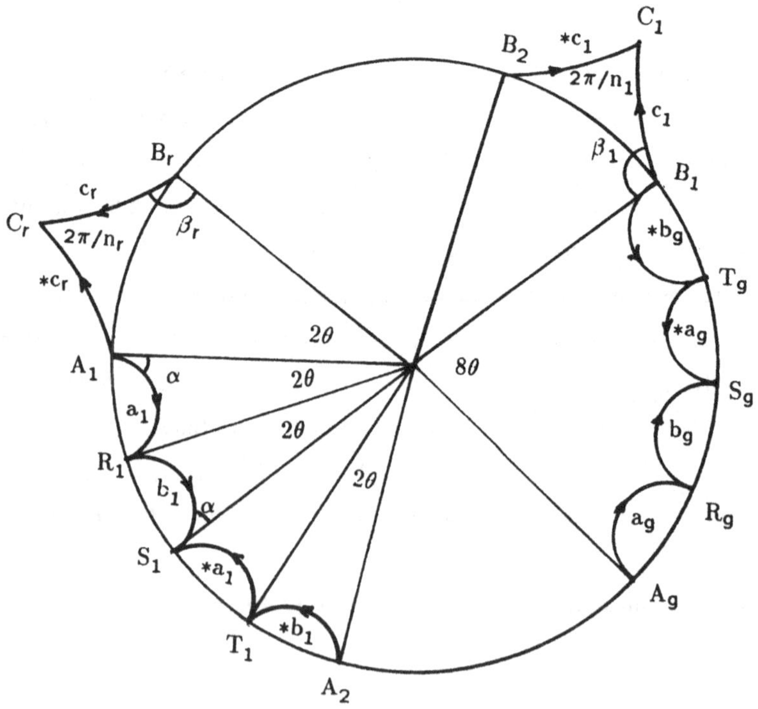

The boundary cycle for the polygon is given by

$$a_1 b_1 * a_1^{-1} * b_1^{-1} \dots a_g b_g * a_g^{-1} * b_g^{-1} c_1 * c_1^{-1} \dots c_r * c_r^{-1}$$

The edge cycles are

$$c_1^{-1} * c_1,\ c_2^{-1} * c_2\ ,\ \dots\ ,\ c_r^{-1} * c_r$$

$$a_1 * b_1^{-1} * a_1^{-1} b_1 \dots a_g * b_g^{-1} * a_g^{-1} b_g c_1 c_2 \dots c_r$$

The sum of the inner angles for the cycle of initial vertices is

$$f(r) = 8g\alpha + 2\sum_{i=1}^{r} \beta_i$$

If $f(r) = 2\pi$, then Poincaré's theorem assures us that the side transformations generate a Fuchsian group with the given signature. Let vary r through the interval $]0,\infty[$ to see that this can be achieved. The angles α and β_j, $j = 1,\dots,r$ can be analysed through the "alternative cosine formula" III.5.2

$$\cosh r = \frac{\cos \beta_j \cos \theta + \cos \pi/n_j}{\sin \beta_j \sin \theta} \quad ; \quad \cosh r = \cot \alpha \cot \theta$$

From this it follows rather immediately that

$$\lim_{r\to\infty} \alpha = 0 \ , \quad \lim_{r\to\infty} \beta_j = 0 \qquad ; j = 1,..,r$$

which gives $\lim_{r\to\infty} f(r) = 0$. In order to investigate the limit for $r\to 0$ let us rewrite the two cosine formulas as follows

$$\cosh(\theta + \beta_j) = \cos(\pi - \pi/m_j) + \sin \theta \sin \beta_j (\cosh r - 1)$$

$$\cosh \alpha = \cosh r \cot(\pi/2 - \theta)$$

It is now rather obvious that

$$\lim_{r\to 0} \alpha = \tfrac{\pi}{2} - \theta \quad , \quad \lim_{r\to 0} \beta_j = \pi - \tfrac{\pi}{n_j} - \theta$$

We conclude from the formula for f(r) that

$$\lim_{r\to 0} f(r) = 2\pi[\ 2g - 2 - \sum_{i=1}^{r} (1 - \tfrac{1}{n_i})] + 2\pi > 2\pi$$

This conclude the construction of our polygon. □

We shall now complete our investigations by showing that the abstract group structure of our Fuchsian group can be recovered from the signature.

Theorem 7.6 Let Γ be a Fuchsian group with H^2/Γ a compact surface of genus g. If $n_1,\ldots,n_r$ denotes the orders of the conjugacy classes of elliptic subgroups of Γ, then Γ admits a presentation with generators

$$\alpha_1,\ldots,\alpha_g,\beta_1,\ldots,\beta_g,\delta_1,\ldots,\delta_r$$

and relations ($[\alpha,\beta] = \alpha\beta\alpha^{-1}\beta^{-1}$)

$$\delta_1\ldots\delta_r\prod_{i=1}^{g}[\alpha_i,\beta_i] = \iota\ ,\quad \delta^{n_1} = \iota\ ,\ldots,\ \delta^{n_r} = \iota$$

Proof Let us use 6.5 to choose a fundamental domain with boundary cycle

$$\prod_{k=2}^{r}(d_k * d_k^{-1})\prod_{i=1}^{g}(a_i * b_i^{-1} * a_i^{-1}\ b_i)$$

The edge cycles are

$$d_2\ldots d_r\prod_{i=1}^{g}(a_i b_i * a_i^{-1} * b_i^{-1})\ ,\ d_2^{-1},\ldots,\ d_r^{-1}$$

Let us use Greek letters for the corresponding side transformations to get the presentation of Γ with a complete set of relations

$$(\delta_2\ldots\delta_r\prod_{i=1}^{g}[\alpha_i,\beta_i])^{n_1} = \iota\ ,\quad \delta^{n_2} = \iota\ ,\ldots,\ \delta^{n_r} = \iota$$

Upon introducing the auxiliary transformation δ_1 given by

$$\delta_1 = \delta_2\ldots\delta_r\prod_{i=1}^{g}[\alpha_i,\beta_i]$$

we can bring the presentation of Γ into the announced form □

Analogous results can be derived by the same method for the general group Γ with H^2/Γ compact, see [Wilkie] and [Macbeath]. For a different approach see [Zieschang, Vogt, Caldeway].

VII EXERCISES

EXERCISE 2.1 Let p and q be ordinary integers which satisfy the relation

$$(p-2)(q-2) > 4$$

1° Show that there exists a regular p-gon Δ with angles $2\pi/q$.

2° Show that there exists a tesselation $\mathcal{T}_{p,q}$ of H^2 consisting of regular p-gons with angles of measure $2\pi/q$, such that any two of the p-gons have empty intersection or have an edge in common. Hint: Consider the triangle group G(2,p,q) and its canonical tesselation.

3° Let ρ denote a rotation of order p about the centre of Δ and let σ denote a rotation of order q with respect to a vertex of Δ. Show, that the group of even symmetries of the tesselation $\mathcal{T}_{p,q}$ is generated by ρ and σ, and that

$$\rho^p = \sigma^q = (\rho\sigma)^2 = \iota$$

is a complete set of relations for the group of even symmetries of $\mathcal{T}_{p,q}$.

4° Let us consider integers p and q such that

$$(p-2)(q-2) \geq 4$$

Show that the universal group generated by R and S subject to the relations

$$R^p = S^q = (RS)^2 = \iota$$

is infinite. Hint: When $(p-2)(q-2) = 4$ use Euclidean geometry.

EXERCISE 2.2 Let Δ denote a regular p-gon with angles of measure $2\pi/q$ with the notation of the previous exercise.

1° Show that the group $D(\Delta)$ generated by reflections in the sides of Δ is discrete if and only if q is an even integer.

2° Suppose q is even. Show that the group $D(\Delta)$ is a normal subgroup of the symmetry group of $\mathcal{T}_{p,q}$, and identify $D(\Delta)$ with a normal subgroup of G(p,q,2) of index 2p.

EXERCISE 2.3 Let us consider integers p and q such that

$$(p-2)(q-2) \geq 4$$

Show that the universal group generated by R and S subject to the relations

$$R^p = S^q = (RS)^2 = \iota$$

is infinite. Hint: When $(p-2)(q-2) > 4$ use hyperbolic geometry, but when $(p-2)(q-2)$ use Euclidean geometry.

EXERCISE 2.4 Let Δ be an r-gon in the hyperbolic plane whose vertices has angles of measures

$$\angle P_i = \pi/n_i \qquad ; \quad n_i \in \mathbb{N},\ n_i \geq 2,\ i=1,...,r$$

1° Show that reflections in the sides of Δ generate a discrete group $D(\Delta)$ with Δ as fundamental domain.

2° Show that "Dyck's group" $D^+(\Delta)$ can be generated by $\rho_1,...,\rho_r$ subject to

$$\rho_1^{\ n_1} = \ ... \ = \rho^{n_r} = \rho_1\rho_2...\rho_r = \iota$$

3° Show that the genus of the Riemann surface $H^2/D^+(\Delta)$ is zero.

EXERCISE 2.5 Let $n \geq 2$ be an integer and let Δ be a regular 2n-gon with angles of measure π/n. Let us consider a side pairing on Δ with boundary cycle

$$a_1 {*a_1}\ a_2 {*a_2} \ ...\ a_n {*a_n}$$

For $i = 1,...,n$ we let M_i resp. $*M_i$ denote the midpoint of the edge a_i resp. $*M_i$. Show that the side transformation α_i is glide transformation along $M_i{*M_i}$.

2° Show that $\alpha_1,...,\alpha_n$ generate a discrete group G with fundamental domain Δ and that a complete set of relations is

$$\alpha_n^{\ 2}\alpha_{n-1}^{\ 2} ... \ \alpha_1^{\ 2} = \iota$$

3° Identify the group G with a subgroup of index 4n in the triangle group G(2,2n,2n).

EXERCISE 2.6 Let $g \geq 2$ be an integer and let Δ be a regular 4g-gon with angles of measure $\pi/2g$. Let us consider the side pairing on Δ which identifies each side with the opposite side of the polygon by means of parallel translation along the midpoints of the two sides. Let us enumerate the sides of Δ such that the boundary cycle reads

$$a_1\ b_1\ a_2\ b_2\ ...\ a_g\ b_g\ {*a_1^{-1}}{*b_1^{-1}}\ ...\ {*a_g^{-1}}{*b_g^{-1}}$$

1° Show that $\alpha_1,\beta_1,....,\alpha_g,\beta_g$ generates a discrete group G with fundamental domain Δ and that a complete set of relations is

$$\alpha_1\beta_1^{-1}\alpha_2\beta_2^{-1}\ldots\ \alpha_n\beta_n^{-1} = \beta_n^{-1}\alpha_n\ \ldots\ \beta_1^{-1}\alpha_1$$

2° Show that the Riemann surface H^2/G has genus g.

3° Identify the group G with a subgroup of index 8g in the triangle group G(2,4g,4g).

EXERCISE 2.7 Let Δ be the regular 8-gon with angles of measure $\pi/4$. Consider the side pairing with boundary cycle

$$a\ b\ *a^{-1}*c^{-1}*d^{-1}\ c\ *b^{-1}d$$

Show that the corresponding Riemann surface has genus 2 and a fundamental group generated by α, β, γ, δ subject to the relation

$$\delta^{-1}\gamma\ \delta\ \beta\ \alpha^{-1}\ \beta^{-1}\ \gamma^{-1}\ \alpha = \iota$$

EXERCISE 5.1 Show that the commutator subgroup [Γ,Γ] of the modular group Γ has the fundamental domain $\Delta \cup \tau\Delta \cup \tau^2\Delta \cup \tau^3\Delta \cup \tau^4\Delta \cup \tau^5\Delta$ with side transformations τ^6, $\tau^3\sigma$, $\tau^4\sigma\tau^{-1}$, $\tau^5\sigma\tau^{-2}$. Work out the edge cycles and apply Poincaré's theorem.

EXERCISE 5.2 1° Show that the subgroup Γ^2 of index 2 in the modular group Γ has a branch scheme (2,1,2), and that this is the only normal subgroup of Γ with a branch scheme of the form (2,1,n).

2° Show that the normal subgroup Γ^3 of Γ of index 3 has a branch scheme (1,3,3) and that this is the only normal subgroup of finite index with a branch scheme of the form (1,3,n).

3° Show that no normal subgroup of finite index of Γ has a branch scheme of the form (1,1,n), except for Γ itself.

EXERCISE 5.3 Let Δ denote the hyperbolic triangle ∞,0,1 with side transformations σ, $\rho^{-1}\sigma\rho$, $\rho\sigma\rho^{-1}$ where $\rho = \tau\sigma$. Show that Poincaré's theorem applies and deduce that Γ^3 is the free product of the three cyclic subgroups generated by

$$\sigma,\ \rho^{-1}\sigma\rho,\ \rho\sigma\rho^{-1}$$

EXERCISE 5.4 1° Show that Γ^3 is the smallest normal subgroup of Γ containing ρ.

2° Show that the subgroup Γ^2 of Γ of index 2 is the smallest normal subgroup of Γ containing σ.

3° Show that Γ^2 and Γ^3 are the only non-trivial normal subgroups of Γ which contains elements of finite orders.

EXERCISE 5.5 Let Π denote a normal subgroup of finite index μ of Γ with a branch scheme of the the form (2,3,n).

1° Show that the corresponding compact Riemann surface has genus 0 for the following values of (μ,n) only

$$(6,2) \quad (12,3) \quad (24,4) \quad (60,5)$$

2° Show that the table above is realised by $\Gamma(2)$, $\Gamma(3)$, $\Gamma(4)$, $\Gamma(5)$

EXERCISE 5.6 Verify the following calculation of genus of $\Gamma(n)$

n	1	2	3	4	5	6	7	8	9	10	11	12
genus	0	0	0	0	0	1	3	5	10	13	26	25

VIII HYPERBOLIC 3-SPACE

In the first two sections of this chapter we return to the generality of chapter I and study the exterior algebra over an oriented n–dimensional real vector space E equipped with a non-singular quadratic form. The main issue is parametrisation of subspaces of E of given Sylvester type by exterior vectors.

This construction is particularly satisfactory where our space is an oriented Minkowski space M, i.e. is a four-dimensional space of Sylvester type (−3,1). In this case the space $\wedge^2 M$ has a natural complex structure and a complex inner product familiar from special relativity. We find that a 2-vector $z \in \wedge^2 M$ is decomposable (primitive) if and only if $Im[\langle z,z\rangle] = 0$. A decomposable 2-vector $z \neq 0$ determines a plane in M whose Sylvester type is determined by the sign of the norm $\langle z,z\rangle$. In particular we find a one to one correspondence between oriented geodesics in H^3 and 2-vectors $z \in \wedge^2 M$ with $\langle z,z\rangle = -1$. Our general discussion of the geometry of H^3 is based on this fact.

So far, our discussion has been based on a general Minkowski space. In section VIII.6 we introduce a specific Minkowski space, namely the space M of 2×2 Hermitian matrices. This space is, in a natural way, a linear representation of the group $Sl_2(\mathbb{C})$ and we deduce an isomorphism

$$PSl_2(\mathbb{C}) \;\; \simeq \;\; Lor^+(M)$$

where $Lor^+(M)$ is the group of special Lorentz transformations of M. We make a definite choice of a sheet of the hyperbola in M and call this the Hermitian model of H^3, compare [Shafarevich p.149] and [Manin].

Beyond its own intrinsic interest, the Hermitian model has a trace formula not available in general: Let $S \in Sl_2(\mathbb{C})$ act on H^3 as the product of half-turns α and β in two geodesics h and k with normal vectors H and K. Then

$$tr^2(S) = 4\langle H,K\rangle^2$$

This leads to a classification theorem: the conjugacy classes in the special Lorentz group are in one to one correspondence with classes of pairs of geodesics in H^3.

The Hermitian model has its origin in physics. We shall make use of another tool with its origin in physics, the Dirac algebra. For readers familiar with the formalism of Clifford algebra we can add that Dirac's algebra is a concrete realisation of the Clifford algebra of M.

The trace formula above makes it clear that geodesics have an important role to play in the study of even isometries of H^3. Geodesics occur in bundles as we shall see in section VIII.5. In the final section of the book we shall illustrate how the geometric notion of bundles of geodesics in H^3 applies to fixed point theorems.

VIII.1 EXTERIOR ALGEBRA

In this section we consider a vector space E of dimension n over $\mathbb{R}$ equipped with a non-singular quadratic form. We shall be concerned with the extra structure induced by the form on the exterior algebra $\wedge E$. For $p \in \mathbb{N}$ we can introduce a symmetric bilinear form on $\wedge^p E$ by the formula

1.1 $$<u_1 \wedge \dots \wedge u_p, v_1 \wedge \dots \wedge v_p> = det_{ij} <u_i, v_j>$$

A vector $e \in E$ gives rise to an operator $\lfloor e$ on $\wedge E$ given by

1.2 $$u_1 \wedge \dots \wedge u_p \lfloor e = \sum_i (-1)^{i-1} <u_i, e> u_1 \wedge \dots \wedge \hat{u}_i \dots \wedge u_p$$

It is easily verified that our operation satisfies the general rule

1.3 $$(u \wedge v)\lfloor e = (u \lfloor e) \wedge v + (-1)^r u \wedge (v \lfloor e) \qquad ; u \in \wedge^r E, v \in \wedge^s E, \ e \in E$$

Observe that the operator $\lfloor a$ on $\wedge E$ has square 0. It follows that we can extend the operator $\lfloor$ to all of $\wedge E$ making $\wedge E$ into a <u>right module</u> over the ring $\wedge E$

1.4 $$u \lfloor (v \wedge w) = (u \lfloor v) \lfloor w \qquad ; u, v, w \in \wedge E$$

It is now time to introduce the key formula

1.5 $$\boxed{u \lfloor v = \langle u,v \rangle} \qquad ; u,v \in \wedge^p E$$

Proof Let us proceed by induction on p. Consider two decomposable vectors $u = u_1 \wedge \dots \wedge u_p$ and $v = v_1 \wedge \dots \wedge v_p$. Using formulas 1.2 and 1.3 we get that

$$u \lfloor v = \left(\sum_i (-1)^{i+1} \langle u_i, v_1 \rangle u_1 \wedge \dots \wedge \hat{u}_i \dots \wedge u_p\right) \lfloor v_2 \wedge \dots \wedge v_p =$$
$$\sum_i (-1)^{i+1} \langle u_i, v_1 \rangle \langle u_1 \wedge \dots \wedge \hat{u}_i \dots \wedge u_p, v_2 \wedge \dots \wedge v_p \rangle = \langle u,v \rangle$$

where we have used expansion of the determinant after the first row. □

Let us combine the two previous formulas to get that

1.6 $$\langle u, v \wedge w \rangle = \langle u \lfloor v, w \rangle \qquad ; u \in \wedge^{p+q}E,\ v \in \wedge^p E,\ w \in \wedge^q E$$

Orientations Let us recall the algebra underlying the concept of orientation of our n-dimensional real vector space E. By a <u>frame</u> in E we understand an ordered basis $e_1,\dots,e_n$ for E. The frames make up a subset $Fr(E)$ of E^n on which the group $Gl(E)$ acts transitively. The group $Gl^+(E)$ of transformations with positive determinant acts with two orbits on $Fr(E)$, the <u>orientations</u> of E.

In the rest of this section we shall assume that an orientation of E has been singled out. A frame $e_1,\dots,e_n$ is said to be <u>positively oriented</u> if it represents the preferred orientation. In particular, if we depart from a positively oriented orthonormal basis $e_1,\dots,e_n$ for E, the <u>orientation vector</u> $or = e_1 \wedge \dots \wedge e_n$ is independent of the positively oriented orthonormal basis considered. It is now time to introduce the <u>Hodge star operator</u> $*$ in the exterior algebra of the oriented space E of Sylvester type $(-s,t)$. To a p-vector v we shall associate the $(n-p)$-vector $*v$ given by

1.7 $$* \, v = (-1)^s \, or \lfloor v \qquad ; \; v \in \wedge^p E$$

By definition $*1 = (-1)^s or$. Simple evaluation gives us

1.8 $$*or = 1$$

The $*$ operator satisfies the formula

1.9 $$\boxed{** \, v = (-1)^{p(n-p)+s} v} \qquad ; \; v \in \wedge^p E$$

Proof Let us observe that $\wedge^p E$ is spanned by p-vectors of the form $e_1 \wedge \ldots \wedge e_p$ where $e_1,\ldots,e_p$ are part of a positively oriented orthonormal basis $e_1,\ldots,e_p,\ldots,e_n$ for E. Simple direct calculation yields

1.10 $$(-1)^s * \, e_1 \wedge \ldots \wedge e_p = \prod_{i=1}^{p} \langle e_i,e_i \rangle \; e_{p+1} \wedge \ldots \wedge e_n$$

Let us remark that the operator $*$ changes sign if the orientation is changed; the orientation must be changed $(n-p)p$ times to make $e_{p+1},\ldots,e_n,e_1,\ldots,e_p$ positively oriented. Thus we get from 1.7 that

$$(-1)^s * \, e_{p+1} \wedge \ldots \wedge e_n = (-1)^{n(n-p)} \prod_{i=p+1}^{n} \langle e_i,e_i \rangle \; e_1 \wedge \ldots \wedge e_p$$

Apply 1.7 once more and observe that $\prod_{i=1}^{n} \langle e_i,e_i \rangle = (-1)^s$. □

Consider an $(n-p)$-vector w and a p-vector u and let us write $w \wedge u = \lambda \, or$ with $\lambda \in \mathbb{R}$. The constant λ can be determined by formula 1.6 :

$$\langle or, \, w \wedge u \rangle = \langle or \lfloor w, \, u \rangle = (-1)^s \langle *w,u \rangle$$

Taking into account that $\langle or,or \rangle = (-1)^s$ we get that $\lambda = \langle *w,u \rangle$. Thus

$$w \wedge u = \langle *w,u \rangle \, or$$

Let us apply the Hodge operator and use 1.8 to get that

1.11 $$*(w \wedge u) = <*w,u> \qquad ; w \in \wedge^{n-p}E,\ u \in \wedge^{p}E$$

In particular take $w = *v$, $v \in \wedge^{p}E$, and use 1.9 to get that

1.12 $$<u,v> = (-1)^s*(u \wedge *v) \qquad ; u,v \in \wedge^{p}E$$

A combination of 1.11 and 1.12 gives that

1.13 $$<u,v> = (-1)^s<*u,*v> \qquad ; u,v \in \wedge^{p}E$$

Complex structure

Let M be an oriented Minkowski space, i.e. a space of Sylvester type $(-3,1)$. The operator $*$ on $\wedge^2M$ satisfies $*^2 = -\iota$ and is self adjoint as follows from 1.13

$$<u,*v> = <*u,v> \qquad ; u,v \in \wedge^2M$$

This allows us to introduce a complex structure on $\wedge^2E$ through the formula

1.14 $$(x + iy)\,v = xv + y*v \qquad ; x,y \in \mathbb{R},\ v \in \wedge^2M$$

and a symmetrical $\mathbb{C}$-valued bilinear form

1.15 $$<u,v>_c = <u,v> - i<*u,v> \qquad ; u,v \in \wedge^2M$$

The subscript c will be omitted when no confusion is possible. Observe that

$$<iu,v>_c = <*u,v>_c = <*u,v> + i<u,v> = i(<u,v> - i<*u,v>) = i<u,v>_c$$

In conclusion we have given $\wedge^2E$ the structure of a three-dimensional vector space over $\mathbb{C}$ with a complex quadratic form. Let us look at this in terms of a positively oriented orthonormal basis $\sigma_0,\sigma_1,\sigma_2,\sigma_3$ for our Minkowski space with

$$<\sigma_0,\sigma_0> = 1\ ,\quad <\sigma_i,\sigma_i> = -1 \qquad ; i = 1,2,3$$

We find by explicit calculations that

1.16 $$*(\sigma_1 \wedge \sigma_2) = \sigma_3 \wedge \sigma_0\ ,\ *(\sigma_2 \wedge \sigma_3) = \sigma_1 \wedge \sigma_0\ ,\ *(\sigma_3 \wedge \sigma_1) = \sigma_2 \wedge \sigma_0$$

It is now straightforward to check that $\sigma_1 \wedge \sigma_2$, $\sigma_2 \wedge \sigma_3$, $\sigma_3 \wedge \sigma_1$ form a C-linear basis for $\wedge^2 E$; in fact an orthonormal basis for the complex inner product.

Volume form

For an oriented Minkowski space M as above we can introduce the volume form

1.17 $$vol(a,b,c,d) = *(a \wedge b \wedge c \wedge d) \qquad ; a,b,c,d \in M$$

This is an alternating 4-form which takes the value 1 on a positively oriented orthonormal basis as follows from 1.8. Conversely, an alternating 4-form which takes the value 1 on some positively oriented orthonormal basis for M must be the orientation form. We ask the reader to use 1.12 to show that

1.18 $$\langle a \wedge b, c \wedge d\rangle_c = \langle a,c\rangle\langle b,d\rangle - \langle a,d\rangle\langle b,c\rangle - i\, vol(a,b,c,d)$$

for any four vectors a,b,c,d $\in$ M.

The reader is referred to [Bourbaki] for the exterior algebra. Note that we have used the sign conventions which makes $\wedge E$ into a right module over itself, as opposed to later editions of Bourbaki.

VIII.2 GRASSMANN RELATIONS

Let us continue our study of an oriented vector space E of dimension n over $\mathbb{R}$ equipped with a non-singular quadratic form. We shall try to decide when a given p-vector v is pure, i.e. of the form $v = e_1 \wedge e_2 \ldots \wedge e_p$, where $e_1,\ldots,e_p$ are linearly independent in E.

Quite generally, let us associate with a non-zero $v \in \wedge^p E$ the smallest subspace E_v of E such that $v \in \wedge^p E_v$. Note that

2.1 $$dim\, E_v \geq p \qquad ; v \in \wedge^p E,\ v \neq 0$$

and that equality holds if and only if v is a pure p-vector.

Proposition 2.2 The subspace E_v of E associated with a p-vector v is

$$E_v = Im(v\lfloor)\ , \quad v\lfloor : \wedge^{p-1}E \to E \qquad ; x \mapsto v\lfloor x$$

The subspace of E orthogonal to E_v is given by

$$E_v^{\perp} = Ker(v\lfloor)\ , \quad v\lfloor : E \to \wedge^{p-1}E \qquad ; e \mapsto v\lfloor e$$

Proof We shall take our point of departure in the identity

$$\langle v\lfloor x, y\rangle = v\lfloor x \lfloor y = (-1)^{p-1} v\lfloor y\lfloor x = (-1)^{p-1}\langle x, v\lfloor y\rangle \qquad ; x \in \wedge^{p-1}E,\ y \in E$$

Using that the inner product on $\wedge^{p-1}E$ is non-singular we get that

$$Im(v\lfloor)^{\perp} = Ker(v\lfloor)$$

The inclusion $E_v^{\perp} \subseteq Ker(v\lfloor)$ follows by direct verification using 1.2. It remains to prove $E_v \subseteq Ker(v\lfloor)^{\perp}$ or what amounts to the same thing, $E_v \subseteq e^{\perp}$ for all $e \in Ker(v\lfloor)$, $e \neq 0$. Choose a basis $e_1,\ldots,e_n$ for E where $e_2,\ldots,e_n \in e^{\perp}$ and write

$$v = \sum_H a_H\, e_H \qquad ; \ H: [1,p] \to [1,n]$$

where H runs through all strictly increasing maps in the range indicated above and $e_H = e_{H(1)} \wedge \ldots \wedge e_{H(p)}$. The relation $v\lfloor e = 0$ can be written

$$0 = \sum_H a_H \langle e_{H(1)}, e\rangle\, e_{H(2)} \wedge \ldots \wedge e_{H(p)}$$

which gives $a_H = 0$ whenever $H(1) = 1$. Thus $E_v \subseteq e^{\perp}$ as required. □

Proposition 2.3 If the vector $w \in \wedge^p E$ is pure, then the vector $*w \in \wedge^{n-p} E$ is pure and $E_{*w} = E_w{}^{\perp}$.

Proof From 1.3 it follows that for $u \in \wedge^r E$ and $v \in \wedge^s E$ we have the formula

2.4 $$(u \wedge v) \lfloor e_1 \wedge \ldots \wedge e_p = u \lfloor (e_1 \wedge \ldots \wedge e_p) \wedge v \quad ; e_i \in E,\ v \lfloor e_i = 0,\ i = 1, \ldots, p$$

Let us consider a pure vector $w = e_1 \wedge \ldots \wedge e_p$ and let E_w denote the space spanned by $e_1, \ldots, e_p$. Pick a basis $f_1, \ldots, f_n$ for E with $f_{p+1}, \ldots, f_n \in E_w{}^{\perp}$ and apply 2.5 to get
$$(f_1 \wedge \ldots \wedge f_n) \lfloor e_1 \wedge \ldots \wedge e_p = \langle f_1 \wedge \ldots \wedge f_p, e_1 \wedge \ldots \wedge e_p \rangle f_{p+1} \wedge \ldots \wedge f_n$$
Observe that we can arrange for $or = f_1 \wedge \ldots \wedge f_n$ and the result follows. □

Corollary 2.5 Any non-zero vector $w \in \wedge^{n-1} E$ is pure.

Theorem 2.6 A non-zero $z \in \wedge^p E$ is pure if and only if the composite map
$$\wedge^{p-1} E \xrightarrow{z\lfloor} E \xrightarrow{z\wedge} \wedge^{p+1} E$$
is zero. When z is pure, the sequence is exact.

Proof Let us write down the general formula

2.7 $$*(z \wedge e) = or \lfloor (z \wedge e) = (or \lfloor z) \lfloor e = *z \lfloor e \qquad ; e \in E$$

to see that $z \wedge e = 0$ if and only if $e \in Ker(*z\lfloor)$. We conclude from 2.2 that the image of the first map is M_z while the kernel of the second map is $M_{*z}{}^{\perp}$. Thus
$$(z \wedge) \circ (z\lfloor) = 0 \quad \Leftrightarrow \quad E_z \text{ is orthogonal to } E_{*z}$$
If z is pure, we have seen in 2.3 that E_z is orthogonal to E_{*z}. Conversely, if E_z is orthogonal to E_{*z} we conclude from I.1.7 that
$$dim\ E_z + dim\ E_{*z} \leq n$$
Since in general $dim\ E_z \geq p$ and $dim\ E_{*z} \geq n - p$ we get $dim\ E_z = p$. This means that z is a pure p-vector. □

Let us now specify our theory to a Minkowski space M. Our main concern will be the classification of two-dimensional subspaces of M by 2-vectors.

Theorem 2.8 A non-zero 2-vector $w \in \wedge^2 M$ is pure if and only if $<w,w>_c \in \mathbb{R}$. A pure 2-vector w gives rise to an exact sequence

$$M \xrightarrow{w\lfloor} M \xrightarrow{*w\lfloor} M$$

For a pure 2-vector w the plane $M_w = Ker(w\wedge: M \to \wedge^3 M)$ has Sylvester type (−1,1), (−1,0),(−2,0) depending on whether <w,w> is negative, zero or positive.

Proof The composite of the two maps we have just written is zero if and only if w is pure as follows from 2.6 and 2.7. The first statement will follow once we have established the formula

2.9
$$\boxed{*w\lfloor(w\lfloor e) = -\tfrac{1}{2}<*w,w>e = \tfrac{1}{2}Im[<w,w>_c]\, e} \qquad ; e \in M$$

We shall be content with a particular case, namely, with the notation of 1.16
$w = (x+iy)\sigma_1 \wedge \sigma_2 = x\sigma_1 \wedge \sigma_2 + y\sigma_3 \wedge \sigma_0$, $*w = iw = -y\sigma_1 \wedge \sigma_2 + x\sigma_3 \wedge \sigma_0$
observe that $Im[<w,w>_c] = Im[(x+iy)^2] = 2xy$. Finally check through the formula as e varies through $\sigma_0,\sigma_1,\sigma_2,\sigma_3$. An alternative proof can be given using the technique from VIII.7.

Let us investigate the general subspace E of M of dimension 2. Pick an orthonormal basis e,f for E and calculate the norm of the 2-vector $e\wedge f$

$$<e\wedge f, e\wedge f> = <e,e><f,f> - <e,f>^2$$

The discriminant lemma I.6.3 shows that the norm is negative, zero, positive depending on the Sylvester type as stated. □

Let us end this section by the observation that an operator on Minkowski space M of the form $z\lfloor$, $z \in \wedge^2 M$, is skew adjoint

$$<z\lfloor e,f> = z\lfloor e\lfloor f = -\, z\lfloor f\lfloor e = -<e,z\lfloor f>$$

VIII.3 NORMAL VECTORS

Let H^3 be one of the sheets of the unit hyperbola in Minkowski space M. We shall study the geometry of H^3 by assigning to a geodesic in H^3 a normal vector $N \in \wedge^2 M$. The relative position of two geodesics will be studied through the complex inner product of their normal vectors.

Proposition 3.1 The points of a geodesic curve $\gamma:\mathbb{R}\rightarrow H^3$ span a plane in M of Sylvester type $(-1,1)$. The normal vector

$$\gamma(t) \wedge \gamma'(t) \in \wedge^2 M \qquad ; t \in \mathbb{R}$$

has norm -1 and is independent of $t \in \mathbb{R}$. This establishes a one to one correspondence between oriented geodesics in H^3 and 2-vectors in $\wedge^2 M$ of norm -1.

Proof A geodesic $\gamma:\mathbb{R}\rightarrow H^3$ can be written in standard form

$$\gamma(t) = A \cosh t + T \sinh t \qquad ; t \in \mathbb{R}$$

where A and T are vectors from M of norm 1 and -1. Let us differentiate this to get $\gamma'(t) = A \sinh t + T \cosh t$. A simple calculation gives

$$\gamma(t) \wedge \gamma'(t) = A \wedge T\,(\cosh^2 t - \sinh^2 t) = A \wedge T$$

It follows from 3.6 that the norm of $N = A \wedge T$ is -1.

Conversely, let $N \in \wedge^2 M$ be a vector of norm -1. It follows from 2.8 that we can find vectors A and T in M with $A \wedge T = N$. The discriminant

$$-1 = \langle N,N\rangle = \langle A \wedge T, A \wedge T\rangle = \langle A,A\rangle\langle T,T\rangle - \langle A,T\rangle^2$$

shows us that A and T span a plane of type $(-1,1)$. We may assume that A and T are orthogonal with $A \in H^3$ and $\langle T,T\rangle = -1$. We can use the standard formula above to write down a geodesic curve with normal vector $N = A \wedge T$. □

By a hyperbolic plane we understand a subset of H^3 traced by a linear subspace of M of Sylvester type $(-2,1)$. Let us observe that the linear span of a hyperbolic plane in H^3 is a linear subspace of M of type $(-2,1)$

Proposition 3.2 Two geodesics h and k in H^3 with normal vectors H and K are coplanar, i.e. contained in a hyperbolic plane if and only if $<\mathrm{H},\mathrm{K}>_c \in \mathbb{R}$.

Proof Pick points $A \in h$ and $B \in k$ and unit tangent vectors S at A and T at B such that $H = A \wedge S$ and $K = B \wedge T$. We get from 1.18

$$Im[<\mathrm{H},\mathrm{K}>_c] = -\,vol(\mathrm{A},\mathrm{S},\mathrm{B},\mathrm{T})$$

Thus $<\mathrm{H},\mathrm{K}>_c \in \mathbb{R}$ if and only if A,S,B,T are linearly dependent. Let us observe that a three-dimensional subspace E of M which meets H^3 has type $(-2,1)$, since the two alternative types $(-2,0)$ and $(-1,0)$ are excluded on the grounds that they do not contain vectors of norm 1, compare I.6.3. In conclusion, if $<\mathrm{H},\mathrm{K}>_c \in \mathbb{R}$, then the linear hyperplane E spanned by h and k cuts H^3 in a hyperbolic plane which contains h and k. □

Let us say that two geodesics h and k are perpendicular if they meet in a point A and their tangent vectors at A are orthogonal.

Corollary 3.3 Two geodesics h and k are perpendicular if and only if their normal vectors H and K are orthogonal $<\mathrm{H},\mathrm{K}>_c = 0$.

Proof If $<\mathrm{H},\mathrm{K}>_c = 0$ then h and k are coplanar by 3.2. Observe also that if h and k are perpendicular then they are contained in a hyperbolic plane. Thus the result follows from III.2. □

Theorem 3.4 The action of the special Lorentz group on $\wedge^2\mathrm{M}$ defines an isomorphism of groups

$$Lor^+(\mathrm{M}) \xrightarrow{\sim} SO_3(\mathbb{C})$$

Proof The group $SO_3(\mathbb{C}) = SO(\wedge^2\mathrm{M})$ acts transitively on orthonormal pairs H,K in $\wedge^2\mathrm{M}$, while the group $Lor^+(\mathrm{M})$ acts transitively on pairs h,k of oriented perpendicular geodesics. □

Let ∂H^3 denote the set of isotropic lines in M (points at infinity). We say that $P \in \partial H^3$ is an end for a geodesic h in H^3 if the line S is contained in the linear span of h. It is clear that a geodesic h has two ends, and conversely, given two distinct points A and B of ∂H^3, there exists a unique geodesic in H^3 with ends A and B. Observe also that two geodesics h and k with a common end are coplanar.

Theorem 3.5 Let h and k be two geodesics in H^3. There exists a geodesic g perpendicular to h and k if and only if h and k don't have a common end.

Proof Let H and K be normal vectors for h and k respectively. The vectors H and K are linearly independent over $\mathbb{C}$: from $H = \lambda K$, $\lambda \in \mathbb{C}$, we get, taking norms, that $\lambda^2 = 1$ and therefore $H = K$ or $H = -K$, which means that H and K represent the same unoriented geodesic. Let us introduce the complex vector space $E \subseteq \wedge^2 M$ spanned by H and K and let us form the discriminant

$$\Delta = \langle H,H\rangle_c \langle K,K\rangle_c - \langle H,K\rangle_c^{\,2} = 1 - \langle H,K\rangle_c^{\,2}$$

The complex line $E^{\perp}$ is isotropic if and only if $\Delta = 0$.

If $\Delta \neq 0$, we can find a 2-vector G with norm -1 spanning $E^{\perp}$. The geodesic g with normal vector G is perpendicular to both h and k as follows from 3.3. Conversely, if the geodesic g is perpendicular to h and k, then the normal vector G for g spans the complex line $E^{\perp}$ and it follows that $\Delta \neq 0$

If $\Delta = 0$, then we have $\langle H,K\rangle_c = \pm 1$, and we conclude that h and k coplanar. Thus the result follows from the discussion in III.2. □

VIII.4 PENCILS OF PLANES

Hyperbolic planes occur in certain families called pencils. In this section we shall find all such pencils and demonstrate their significance for simple geometric problems.

Let us recall that a hyperbolic plane P in H^3 is the trace on H^3 of a linear hyperplane E of M of Sylvester type (−2,1): $P = E \cap H^3$. Let there be given an orientation of the supporting hyperplane E. A positively oriented orthonormal basis K,L,M for E defines the normal vector for P

4.1
$$S = K \wedge L \wedge M \in \wedge^3 M$$

The hyperplane E can be recovered from S as the kernel of $S\wedge : M \to \wedge^4 M$. Alternatively, the space E is the kernel of the composite map

$$M \xrightarrow{S\wedge} \wedge^4 M \xrightarrow{*} \wedge^0 M$$

By 2.7 this map is $(*S\lfloor): M \to \mathbb{R}$, so we find that $E = (*S)^{\perp}$. In conclusion

4.2
$$P = H^3 \cap (*S)^{\perp}$$

Proposition 4.3 The normal vector $S \in \wedge^3 M$ of an oriented hyperbolic plane P in H^3 has norm 1. This construction establishes a one to one correspondence between 3-vectors $S \in \wedge^3 M$ of norm 1 and oriented hyperbolic planes P in H^3.

Proof With the notation of 4.1, pick a positively oriented orthonormal basis e_1,e_2,e_3 with norms −1,−1,+1. This gives

$$\langle S,S\rangle = det_{ij}\langle e_i,e_j\rangle = 1$$

The main body of the proposition follows from VIII.2. □

We shall investigate a number of incidence relations in terms of normal vectors.

Proposition 4.4 Let P be a hyperbolic plane with normal vector $S \in \wedge^3 M$. A geodesic k with normal vector $K \in \wedge^2 M$ is contained in P if and only if

$$(*K) \wedge (*S) = 0$$

Proof Let us observe that M_K is contained in P if and only if M_K is orthogonal to $*S$ as follows from 4.2. This means $*S \in {M_K}^{\perp}$ or $*S \in M_{*K}$ as follows from 2.3. But $*S \in M_{*K}$ means $*S \wedge *K = 0$ by 2.8. □

Definition 4.5 A geodesic n is said to be <u>perpendicular</u> to a hyperbolic plane P in H^3 if n intersects P in a point A and all geodesics h in P through A are perpendicular to n.

Proposition 4.6 A geodesic n with normal vector $N \in \wedge^2 M$ is perpendicular to the hyperbolic plane P with normal vector $S \in \wedge M^3$ if and only if $N \wedge *S = 0$.

Proof Suppose that the geodesic n is perpendicular to P. Let us observe that a unit tangent vector T to n at the point A of intersection between P and h is orthogonal to any unit tangent S to P at A. Observe that $<T,A> = 0$ to conclude that T is orthogonal to the linear span E of P. But the vector $*S$ is orthogonal to E too as follows from 4.2. Let us recall from 1.13 that

4.7 $$<*S,*S> = -<S,S> \qquad ; \ S \in \wedge^3 M$$

It follows that $*S$ has norm -1. Thus $T = \pm S$ and

$$N \wedge *S = A \wedge T \wedge *S = 0$$

Conversely, let us assume that the normal vector N to n satisfies $N \wedge *S = 0$. According to 2.8 this means that $*S$ belongs to the linear span of n. From this we can conclude the existence of a point $A \in n$ with $<A,*S> = 0$. The point A is common to n and P and the vector $*S$ is tangent to n at A. It follows from 3.3 that the geodesic n is perpendicular to P. □

Proposition 4.8 Let P and Q be two distinct hyperbolic planes with normal vectors S and T. The plane M_K associated with the pure 2–vector

$$K = *(*S \wedge *T)$$

is the intersection of the hyperplanes E and F in M spanned by P and Q.

Proof From the assumption that $P \neq Q$ it follows that $*S$ and $*T$ are linearly independent which implies that $*S \wedge *T \neq 0$. Observe that $*K \wedge *S=0$ to conclude that $*S$ belongs to M_{*K}, compare 2.8. . Since $M_{*K} = M_K{}^{\perp}$ by 2.3, we conclude that $*S$ is orthogonal to M_K or $M_K \subseteq (*S)^{\perp}$. Similarly, we find that $M_K \subseteq (*T)^{\perp}$. For dimension reasons we must have that $M_K = (*S)^{\perp} \cap (*T)^{\perp}$ □

By a <u>pencil</u> $\mathcal{P}$ of hyperbolic planes in H^3 we understand a two-dimensional subspace of $\wedge^3 M$. The actual hyperbolic planes in the pencil are those hyperbolic planes with normal vectors in $\mathcal{P}$. The Sylvester type of $\wedge^3 M$ is (−1,3) as follows from 4.7; thus the possible pencil types are (0,2), (0,1), (−1,1).

Let us consider a fixed pencil $\mathcal{P}$ and pick two distinct hyperbolic planes P and Q in the pencil with normal vectors S and T.

4.9

$\langle S,T\rangle$	Sylvester type	Intersection $P \cap Q$	$\#(\partial P \cap \partial Q)$
$\lvert\langle S,T\rangle\rvert > 1$	(−1,1)	∅, common perp.l.	0
$\lvert\langle S,T\rangle\rvert < 1$	(0,2)	geodesic	2
$\lvert\langle S,T\rangle\rvert = 1$	(0,1)	∅, common end	1

Proof With the notation of 4.8, let us calculate the norm of K

$$\langle K,K\rangle = -\langle *K,*K\rangle = -(\langle *S,*S\rangle\langle *T,*T\rangle - \langle *S,*T\rangle^2) = \langle S,T\rangle^2 - 1$$

If $\langle K,K\rangle > 0$ we find that the discriminant Δ of $\mathcal{P}$ is negative and that $E \cap F = M_K$ has type (0,2). The plane $G = M_K{}^{\perp}$ has type (1,−1) and defines a geodesic perpendicular to P and Q.

If $\langle K,K\rangle < 0$, we have $\Delta > 0$ and the plane $E \cap F$ has type (−1,1). It follows that $P \cap Q$ is a geodesic.

If $< K,K> = 0$, we have $\Delta = 0$, and the plane $E \cap F$ has type $(-1,0)$. It follows that P and Q have precisely one common end. □

Corollary 4.10 Let n be a geodesic in H^3. Through any point $A \in H^3$ there passes a unique geodesic perpendicular to n. There is a unique hyperbolic plane perpendicular to n with a given end $Q \in \partial H^3$.

Proof Let us introduce the following subset of $\wedge^3 M$

$$\mathcal{A} = \{ S \in \wedge^3 M \mid <A,*S> = 0 \}$$

This is a three-dimensional subspace of $\wedge^3 M$ of type $(0,3)$. The pencil $\mathcal{N}$ of hyperbolic planes perpendicular to n has type $(-1,1)$, and it follows that $\mathcal{N} \cap \mathcal{A}$ has dimension 1, spanned by a vector of norm 1. Introduce the set

$$\mathcal{Q} = \{ S \in \wedge^3 M \mid <Q,*S> = 0 \}$$

This is a three-dimensional subspace of $\wedge^3 M$ of type $(0,2)$ and it follows that $\mathcal{N} \cap \mathcal{Q}$ has dimension 1, spanned by a vector of norm 1. □

Proposition 4.11 Let $\mathcal{P}$ be a pencil of hyperbolic planes. Through each point A of H^3 there passes at least one plane from the pencil. For pencils of types $(-1,1)$ or $(0,1)$ there will pass precisely one plane from the pencil through A.

Proof Let us consider a pencil $\mathcal{P}$ of type $(0,1)$. The condition for the hyperbolic plane P with normal vector S to pass through the point A is $<*S,A> = 0$. Thus we are lead to consider the intersection $A^{\perp} \cap *\mathcal{P}$. The plane $*\mathcal{P}$ in M has type $(-1,0)$ while $A^{\perp}$ has type $(-3,0)$. It follows that the intersection is one-dimensional, generated by a vector S of norm -1. We leave it to the reader to treat the remaining two types of pencil. □

VIII.5 BUNDLES OF GEODESICS

By a pencil $\mathfrak{P}$ of geodesics in H^3 we understand a two-dimensional linear subspace of $\wedge^2$M consisting of 2-vectors with real norm and spanned by 2-vectors of norm -1. The actual geodesics in the pencil are those geodesics whose normal vectors lie in $\mathfrak{P}$.

Proposition 5.1 The geodesics in a pencil $\mathfrak{P}$ are contained in a hyperbolic plane. Conversely, the normal vectors of two geodesics span a pencil if and only if the two geodesics are coplanar, i.e. is contained in a hyperbolic plane.

Proof Let us pick two distinct geodesics h and k from the pencil and let H and K be normal vectors for these. Observe that the formula for polarisation I.1.2 shows that $\langle H,K \rangle \in \mathbb{R}$ since H, K and H + K all have real norms; it follows from 3.2 that h and k are contained in the same linear hyperplane N of M of type (−2,1). It follows that $\mathfrak{P} \subseteq \wedge^2$N. □

Definition 5.2 By a bundle $\mathfrak{B}$ of geodesics in H^3 we understand a three-dimensional linear subspace of $\wedge^2$M consisting of 2-vectors with real norm, and spanned by 2-vectors of norm -1. The actual geodesics in the bundle are the geodesics with normal vectors in $\mathfrak{B}$.

Geodesics through a point Let us fix a point A of H^3 and introduce the following set of 2-vectors

$$\mathcal{A} = \{ X \in \wedge^2 M \mid A \wedge X = 0 \}$$

It is easy to check that exterior multiplication $\wedge$ A defines an exact sequence

5.3 $$0 \to \wedge^0 M \to \wedge^1 M \to \wedge^2 M \to \wedge^3 M \to \wedge^4 M \to 0$$

From this it follows that $\mathcal{A} = Im(\wedge A: \wedge^1 M \to \wedge^2 M)$. This description reveals that $\mathcal{A}$ has dimension 3 and consists of 2-vectors with real norm, 1.18. From

$$<A \wedge X, A \wedge Y> = <X,Y> \qquad ; X,Y \in T_A(H^3)$$

it follows that $\mathcal{A}$ is isometric to the tangent space $T_A(H^3)$, in particular the Sylvester type is $(-3,0)$. An oriented geodesic curve k has normal vector $K \in \mathcal{A}$ if and only if k passes through A, compare 2.8. In resume, we have realised the set of geodesics through A as the geodesics in the bundle $\mathcal{A}$.

Geodesics with a given end

Let us fix an isotropic line in M represented by the isotropic vector Q and introduce the set $\mathbb{Q}$ of 2-vectors

$$\mathbb{Q} = \{ X \in \wedge^2 M \mid X \wedge Q = 0 \}$$

From the exact sequence 5.3 it follows that $\mathbb{Q} = Im(\wedge Q: \wedge^1 M \to \wedge^2 M)$. This description reveals that $\mathcal{A}$ has dimension 3 and consists of 2-vectors with real norm, 1.18. Let us pick three linearly independent vectors A,B,C in M with $<A,Q> = <B,Q> = 0$ and $<C,Q> = 1$. From

$$<Q \wedge X, Q \wedge Y> = -<Q,X><Q,Y> \qquad ; X,Y \in M^3$$

we find that $Q \wedge A$, $Q \wedge B$, $Q \wedge C$ form an orthonormal basis with norms 0, 0, -1. An oriented geodesic k has normal vector $K \in \mathbb{Q}$ if and only if Q belongs to the plane spanned by k, compare 2.8. To summarise, we have realised the set of geodesics with end $\mathbb{R}Q$ as the set of geodesics in the bundle $\mathbb{Q}$.

Geodesics perpendicular to a plane

Let us consider a hyperbolic plane P in H^3 with normal vector $S \in \wedge^3 M$ and let us introduce the following set of 2-vectors

$$\mathcal{N} = \{ X \in \wedge^2 M \mid X \wedge *S = 0 \}$$

A look at the exact sequence 5.3 reveals that $\mathcal{N}$ consists of vectors of real norm and that its dimension is 3. In order to calculate the Sylvester type , observe that $<*S,*S> = -1$ and that $X \mapsto X \wedge *S$ defines an anti-isometry between $(*S)^{\perp}$ and $\mathcal{N}$. In particular, the Sylvester type of $\mathcal{N}$ is $(-1,2)$. It follows from 4.6 that a geodesic n has normal vector N in $\mathcal{N}$ if and only if n is perpendicular to P.

Geodesics in a plane Let us consider a hyperbolic plane P with given normal vector $S \in \wedge^3 M$ and form the following set of 2-vectors

$$\mathcal{P} = \{ X \in \wedge^2 M \mid *X \wedge *S = 0 \}$$

From the previous case we find that $\mathcal{P}$ consists of vectors of real norm and that its dimension is 3 of Sylvester type $(-2,1)$. A geodesic k has normal vector in $\mathcal{P}$ if and only if $k \subseteq P$ as follows from 4.4.

Let us summarise our findings in table 5.4 below. The heading "Pencils" indicates the possible Sylvester types of pencil in the bundle.

5.2

Bundle	Sylvester type	Pencils
Geodesics through a point	$(-3,0)$	$(-2,0)$
Geodesics with a given end	$(-1,0)$	$(-1,0)$
Geodesics perpendicular to a plane	$(-1,2)$	$(-1,1)$
Geodesics in a plane	$(-2,1)$	$(-1,1)$
		$(-1,0)$ $(-2,0)$

Let us notice that the Hodge $*$ transforms spaces of type $(-3,0)$ or $(-1,0)$ into spaces of type $(0,3)$ or $(0,1)$; however, such spaces will not contain 2-vectors of norm -1. We have, in fact, exhausted the possible Sylvester types:

Lemma 5.5 A three-dimensional subspace B of $\wedge^2 M$ consisting of 2-vectors with real norm has Sylvester type from the following list

$$(0,3)\ (-3,0)\ (0,1)\ (-1,0)\ (-1,2)\ (1,-2)$$

Proof We must rule out the four types $(0,2)$, $(-2,0)$, $(-1,1)$ and $(0,0)$. In the first three cases it is easily checked that an orthonormal real basis forms a complex basis as well. These three types can now be ruled out on the grounds that the complex inner product on $\wedge^2 M$ is non-singular.

In order to rule out the fourth case, observe that a hypothetical three-dimensional subspace consisting of 2-vectors of norm 0 would span a complex subspace E isotropic for the complex inner product, i.e $E \subseteq E^{\perp}$. Since the inner product on $\wedge M^2$ is non-singular we get that $dim_{\mathbb{C}} E = 1,\ 0$, which is a contradiction. □

Theorem 5.6 Table 5.4 is a complete list of bundles of geodesics in hyperbolic 3-space H^3.

Proof By theorem 3.4 it suffices to prove that $SO_3 = SO(\wedge^2 M)$ acts transitively on the bundles of given type.

Let us treat type $(-2,1)$ separately. A bundle is specified by a sequence e_1, e_2, e_3 of orthogonal 2-vectors of norms $-1, -1, 1$. It is clear that $O_3 = O(\wedge^2 M)$ acts transitively on such sequences. An orthogonal transformation has $det = \pm 1$, but the sign can be changed by permuting the first two vectors leaving the third invariant. Types $(-3,0)$ and $(-1,2)$ can be treated in a similar fashion.

Let us consider a bundle E of type $(-1,0)$ and let us prove that $E \cap iE \neq 0$. Pick an orthogonal basis e_1, e_2, e_3 with norms $-1,\ 0,\ 0$ to see that $E \cap iE = 0$ is equivalent to $E + iE = M$, which implies that e_1, e_2, e_3 are linearly independent over C; in particular we find that $(Ce_2 + Ce_3)^{\perp} = Ce_2 + Ce_3$ contradicting that our complex form is non-singular. Let us show that $E \cap iE \subseteq E^{\perp}$: starting with

$$<e,z> = -i<ie,z> \qquad ; z \in E\ ,\ e \in E \cap iE$$

and the observation $ie \in E$ we conclude that $<e,z> = 0$ as required. This shows that we can construct a real basis for E as follows: pick any vector $e \in E$ with norm -1 and a non-zero vector $f \in E \cap iE$ to get the basis e, f, if. This shows how to recover E from a pair of orthogonal vectors e,f of norms 1, 0. It follows from Witt's theorem I.2.4 that O_3 acts transitively on such data. To see that SO_3 acts transitively, let us consider an orthonormal basis e_1, e_2, e_3 and put $e = e_1$, $f = e_2 + ie_3$; observe that we can permute e_2 and e_3 keeping e_1 fixed. □

Let us emphasise that the fact that geometrically different pencils have different Sylvester types amplifies statements like "through a given point there passes at most one geodesic perpendicular to a given geodesic". As an illustration let us apply this to the theorem of Whitehead, Smid and Van der Waerden, see [Coxeter[1]] § 4.

Whitehead's theorem 5.7 If there are five distinct hyperbolic planes containing pairs of four given geodesics, the remaining pair also consists of two coplanar geodesics (a more precise statement is given in the proof).

Proof Let the four geodesics be a, b, c, d. In the graph below we have represented each of the five hyperbolic planes by a bar. The problem is to show that the geodesics b and d are coplanar.

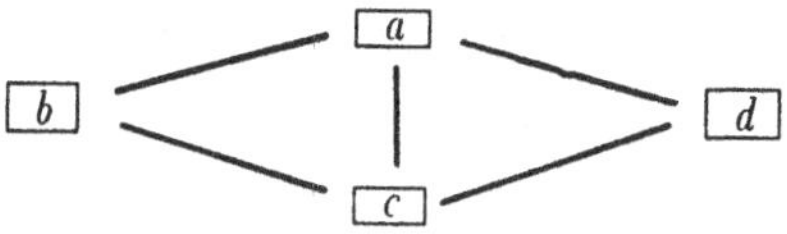

The assumptions may be restated: the triples a, b, c and a, c, d generate bundles $\mathfrak{B}$ and $\mathfrak{D}$ of Sylvester type $\neq (-2,1)$. Observe that $\mathfrak{B}$ and $\mathfrak{D}$ have the pencil generated by a, c in common. It follows from 5.8 below that $\mathfrak{B} = \mathfrak{D}$. □

Lemma 5.8 Let $\mathfrak{B}$ and $\mathfrak{D}$ be bundles of geodesics both of type $\neq (-2,1)$. If $\mathfrak{B}$ and $\mathfrak{D}$ have a common pencil, then $\mathfrak{B} = \mathfrak{D}$.

Proof Inspection of 5.4 reveals that $\mathfrak{B}$ and $\mathfrak{D}$ must be of the same type in order to have a common pencil. We ask the reader to go through the three cases. □

VIII.6 HERMITIAN MATRICES

In this section we shall introduce the space M of 2×2 Hermitian matrices and show that the determinant is a quadratic form on M of type $(-3,1)$. In fact, this realisation of Minkowski space has been used in physics for a long time.

Let us fix the notation with respect to some basic operators on the algebra $M_2(\mathbb{C})$ of 2×2 complex matrices. The cofactor matrix of A is denoted $A^{\check{}}$

6.1
$$\begin{bmatrix} a & b \\ c & d \end{bmatrix}^{\check{}} = \begin{bmatrix} d & -b \\ -c & a \end{bmatrix} \qquad ; a,b,c,d \in \mathbb{C}$$

The basic formula for manipulations with this symbol is

6.2
$$A\,A^{\check{}} = A^{\check{}}A = \iota\, det\, A \qquad ; A \in M_2(\mathbb{C})$$

Our symbol is an anti-automorphism in the sense that

6.3
$$(A\,B)^{\check{}} = B^{\check{}}A^{\check{}} \qquad ; A, B \in M_2(\mathbb{C})$$

On the space $M_2(\mathbb{C})$ we shall consider the symmetric bilinear form

6.4
$$<A,B> = \tfrac{1}{2} tr(A\,B^{\check{}}) \qquad ; A, B \in M_2(\mathbb{C})$$

From the fact that $A\,A^{\check{}} = \iota\, det\, A$ it follows at once that

6.5
$$<A,A> = det\, A \qquad ; A \in M_2(\mathbb{C})$$

Polarisation of formula 6.2 gives us the fundamental formula

6.6
$$\boxed{A\,B^{\check{}} + B\,A^{\check{}} = <A,B>\, 2\iota} \qquad ; A,B \in M_2(\mathbb{C})$$

The second operator of importance is the <u>star</u> $A^* = {}^T\bar{A}$ or complex transposed. A matrix which is invariant under the star operator is called <u>Hermitian</u>. The space of Hermitian matrices is denoted

$$M = \{ X \in M_2(\mathbb{C}) \mid X^* = X \}$$

Quite explicitly, a Hermitian matrix looks like this

$$\begin{bmatrix} a & b \\ \bar{b} & d \end{bmatrix} \qquad ; \ a, d \in \mathbf{R}, \ b \in \mathbb{C}$$

From this description it follows that the inner product takes real values on M. In particular, the determinant defines a real quadratic form on M. An orthonormal basis for M is formed by the

6.7 Pauli matrices

$$\sigma_0 = \begin{bmatrix} 1 & 0 \\ 0 & 1 \end{bmatrix}, \ \sigma_1 = \begin{bmatrix} 0 & 1 \\ 1 & 0 \end{bmatrix}, \ \sigma_2 = \begin{bmatrix} 0 & -i \\ i & 0 \end{bmatrix}, \ \sigma_3 = \begin{bmatrix} 1 & 0 \\ 0 & -1 \end{bmatrix}$$

which have norms $1, -1, -1, -1$ respectively. This shows that M is a Minkowski space. The sheet of the unit hyperbola in M consisting of <u>positive definite</u> forms ($a_{11} > 0$) will be our <u>Hermitian model</u> of H^3. The group $Sl_2(\mathbb{C})$ acts on M:

6.8
$$\sigma.X = \sigma X \sigma^* \qquad ; \ \sigma \in Sl_2(\mathbb{C}), \ X \in M$$

These transformations are orthogonal and we conclude from the fact that $Sl_2(\mathbb{C})$ is connected that a transformation of the form 6.8 preserves orientation and the connected components of the unit hyperbola in M. Thus we deduce a morphism[1]

$$Sl_2(\mathbb{C}) \rightarrow Lor_+(M)$$

which we shall demonstrate to be surjective, see 8.5. It is this fine action of $Sl_2(\mathbb{C})$ that makes the Hermitian model of H^3 extremely useful.

---

[1]A different proof of this will be presented during the proof of 8.5. As it happens, it follows from the same source that $Sl_2(\mathbb{C})$ is connected. See also 8.9.

VIII.7 DIRAC'S ALGEBRA

In this section we shall introduce an algebra which has been used in physics by P.M.Dirac[2]. In mathematical terms it is a concrete realisation of the Clifford algebra of the Minkowski space M, [Deheuvels, X.6]. However, the reader is not assumed to be familiar with the formalism of Clifford algebras.

The Dirac algebra D is built over the real vector space $M_2(\mathbb{C}) \oplus M_2(\mathbb{C})$ with multiplication given by means of the automorphism $A \mapsto \tilde{A} = {}^T\bar{A}^{\check{}}$ of $M_2(\mathbb{C})$

7.1 $$(A,B)(C,D) = (AC + B\tilde{D}, AD + B\tilde{C}) \qquad ; A,B,C,D \in M_2(\mathbb{C})$$

We ask the reader to verify that the Dirac algebra D is associative with multiplicative unit $(\iota,0)$. Let us introduce the element $\Psi = (0,\iota)$ and let us identify a matrix $A \in M_2(\mathbb{C})$ with the element $(A,0)$ of the Dirac algebra. Then we can write

$$(A,B) = A + B\Psi \qquad ; A,B \in M_2(\mathbb{C})$$

The multiplication is ruled by the formulas

7.2 $$\Psi^2 = \iota \ , \quad \Psi A = \tilde{A}\Psi \qquad ; A \in M_2(\mathbb{C})$$

The principal involution $X \mapsto X^{\#}$ is the automorphism of D given by

7.3 $$(A + B\Psi)^{\#} = A - B\Psi \qquad ; A,B \in M_2(\mathbb{C})$$

Its eigenspaces relative to 1 and -1 are denoted D_+ and D_- respectively. The principal anti-involution $X \mapsto X^{\check{}}$ of the Dirac algebra is given by

7.4 $$(A + B\Psi)^{\check{}} = A^{\check{}} + \Psi B^{\check{}} \qquad ; A,B \in M_2(\mathbb{C})$$

We ask the reader to check that indeed $(XY)^{\check{}} = Y^{\check{}}X^{\check{}}$, $X,Y \in D$. Our next

---

[2]See A.Messiah, *Quantum Mechanics*, North-Holland Publishing Company, Amsterdam 1961.

formula is a simple transcription of formula 6.6

7.5 $$\boxed{R\Psi\, S\Psi + S\Psi\, R\Psi = 2\langle R,S\rangle \iota} \qquad ; R,S \in M$$

The Minkowski space M is oriented by means of the Pauli matrices 6.7. This has the effect of making $\wedge^2 M$ into a complex vector space with a complex quadratic form, 1.15. Let us introduce the D-valued alternating bilinear form on M

7.6 $$R \times S = \tfrac{1}{2}(R\Psi S\Psi - S\Psi R\Psi) \qquad ; R,S \in M$$

Observe that this symbol is invariant under the principal automorphism but anti-invariant under the principal anti-automorphism. As a consequence we conclude that the symbol takes values in the following subspace of D_+

$$sl_2(\mathbb{C}) = \{X \in M_2(\mathbb{C}) \mid tr\, X = 0\}$$

The space $sl_2(\mathbb{C})$ has a natural complex structure and the determinant defines a complex quadratic form on $sl_2(\mathbb{C})$.

Theorem 7.7 The assignment $A \wedge B \mapsto A \times B$ defines an isometry

$$\wedge^2 M \xrightarrow{\sim} sl_2(\mathbb{C})$$

from the complex vector space $\wedge^2 M$ equipped with the complex quadratic form from 1.14 onto $sl_2(\mathbb{C})$ equipped with the determinant form.

Proof The Pauli matrices $\sigma_0, \sigma_1, \sigma_2, \sigma_3$ form a positively oriented orthonormal basis for M. It follows from 6.6 that

7.8 $$\begin{aligned} &\sigma_r\Psi\sigma_s\Psi + \sigma_s\Psi\sigma_r\Psi = 0 && ;\ r \neq s \\ &(\sigma_0\Psi)^2 = \iota \ , \quad (\sigma_s\Psi)^2 = -\iota && ;\ s=1,2,3 \end{aligned}$$

It is also worth recording that the scalar matrix $i = i\iota$ can be written

7.9 $$i = \sigma_0\Psi\sigma_1\Psi\sigma_2\Psi\sigma_3\Psi = \sigma_0\tilde{\sigma}_1\sigma_2\tilde{\sigma}_3$$

Let us prove that the map $R \wedge S \mapsto R \times S$ is C-linear. According to the formulas 1.16 this is equivalent to the formulas

$$i(\sigma_1 \times \sigma_2) = \sigma_3 \times \sigma_0 \, , \quad i(\sigma_2 \times \sigma_3) = \sigma_1 \times \sigma_0 \, , \quad i(\sigma_3 \times \sigma_1) = \sigma_2 \times \sigma_0$$

These can be verified by means of 7.7 and 7.10. Let us be content with the first

$$\sigma_0 \Psi \sigma_1 \Psi \sigma_2 \Psi \sigma_3 \Psi \, (\sigma_1 \Psi \sigma_2 \Psi) = -\sigma_0 \Psi \sigma_2 \Psi \sigma_3 \Psi \sigma_2 \Psi = \sigma_3 \Psi \sigma_0 \Psi$$

It remains to verify that $\sigma_1 \times \sigma_2$, $\sigma_2 \times \sigma_3$, $\sigma_3 \times \sigma_1$ form an orthogonal basis for $sl_2(\mathbb{C})$. Let us verify that the first two vectors are orthogonal:

$$-\tfrac{1}{2} tr((\sigma_1 \times \sigma_2)(\sigma_2 \times \sigma_3)) = -\tfrac{1}{2} tr(\sigma_1 \Psi \sigma_2 \Psi \sigma_2 \Psi \sigma_3 \Psi) = \tfrac{1}{2} tr(\sigma_1 \Psi \sigma_3 \Psi) = 0$$

The remaining details are left to the reader. □

Let us end this section with some information for readers familiar with the general formalism of Clifford algebras [Deheuvels].

Corollary 7.10 The Dirac algebra D is isomorphic to the Clifford algebra of the quadratic space M of Hermitian matrices.

Proof Let us observe that we have the direct sum decompositions

$$D_+ = \mathbb{R} \oplus sl_2(\mathbb{C}) \oplus i\mathbb{R} \quad , \qquad D_- = M\Psi \oplus iM\Psi$$

It follows from this that the $\mathbb{R}$–algebra D is generated by $M\Psi$. On the other hand the real dimension of D is 16 and the dimension of the Clifford algebra is 2^4. □

For the benefit of the reader we should mention that the <u>Dirac</u> <u>matrices</u> refer to the following eight-dimensional real representation of Dirac's algebra

$$A + B\Psi \mapsto \begin{bmatrix} A & B \\ \tilde{B} & \tilde{A} \end{bmatrix} \qquad ; D \to M_4(\mathbb{C})$$

It is known that $D \xrightarrow{\sim} M_2(\mathbb{H})$, compare execise 7.1. It follows that the eight-dimensional representation above is simple.

VIII.8 CLIFFORD GROUP

In this section we see how a certain subgroup of the multiplicative group $U(D)$ of Dirac's algebra D acts on M. Since we have already identified $\wedge^2 M$ with a subspace of D, this will bring, for example, normal vectors of geodesics in H^3 to act on M. We shall apply this to the finer study of even Lorentz transformations, and conclude the section with the main theorem of this chapter.

The map $R \mapsto R\Psi$ identifies the space M with a linear subspace of the Dirac algebra D. The Clifford group Γ is the subgroup of those $G \in U(D)$ which have the following stability property relative to the subspace $M\Psi$ of D

8.1
$$G^{\#}\, M\Psi\, G^{-1} = M\Psi$$

Thus the group Γ acts on M through the formula

8.2
$$(G.X)\Psi = G^{\#}X\Psi G^{-1} \qquad ; X \in M,\ G \in \Gamma$$

Let us observe that $G \in \Gamma$ acts on M as an orthogonal transformation:

$$<G.X,G.X>\iota = (G.X)\Psi(G.X)\Psi = -(G.X)\Psi((G.X)\Psi)^{\#} = <X,X>\iota$$

As an example, let us consider a non-isotropic vector $S \in M$ and the corresponding unit $S\Psi \in \Gamma$. The action of $S\Psi$ on M is reflection τ_S along S:

8.3
$$\boxed{(S\Psi).X = \tau_S(X)} \qquad ; X \in M$$

Proof Let us recall from 7.5 the formulas

$$S\Psi S\Psi = <S,S>\iota \ , \quad S\Psi X\Psi + X\Psi S\Psi = 2<X,S>\iota \qquad ; X \in M$$

Let us multiply the second formula from the right by $(S\Psi)^{-1} = S\Psi<S,S>^{-1}$ to get

$$-S\Psi X\Psi(S\Psi)^{-1} = X\Psi - 2<X,S><S,S>^{-1}S\Psi = \tau_S(X)\Psi \qquad \square$$

Let us identify $Gl_2(\mathbb{C})$ with the group of multiplicative units in D_+. For a matrix $R \in Gl_2(\mathbb{C})$ with real determinant we have $\tilde{R}^{-1} = R^*(\det R)^{-1}$, thus

$$RX\Psi R^{-1} = RX\tilde{R}^{-1}\Psi = RXR^*\Psi(\det R)^{-1} \qquad ; \; X \in M$$

which shows that $R \in \Gamma_+$, where $\Gamma_+ = D_+ \cap \Gamma$. We ask the reader to show that

8.4

$$\Gamma_+ = \{ R \in Gl_2(\mathbb{C}) \mid \det R \in \mathbb{R} \}$$

$$R.X = RX\tilde{R}^{-1} = RXR^*(\det R)^{-1} \qquad ; \; R \in \Gamma_+, \; X \in M$$

Theorem 8.5 The action of $Sl_2(\mathbb{C})$ on M defines an isomorphism of groups

$$PSl_2(\mathbb{C}) \xrightarrow{\sim} Lor^+(M)$$

Proof Let us show that $R \in \Gamma_+$ acts as a Lorentz transformation on M if and only if $\det R$ is positive. To this end we ask the reader to verify the identity

$$tr \begin{bmatrix} a & b \\ c & d \end{bmatrix} \begin{bmatrix} 1 & 0 \\ 0 & 1 \end{bmatrix} \begin{bmatrix} \bar{a} & \bar{c} \\ \bar{b} & \bar{d} \end{bmatrix} \begin{bmatrix} 1 & 0 \\ 0 & 1 \end{bmatrix} = a\bar{a} + b\bar{b} + c\bar{c} + d\bar{d}$$

Let us also remark that $R \in \Gamma_+$ acts trivially on M if and only if R is a real scalar matrix: this follows from the proof of 7.10. Consider $R \in \Gamma_+$ with $\det R > 0$ and decompose the action of R on M into a product of reflections $\tau_1\tau_2...\tau_d$ along vectors $S_1,...,S_d$ of norm -1. It follows that we can find $k \in \mathbb{R}^*$ such that

$$R = k\, S_1\Psi\, S_2\Psi ... S_d\Psi$$

Let us make use of the principal anti-involution 7.4 to get that

$$R\, R\check{} = k^2\, S_1S_1\check{}\, S_2S_2\check{} ...\, S_dS_d\check{} = k^2(-1)^d$$

Using that $\det R = RR\check{}$ is positive we conclude that d is even.

An even Lorentz transformation of M can be written $\tau_A\tau_B\tau_C\tau_D$, where $A,B,C,D \in M$, each of determinant -1. According to 7.5 this is the action of

$$A\Psi B\Psi C\Psi D\Psi = A\tilde{B}C\tilde{D}$$

This allows us to conclude that $Sl_2(\mathbb{C}) \rightarrow Lor_+(M)$ is surjective. □

Definition 8.6 By a <u>half-turn</u> in a geodesic k in H^3 we understand the orthogonal reflection in the linear span of k in the ambient Minkowski space.

Main theorem 8.7 Let $S \in Sl_2(\mathbb{C})$ act on H^3 as the product $\alpha\beta$ of half-turns α and β in geodesics h and k with normal vectors H and K. Then

$$tr^2 S = 4<H,K>^2$$

Any even isometry of H^3 can be decomposed into a product of two half-turns.

Proof Pick a point $A \in h$ and a unit tangent vector T to h at A such that

$$H = A \times T = \tfrac{1}{2}(A\Psi T\Psi - T\Psi A\Psi) = A\Psi T\Psi$$

We conclude from formula 8.3 that $H \in \Gamma_+$ acts on M as $-\alpha$. Observe that the scalar matrix $i = i\iota$ acts as the antipodal map $x \mapsto -x$, $x \in M$, to get that

8.8 $$(iH).X = \alpha(X) \qquad ; \quad X \in M$$

It follows that $(iH)(iK) \in Sl_2(\mathbb{C})$ acts as $\alpha\beta$ on M which gives $S = \pm\, iH\ iK$. Let us recall that $K\check{} = -K$, since $K \in sl_2(\mathbb{C})$. This gives us

$$tr(iH\ iK) = -tr(HK) = tr(HK\check{}) = 2<H,K>$$

The required formula follows by squaring this relation. Conclusion by 8.9. □

Lemma 8.9 Any $S \in Sl_2(\mathbb{C})$ can be written $S = HK$ where $H,K \in sl_2(\mathbb{C})$ with $det\,H = det\,K = -1$.

Proof We ask the reader to use the theory of quadratic forms to find vectors $H,K \in sl_2(\mathbb{C})$ with $<H,H> = <K,K> = -1$ and $2<H,K> = tr(S)$. Observe that $tr^2 S = tr^2 HK$ and conclude from I.9.15 that S and HK are conjugated in $PSl_2(\mathbb{C})$. Upon changing the sign of H we may assume that S and HK are conjugated in $Sl_2(\mathbb{C})$. □

VIII.9 LIGHT CONE

In this section we shall work with the Hermitian model for H^3 and identify its boundary ∂H^3 with $\hat{\mathbb{C}}$ keeping track of the action of $Sl_2(\mathbb{C})$. Once this is done we write down the normal vector $L(A,B)$ for the oriented geodesic with ends $A,B \in \hat{\mathbb{C}}$ and evaluate $<L(A,B),L(P,Q)>$ in terms of the cross ratio $[A,B,P,Q]$. To begin let us assign to $\mathbf{e} \in \mathbb{C}^2$ the following Hermitian matrix

9.1
$$L(\mathbf{e}) = \begin{bmatrix} z\bar{z} & z\bar{w} \\ w\bar{z} & w\bar{w} \end{bmatrix} = \begin{bmatrix} z \\ w \end{bmatrix} \begin{bmatrix} \bar{z} \\ \bar{w} \end{bmatrix}^T \qquad ; \mathbf{e} = \begin{bmatrix} z \\ w \end{bmatrix}$$

Notice that $L(\lambda\mathbf{e}) = \lambda\bar{\lambda}L(\mathbf{e})$, $\lambda \in \mathbb{C}$, and observe the pleasing transformation rule

9.2
$$L(\sigma(\mathbf{e})) = \sigma L(\mathbf{e})\sigma^* \qquad ; \sigma \in Sl_2(\mathbb{C}), \ \mathbf{e} \in \mathbb{C}^2$$

which follows immediately from the decomposition

$$\begin{bmatrix} z\bar{z} & z\bar{w} \\ w\bar{z} & w\bar{w} \end{bmatrix} = \begin{bmatrix} z \\ w \end{bmatrix} \begin{bmatrix} \bar{z} \\ \bar{w} \end{bmatrix}^T \qquad ; z,w \in \mathbb{C}$$

Proposition 9.3 The transformation L induces a $Sl_2(\mathbb{C})$ equivariant bijection from $\hat{\mathbb{C}}$ onto the space *PC(M)* of isotropic lines in M.

Proof The equivariant part of the statement is formula 9.2. The verification of bijectivity is straightforward. Let me take the opportunity to point out that the construction performed here is a special case of the general treatment in I.7. It suffices to identify $\mathbb{C} \oplus \mathbb{R}^2$ with M through the assignment

$$(z,a,b) \mapsto \begin{bmatrix} a & z \\ \bar{z} & b \end{bmatrix} \qquad ; \ z \in \mathbb{C}\, ,\ x,y \in \mathbb{R}$$

Specification of I.7.5 gives another proof of the isomorphism $M\ddot{o}b(\mathbb{C}) \xrightarrow{\sim} Lor(M)$. □

Circles Let us consider a vector $N \in M$ of norm -1, quite explicitly

$$N = \begin{bmatrix} a & b \\ \bar{b} & d \end{bmatrix} \qquad ; b \in \mathbb{C},\ a,d \in \mathbb{R}\ ,\ ad - b\bar{b} = -1$$

This determines a hyperbolic plane $P = H^3 \cap N^{\perp}$. By the <u>boundary</u> ∂P of P we understand the set of isotropic lines contained in the linear hyperplane $N^{\perp}$ spanned by P. If we take advantage of the isomorphism $\hat{\mathbb{C}} \xrightarrow{\sim} PC(M)$ we get the following equation for ∂P

$$tr \begin{bmatrix} a & b \\ \bar{b} & d \end{bmatrix}^{\vee} \begin{bmatrix} z \\ w \end{bmatrix} \begin{bmatrix} \bar{z} \\ \bar{w} \end{bmatrix}^{T} = 0 \qquad ; \begin{bmatrix} z \\ w \end{bmatrix} \in \hat{\mathbb{C}}$$

an equation which may be written

9.4
$$\begin{bmatrix} \bar{z} \\ \bar{w} \end{bmatrix}^{T} \begin{bmatrix} d & -b \\ -\bar{b} & a \end{bmatrix} \begin{bmatrix} z \\ w \end{bmatrix} = 0 \qquad ; \begin{bmatrix} z \\ w \end{bmatrix} \in \hat{\mathbb{C}}$$

which is the general equation of a circle in $\hat{\mathbb{C}}$.

Pair of ends An oriented geodesic k in H^3 can be specified through its ends which we may consider as an ordered pair (P,Q) of distinct points of $\hat{\mathbb{C}}$. Let us write $P = \left[\begin{smallmatrix} z \\ w \end{smallmatrix}\right]$ and $Q = \left[\begin{smallmatrix} u \\ v \end{smallmatrix}\right]$ and show that the normal vector N of k is given by

9.5
$$\mathcal{L}(P,Q) = \frac{2}{|D|^2}\, \mathcal{L}\left(\begin{smallmatrix} z \\ w \end{smallmatrix}\right) \wedge \mathcal{L}\left(\begin{smallmatrix} u \\ v \end{smallmatrix}\right) \qquad ; D = det \begin{bmatrix} z & u \\ w & v \end{bmatrix}$$

Let us transform this through the correspondence $\wedge^2 M \xrightarrow{\sim} sl_2(\mathbb{C})$ to get

9.6
$$\mathcal{L}(P,Q) = \frac{1}{zv - wu} \begin{bmatrix} zv+wu & -2zu \\ 2wv & -zv-wu \end{bmatrix} \qquad ; P = \begin{bmatrix} z \\ w \end{bmatrix},\ Q = \begin{bmatrix} u \\ v \end{bmatrix}$$

This matrix is easily seen to have determinant -1. Let us recall from 8.11 that the matrix $iL(\mathrm{P,Q})$ represents a Möbius transformation which represents a half-turn with respect to the geodesic with ends P and Q. It follows from 9.2 and 9.5 that the symbol $L(\mathrm{P,Q})$ obeys the transformation rule

9.7 $$\sigma L(\mathrm{P,Q})\sigma^{-1} = L(\sigma(\mathrm{P}),\sigma(\mathrm{Q})) \qquad ; \ \sigma \in Sl_2(\mathbb{C})$$

We shall evaluate the complex inner products of 2-vectors of the type 9.5 in terms of the cross ratio on $\hat{\mathbb{C}}$. Recall from I.9.8 that for four distinct points A,B,P,Q of $\hat{\mathbb{C}}$ the cross ratio $[\mathrm{A,B,C,D}] \neq 0,\ 1,\ \infty$.

9.8 $$\boxed{<L(\mathrm{A,B}),L(\mathrm{P,Q})> = \frac{[\mathrm{A,B,P,Q}]+1}{[\mathrm{A,B,P,Q}]-1}}$$

Proof According to 9.7 and I.9.7 both sides of the formula are invariant under $Sl_2(\mathbb{C})$. From the fact, I.9.3, that the action of $Sl_2(\mathbb{C})$ on $\hat{\mathbb{C}}$ is triply transitive it follows that it suffices to verify the formula in the case where $\mathrm{A} = \infty$, $\mathrm{B} = 0$, $\mathrm{P} = 1$, $\mathrm{Q} = \mathrm{q}$ with $\mathrm{q} \in \mathbb{C}$, $\mathrm{q} \neq 0,\ 1$. Recall from I.9.9 that $[\infty,0,1,\mathrm{Q}] = \mathrm{q}$. We find that

$$L(\infty,0) = \begin{bmatrix} 1 & 0 \\ 0 & -1 \end{bmatrix}, \qquad L(1,\mathrm{Q}) = \frac{1}{1-\mathrm{q}}\begin{bmatrix} 1+\mathrm{q} & -2\mathrm{q} \\ 2 & -1-\mathrm{q} \end{bmatrix}$$

which gives $\frac{1}{2} tr\, L(\infty,0)\breve{}\,L(1,\mathrm{Q}) = \dfrac{\mathrm{q}+1}{\mathrm{q}-1}$ as required. □

Corollary 9.9 Let A,B,P,Q be four distinct points of the Riemann sphere $\hat{\mathbb{C}}$. The geodesics h and k in H^3 with ends A,B and P,Q are coplanar if and only if $[\mathrm{A,B,P,Q}] \in \mathbb{R}$. The two geodesics h and k will intersect if and only if $[\mathrm{A,B,P,Q}]$ is negative; h and k are perpendicular if and only if $[\mathrm{A,B,P,Q}] = -1$.

Proof The transformation $\mathrm{q} \mapsto \dfrac{\mathrm{q}+1}{\mathrm{q}-1}$ is an involution which preserves $\hat{\mathbb{R}}$. For the remaining details see III.10.8. □

VIII.10 QUATERNIONS

Poincaré's half-space model of hyperbolic 3-space is the open subspace $\mathbb{C}\times]0,+\infty[$ of $\mathbb{C}\times\mathbb{R}$ equipped with the global metric ρ given by the formula II.7.2

10.1 $$\cosh \rho(R,S) = 1 + \frac{|R-S|^2}{2rs} \qquad ; \ R = (z,r),\ S = (w,s)$$

The group $Sl_2(\mathbb{C})$ acts as isometries on this model: if we agree to view the point R above as a quaternion $R = x + iy + jr$, then the action is given by the formula

10.2 $$\sigma(R) = (aR+b)(cR+d)^{-1} \qquad ; \ \sigma = \begin{bmatrix} a & b \\ c & d \end{bmatrix}$$

see, for example, [Beardon] or [Fenchel].

We can give an $Sl_2(\mathbb{C})$-equivariant isometry from the Poincaré model to the Hermitian model of H^3 by assigning to the point R from Poincaré's half-space the complex 2×2 matrix

10.3 $$F(R) = \frac{1}{r}\begin{bmatrix} z\bar{z}+r^2 & z \\ \bar{z} & 1 \end{bmatrix} \qquad ; \ R = z + jr$$

Direct computation gives us

$$\tfrac{1}{2}\, tr\, F(R)^{\check{}}\, F(S) = 1 + \frac{|S-R|^2}{2rs} \qquad ; \ R = z+jr\ ,\ S = w+js$$

which shows that F is an isometry as required, compare II.4.1, 7.5. and 10.1. The relevant formula for $Sl_2(\mathbb{C})$-equivariance

10.4 $$F(\sigma(R)) = \sigma\ F(R)\ \sigma^* \qquad ; \ \sigma \in Sl_2(\mathbb{C})$$

is left to the reader.

VIII.11 FIXED POINT THEOREMS

In this section we shall apply geometry to prove some fixed point theorems for subgroups of $PSl_2(\mathbb{C})$. Let us first recall that $\sigma \in PSl_2(\mathbb{C})$, $\sigma \neq \iota$, has one or two fixed points on ∂H^3. In the case of two fixed points, the geodesic through these is stable under σ and is called the axis of σ. We shall analyse σ in terms of theorem 8.7 on half-turns. Details of some of the arguments can be found in the exercises.

Elliptic transformation $tr^2\sigma \in [0,4[$. The transformation σ has two fixed points on ∂H^3 and all points of the axis are fixed points. The transformation σ can be realised as a product of half-turns in two geodesics through a given point of the axis, both perpendicular to the axis. All planes perpendicular to the axis are invariant under σ. Elliptic transformations with a fixed axis h can be realised as the product of two reflections in planes belonging to the pencil of planes through h.

Parabolic transformation $tr^2\sigma = 4$. The transformation σ has precisely one fixed point ∞ on ∂H^3 and is the product of two half-turns in geodesics through ∞. The hyperbolic plane P spanned by these two geodesics is an invariant plane. The transformation σ can be realised as the product of two reflections in planes belonging to the pencil of planes through ∞ perpendicular to P.

The parabolic transformations (including ι) fixing a given point $\infty \in \partial H^3$ make up the commutator group of the subgroup of $PSl_2(\mathbb{C})$ fixing ∞. This can be seen by explicit computations, compare the proof of I.9.13.

Hyperbolic transformation $tr^2\sigma \in]4,\infty[$. The transformation has two fixed points on ∂H^3 but no fixed points on H^3. The transformation leaves any plane containing the axis invariant. The transformation can be realised as the

product of half-turns in two geodesics contained in a fixed invariant plane perpendicular to the axis. A hyperbolic transformation with fixed axis h can be realised as a product of two reflections in planes belonging to the pencil of planes perpendicular to h.

Strictly loxodromic transformation $tr^2\sigma \notin [0,\infty]$. The transformation has two fixed points on ∂H^3 but none on H^3. The transformation can be realised as a product of half-turns in geodesics perpendicular to the axis but not contained in a plane. A transformation which is either strictly loxodromic or hyperbolic is called loxodromic.

The group of transformations fixing two given points A and B of ∂H^3 is commutative, being the product of the group of rotations with axis AB and the group of parallel translations along the geodesic AB. This group has index 2 in the group of transformations stabilising the geodesic with ends A and B.

Elliptic fixed point theorem 11.1 A subgroup G of $PSl_2(\mathbb{C})$ consisting of elliptic transformations has a fixed point on H^3.

Proof Let us consider two distinct non-trivial members σ and τ of G and show that their axes h and k intersect. To this end we shall rule out two alternatives: common end and common perpendicular.

When h and k have a common end, the commutator $\sigma\tau\sigma^{-1}\tau^{-1}$ is parabolic and we conclude that σ and τ commute. The fixed point set h for σ on H^3 is stable under τ as follows from the "normaliser principle" IV.2.6. It follows that $h=k$ or (when τ is a half-turn) h is perpendicular to k.

When h and k have a common perpendicular, let $A \in h$ and $B \in k$ be points such that the geodesic l through A and B is perpendicular to h and k. Pick geodesics a through A and b through B such that $\sigma = \alpha\lambda$ and $\tau = \lambda\beta$ where α,β,λ are half-turns in the geodesics a,b,l. Observe that $\sigma\tau = \alpha\beta$ and conclude from 8.7 that a and b intersect. This implies that the plane P spanned by a and b is perpendicular to the axes h and k for σ and τ. We conclude from the proof of IV.2.8 that A = B and therefore $h = k$.

Since any two axes of rotation are coplanar, we conclude that the axes of rotation are members of some bundle $\mathcal{H}$. Since any two axes intersect, we conclude from 5.4 that all axes pass through the same point of H^3. □

Definition 11.2 A subgroup G of $PSl_2(\mathbb{C})$ is called <u>elementary</u> if G leaves a geodesic invariant or fixes a point of H^3 or a point of the boundary ∂H^3.

Corollary 11.3 A subgroup G of $PSl_2(\mathbb{C})$ with a finite orbit on H^3 or ∂H^3 is elementary.

Proof Suppose that G has a finite orbit on H^3 of order $n \geq 1$. Any non-trivial $\gamma \in G$ acts as a permutation on the set $\mathcal{O}$ on n elements and we find that $\gamma^{n!}$ has a fixed point on H^3. From this it follows that $\gamma^{n!}$ is a elliptic (or ι) and we conclude that γ is a elliptic: the class of loxodromic (resp. parabolic elements) is stable under n!-powers. It follows from 11.1 that G has a fixed point on H^3.

Suppose that G has an orbit on ∂H^3 consisting of $n \geq 1$ elements. In case $n = 1$, the group has a fixed point on ∂H^3. When $n = 2$, the geodesic through the two point orbit is stable. When $n \geq 3$, we find that all elements of G have finite order and the elliptic fixed point theorem tells us that G has a fixed point on H^3. □

Corollary 11.4 A non-elementary subgroup G of $PSl_2(\mathbb{C})$ contains a loxodromic transformation.

Proof Let us assume that G does not contain a loxodromic element. According to the elliptic fixed point theorem, the group G contains a parabolic element γ, fixing ∞ say. For any element $\sigma \in G$ we have that

$$tr^2\,(\gamma^n\sigma) \in [0,4] \qquad ; n \in \mathbb{N}$$

In order to exploit this we put

$$\gamma = \begin{bmatrix} 1 & 1 \\ 0 & 1 \end{bmatrix}, \quad \sigma = \begin{bmatrix} a & b \\ c & d \end{bmatrix} \qquad ; \text{a,b,c,d} \in \mathbf{C},\ \text{ad}-\text{bc} = 1$$

Direct calculation gives us

$$tr^2(\gamma^n\sigma) = (\text{a} + \text{d} + \text{nc})^2 \in [0,4] \qquad ; \text{n} \in \mathbb{N}$$

But this is only possible if $c = 0$, which means that $\sigma(\infty) = \infty$. This being true for all $\sigma \in G$ makes G elementary contrary to the hypothesis. □

Invariant disc theorem 11.5 A non-elementary subgroup G of $PSl_2(\mathbf{C})$ fixes an oriented hyperbolic plane if and only if G doesn't contain strictly loxodromic elements.

Proof Suppose first that G doesn't contain strictly loxodromic elements. Let us show that any two hyperbolic transformations $\mu,\nu \in G$ have coplanar axes. To this end we may assume that the axes m and n have a common perpendicular c. Let us write $\mu = \alpha\gamma$ and $\nu = \gamma\beta$ where γ is a half-turn with respect to c, α is a half-turn with respect to a geodesic a perpendicular to m and β is a half-turn with respect to a geodesic b perpendicular to n. From the fact that none of the transformations $\alpha\gamma$, $\gamma\beta$, $\alpha\beta$ are strictly loxodromic it follows that a,b,c are pairwise coplanar, compare 8.7. It follows from 5.4 that we can find a plane P such that either a,b,c are perpendicular to P or all lie in the same plane P. In both cases we conclude that m and n lie in P.

Let us remark that G contains a hyperbolic transformation μ with axis m say, 11.4. For $\sigma \in G$ we find that $\sigma\mu\sigma^{-1}$ is hyperbolic with axis $\sigma(m)$. It follows that we can find another hyperbolic transformation $\nu \in G$ with axis $n \neq m$. At this point we ask the reader to use 5.4 to find a plane R containing the axes of all hyperbolic elements of G. Observe that R is stable under G.

It remains to prove that G preserves the orientation of R. To this end we must prove that the group $\Gamma \subseteq Isom(\text{R})$ induced by G consists of even transformations only. Assume for a moment that this is not the case. Taking into account that Γ is non-elementary, it follows from IV.2.9 that we can find a glide reflection $\gamma \in \Gamma$ with axes g, say. Let us write $\gamma = \tau_R \kappa_R$ where τ is restriction to R of a translation τ in H^3 along g and κ_R is restriction to R of reflection κ in the hyperbolic plane K perpendicular to R through g. Let ρ denote reflection in R and conclude that $\tau\kappa\rho \in \Gamma$. This is a contradiction since $\tau\kappa\rho$ is strictly loxodromic.

Conversely, let R be an oriented plane fixed by G. An element $\gamma \in G$ induces a transformation of the plane R which is even. It follows that γ is the product of two reflections in planes perpendicular to R. □

Proposition 11.6 Any solvable subgroup G of $PSl_2(\mathbb{C})$ is elementary. More generally, a subgroup G of $PSl_2(\mathbb{C})$ with a composition series

$$\{\iota\} = G_0 \subseteq G_1 \subseteq G_2 \subseteq \ldots \subseteq G_{n-1} \subseteq G_n = G$$

with G_1 non-trivial abelian, is elementary.

Proof Let us start by assuming that G_1 contains a parabolic element σ, say, with fixed point $S \in \partial H^3$. Since S is the only fixed point for σ on ∂H^3, we conclude from the "normaliser principle", IV.2.6, that the fixed point set for G_1 on ∂H^3 is $\{S\}$. Multiple applications of the normaliser principle show that G fixes S.

Let us assume that G_1 contains an element $\kappa \neq \iota$ with two distinct fixed points A and B on ∂H^3 and let us prove that the geodesic k with ends A and B is fixed by G_1. When $\sigma^2 = \iota$ (half-turn) we find that the set of fixed points for σ on H^3 is k and we conclude from the "normaliser principle" that k is stabilised by G_1. When $\kappa^2 \neq \iota$ we find that k is the only geodesic fixed by κ: if κ fixes the geodesic h with ends C and D, we find that κ^2 has fixed points A,B,C,D on ∂H^3 ; it follows that $h = k$. We can apply the "normaliser principle" to the action of G_1 on the set of geodesics in H^3 and conclude that G_1 fixes k.

If k is the only geodesic fixed by G_1, it follows from the "normaliser principle" that G fixes k. Thus we may assume that G_1 fixes two geodesics k and l. It follows from the argument above that G_1 consists of half-turns. By inspection of a single half-turn, we conclude that k and l have a common perpendicular h. It follows that G_1 fixes the two perpendicular geodesics h and k.

The geodesic m perpendicular to h and k is fixed as well and we conclude that G_1 is a subgroup of the group (of order 4) generated by half-turns in h and k. If $\#G_1 = 4$ we find that G_1 has precisely one fixed point on H^3 and it follows from the "normaliser principle" that G has a fixed point on H^3.

We have reduced the problem to the case where $\#G_1 = 2$. In this case G_1 is generated by a half-turn in a geodesic n which is stabilised by G_2 by the "normaliser principle". We can now repeat the arguments of the last paragraph and conclude that G is elementary or $\#G_2 = 2$. We leave it to the reader to conclude the proof by a simple induction on n. □

Theorem 11.7 The group $PSl_2(\mathbb{C})$ is simple, i.e. the group contains no non-trivial normal subgroup.

Proof A non-trivial element σ of $PSl_2(\mathbb{C})$ can be written as a product $\sigma = \alpha\beta$ of half-turns in geodesics a and b respectively. Let us choose an end A for a and an end B $\neq$ A for b. We can write $\alpha\beta = (\alpha\gamma)(\gamma\beta)$ where γ is a half-turn in the geodesic AB. This provides a representation of σ as a product of two parabolic transformations. It is important to recall that any two parabolic transformations are conjugated in $PSl_2(\mathbb{C})$, compare I.9.13.

Let us show that a normal subgroup N $\neq \{\iota\}$ contains a parabolic transformation. If not, we can consider an element ν with two fixed points A and B on ∂H^3. Pick a parabolic transformation α with fixed point A and observe that the commutator $\alpha^{-1}\nu\alpha\nu^{-1} \neq \iota$ is parabolic and belongs to N. □

At this point we refer the reader to [Beardon] for more information, in particular Jørgensen's inequality.

VIII EXERCISES

EXERCISE 2.1 1° Suppose that $dim\,E = 4$. For a non-zero vector $w \in \wedge^2 E$ show that $dim\,E_w = 2$ or $dim\,E_w = 4$. Hint: Rule out the case $dim\,E_w = 3$ by means of proposition 2.5.

2° Let w be a non-pure 2-vector as above. Show that $E_w = E_{*w}$.

EXERCISE 2.2 Let $Z: M \to M$ be a skew adjoint operator on Minkowski space M. Show that there exists a unique 2-vector $z \in \wedge^2 M$ such that $z\lfloor = Z$. Hint: A skew adjoint operator Z defines an alternating 2-form $\langle Z(x), y\rangle$, $x,y \in M$. Show that there is a unique $z \in \wedge^2 M$ with $\langle z, x \wedge y\rangle = \langle Z(x), y\rangle$, $x,y \in M$.

EXERCISE 4.1 Let P and Q be distinct hyperbolic planes in H^3. Show that there exists at most one geodesic h perpendicular to both P and Q. Hint: Use proposition 4.6.

EXERCISE 6.1 1° Prove the general formulas

$$U + U^{\check{}} = \iota\, tr U \qquad ; U \in M_2(\mathbb{C})$$

$$tr(UV) + tr(U^{\check{}}V) = tr U\, tr V \qquad ; U,V \in M_2(\mathbb{C})$$

2° Apply these formulas twice and deduce that

$$tr UV + tr U^{\check{}}V^{\check{}} = 2\, tr U\, tr V - 2\, tr(UV^{\check{}}) \qquad ; U,V \in M_2(\mathbb{C})$$

EXERCISE 6.2 Consider the complex multilinear form on $M_2(\mathbb{C})$

$$vol(A,B,C,D) = -\tfrac{i}{4}\, tr(AB^{\check{}}CD^{\check{}} - A^{\check{}}BC^{\check{}}D) \qquad ; A,B,C,D \in M_2(\mathbb{C})$$

1° Show that the form changes its sign under the permutation

$$(A,B,C,D) \mapsto (D,A,B,C).$$

2° Show that $vol(A,B,C,D)$ changes sign under the permutation of A and B.

3° Show that the form vol is alternating. Hint: Show that the group of permutations of the letters A,B,C,D is generated by the permutations from 2° and 3°.

4° Show that vol takes the value 1 on the Pauli matrices $\sigma_0, \sigma_1, \sigma_2, \sigma_3$.

EXERCISE 6.3 For matrices A,B $\in M_2(\mathbb{C})$ put $A \times B = \frac{1}{2}(AB^{\vee} - BA^{\vee})$. With the notation of exercise 6.2. show that for all $A,B,C,D \in M_2(\mathbb{C})$.

$$\boxed{<A \times B, C \times D> = <A,C><B,D> - <A,D><B,C> - i\ vol(A,B,C,D)}$$

Hint: First observe that the formula is valid for $A,B,C,D \in M$. Second, observe that the formula is $\mathbb{C}$-linear in all variables. Third, observe that the complex vector space $M_2(\mathbb{C})$ is spanned by M.

EXERCISE 7.1 Show that Dirac's algebra D is isomorphic to $M_2(\mathbb{H})$. Hint: Identify the field of quaternions $\mathbb{H}$ with the ring of invariants under the involution $X \mapsto \tilde{X}$ of $M_2(\mathbb{C})$. Observe that the scalar matrices $1, i$ form a basis for $M_2(\mathbb{C})$ as a vector space over $\mathbb{H}$ and define a morphism of rings $F: M_2(\mathbb{C}) \to M_2(\mathbb{H})$ by

$$F(X) = \begin{bmatrix} H & -K \\ K & H \end{bmatrix} \qquad ; \ X = H + iK \ , \ H,K \in \mathbb{H}$$

Extend the definition of F to Dirac's algebra D by the convention $F(\Psi) = \begin{bmatrix} 0 & 1 \\ 1 & 0 \end{bmatrix}$

EXERCISE 7.2 1° Given $D \in sl_2(\mathbb{C})$. View D as a 2-vector on M and show that

$$D \lfloor X = \tfrac{1}{2}(DX - X\tilde{D}) \qquad ; X \in M$$

2° Verify the identity VIII.3.2 $\quad D \lfloor iD \lfloor X = Im[<D,D>]\ X \qquad ; X \in M$

3° Verify the identity $\quad D \lfloor D \lfloor X + iD \lfloor iD \lfloor X = -DX\tilde{D} \qquad ; X \in M$

4° Consider $R \in sl_2(\mathbb{C})$ whose determinant is real and different from zero. Show that

$$DX\tilde{D}^{-1} = (det\ D)(\ D \lfloor D \lfloor X + iD \lfloor iD \lfloor X) \qquad ; X \in M$$

EXERCISE 7.3 Consider the following alternating C-valued 3-form on M

$$\tfrac{1}{3!} \textstyle\sum_\sigma sign(\sigma)\ A_{\sigma(1)} \Psi A_{\sigma(2)} \Psi A_{\sigma(3)} \Psi \qquad ; A_1, A_2, A_3 \in M$$

where the sum is over the group of permutations Σ_3.

1° Show that this form identifies $\wedge^3 M$ and $N\Psi$ where N is the space of anti–Hermitian matrices, $X^* = -X$.

2° Identify the Hodge star $*: \wedge^1 M \to \wedge^3 M$ with multiplication by $i: M \to N$.

3° Identify the exterior product $\wedge^2 M \times \wedge^1 M \to \wedge^3 M$ with the pairing

$$\tfrac{1}{2}(WX - XW^*) \qquad ; \ W \in sl_2(\mathbb{C}),\ X \in M$$

EXERCISE 7.4 Let $\mathcal{M}$ denote the set of matrices $X \in Gl_2(\mathbb{C})$ with $X^{\smile} = -\bar{X}$.

$$\begin{bmatrix} a & b \\ c & -\bar{a} \end{bmatrix} \qquad ; \; a,d \in \mathbb{R} \, , \, b \in \mathbb{C}$$

1° Show that the determinant makes this into a Minkowski space. Hint: Use the transformation $T{:}\mathcal{M} \to M$ given by

$$\begin{bmatrix} a & b \\ c & -\bar{a} \end{bmatrix} \mapsto \begin{bmatrix} -b & a \\ \bar{a} & c \end{bmatrix}$$

2° Let us consider the following action of the group $Sl_2(\mathbb{C})$ on $\mathcal{M}$

$$\sigma . X = \sigma X \bar{\sigma}^{-1} \qquad ; \sigma \in Sl_2(\mathbb{C}) \, , \, X \in \mathcal{M}$$

Show that $Sl_2(\mathbb{C})$ acts on $\mathcal{M}$ through Lorentz transformations.

3° Verify the formula

$$T(\sigma X \bar{\sigma}^{-1}) = \sigma \, T(X) \, \tilde{\sigma}^{-1} \qquad ; X \in M, \; \sigma \in Sl_2(\mathbb{C})$$

4° Define a multiplication on $M_2(\mathbb{C}) \oplus M_2(\mathbb{C})$ by the formula 7.1 replacing $\tilde{A}$ by $\bar{A}$ and show that the resulting algebra F is the Clifford algebra of $(\mathcal{M}, -det)$.

5° Show that the representation

$$D \to M_4(\mathbb{C}) \quad ; \qquad (A,B) \mapsto \begin{bmatrix} A & B \\ \bar{B} & \bar{A} \end{bmatrix}$$

can be decomposed into two simple representations. Hint: Show that vectors of the form $(z_1,z_2,\bar{z}_1,\bar{z}_2)$, $z_1,z_2 \in \mathbb{C}$ make up a D-stable subspace of $\mathbb{C}^4$.

6° Conclude that the Clifford algebra of a form of type $(-1,3)$ is $M_4(\mathbb{R})$.

EXERCISE 7.5 1° Define a multiplication on $M_2(\mathbb{C}) \oplus M_2(\mathbb{C})$ by the formula 7.1 replacing $\tilde{A}$ by A in the definition 7.1 and show that the resulting $\mathbb{C}$-algebra F is the Clifford algebra of $(sl_2(\mathbb{C}), -det)$.

2° Repeat 1° with $\mathbb{C}$ replaced by $\mathbb{R}$.

EXERCISE 8.1 1° Show that the only involutions in $Lor^+(M)$ are the half-turns. Hint: See exercise I.1.1.

2° Find all involutions in the group $Lor(M)$.

EXERCISE 8.2 Show that an even Lorentz transformation σ of M is a product of two plane reflections if and only if $tr^2\sigma \in [0,+\infty[$.

EXERCISE 8.3 For $U \in \Gamma$ show that $UU^{\check{}} \in \mathbb{R}\iota$ and $UU^{\#\check{}} \in \mathbb{R}\iota$. Verify that the <u>spinor norms</u> $N:\Gamma \to \mathbb{R}$ and $S:\Gamma \to \mathbb{R}$ defined by

$$N(U)\iota = UU^{\check{}}\ ,\quad S(U)\iota = UU^{\#\check{}}\iota \qquad ; U \in \Gamma$$

are morphisms of groups.

2° Explain how the spinor norm N was used in the proof of Theorem 8.5.

3° Show that $U \in \Gamma$ acts on M as a Lorentz transformation if and only if the spinor norm is $S(U)$ is positive.

EXERCISE 8.4 Let r denote an oriented geodesic with normal vector R. Show that a half-turn ρ with respect to r acts on M through the formula

$$\rho = R\lfloor\ R\lfloor + *R\lfloor\ *R\lfloor$$

Hint: Consider a positively oriented orthonormal basis A,T,U,V for M where $A \in r$ and T is a positively oriented unit tangent vector to r at A.

EXERCISE 8.5 Verify the following table of involutions.

Lorentz transformation	Γ	$S \in Gl_2(\mathbb{C})$
point reflection	$iS\Psi$	$S \in M,\ detS = 1$
half-turn	iS	$S \in sl_2(\mathbb{C}),\ det\ S = -1$
plane reflection	$S\Psi$	$S \in M,\ det\ S = -1$

EXERCISE 8.6 1° Show that the Clifford group Γ_+ acts on M as orientation preserving transformations and the induced map $\Gamma_+ \to SO(M)$ is surjective.

Hint: Show that Γ_+ is generated by $Sl_2(\mathbb{C})$, ι and $\mathbb{R}^*$.

2° Show that the action of Γ on M defines a surjective map $\Gamma \to O(M)$.

Hint: Observe that $\Gamma = \Gamma_+ \cup \Gamma_+\Psi$

EXERCISE 9.1 Show that $<L(A,B),L(B,C)> = 1 \qquad ;\ A \neq B \neq C$

EXERCISE 9.2 An even isometry σ of H^3 interchanges two distinct points A and B of ∂H^3. Show that σ is a half-turn with respect a geodesic perpendicular to the geodesic with ends A and B.

EXERCISE 9.3 Consider two non-intersecting discs in the complex plane and let the Euclidean line through their centres intersect the bounding circles in the points A,B,C,D in this order.

1° Show that the hyperbolic distance distance d between the corresponding hyperbolic planes is given by

$$tanh^2 \tfrac{1}{2}\mathrm{d} = \frac{\mathrm{AD\ BC}}{\mathrm{AC\ BD}}$$

2° In the case where the two circles are concentric, show that d is the logarithm of the ratio of their radii.

EXERCISE 11.1 1° Let $z = x + iy$ be a complex number. Verify the formula

$$\tfrac{1}{2}\, tr \begin{bmatrix} e^{z} & 0 \\ 0 & e^{-z} \end{bmatrix} = cosh\ x\ cos\ y + i\ sinh\ x\ sin\ y$$

Discuss the action of the matrix on H^3 in the cases $x = 0$ and $y = 0$ separately. In general observe that $e^{z} = e^{x}e^{iy}$.

2° Use this to discuss the general loxodromic transformation.

EXERCISE 11.2 1° Let $\sigma \neq \iota$ be a parabolic transformation of H^3. Show that the set of hyperbolic planes in H^3 invariant under σ form a pencil. Hint: Investigate the special case $\sigma = \begin{bmatrix} 1 & 1 \\ 0 & 1 \end{bmatrix}$, compare I.9.15.

2° Let P be an invariant plane for σ. Show that σ preserves the connected components of $H^3 - \mathrm{P}$.

EXERCISE 11.3 Let $\mathcal{P}$ denote a pencil of hyperbolic planes. Show that the product of three reflections in planes from $\mathcal{P}$ is itself a reflection in a plane from $\mathcal{P}$. Hint: Show that there exists a hyperbolic plane Q whose normal vector is orthogonal to the normal vector of any plane from $\mathcal{P}$.

EXERCISE 11.4 1° Show that an odd isometry σ of H^3 can be decomposed as $\sigma = \kappa\rho$ where κ is a half-turn and ρ is a reflection in a plane. Hint: When $\sigma^2 \neq \iota$, pick a point $A \in \partial H^3$ (distinct from the fixed points of σ^2) and let ρ denote reflection in the plane through $B = \sigma(A)$ perpendicular to the geodesic with ends A and $C = \sigma(B)$. Observe that $\sigma\rho$ interchanges the points B and C and conclude that $\kappa = \sigma\rho$ is a half-turn.

2° Show that an odd isometry σ of H^3 can be decomposed into a product of three plane reflections $\sigma = \alpha\beta\gamma$ where the mirror plane of γ is perpendicular to the mirror planes of α and β. Hint: Consider a decomposition $\sigma = \kappa\rho$ a in 1° and use that the pencil of planes containing the axis k of κ has type (0,2).

APPENDIX: AXIOMS FOR PLANE GEOMETRY

The aim of this appendix is to prove that a metric space which satisfies the incidence axiom and the reflection axiom is isometric to the Euclidean plane or (after rescaling) the hyperbolic plane. For a precise statement see the first page of the introduction. The need for a proof comes from the fact that our axioms are somewhat untraditional. The proof is presented as a systematic series of deductions from the axioms. This will be carried out to the point where we have proved the "Pasch's axiom", see the discussion at the end of the appendix.

Let us agree that a straight geodesic curve in a metric space X is a distance preserving map $\gamma: \mathbb{R} \to X$ and recall that a line in X is the image of a straight geodesic curve, compare II.1. In the following we let X be a plane i.e. a metric space X which satisfies the following two axioms

INCIDENCE AXIOM Through two distinct points of X there passes a unique line. The space X has at least one point.

REFLECTION AXIOM The complement of a given line in X has two connected components. There exists an isometry σ of X which fixes the points of the line, but interchanges the two connected components of its complement.

Sharp triangle inequality A.1

For a point B of the plane not on the line through two distinct points A and C we have that

$$d(A,C) < d(A,B) + d(B,C)$$

Proof

We shall turn the problem upside down and assume that we are given three distinct points A,B,C with $d(A,C) = d(A,B) + d(B,C)$ and proceed to construct a line through all three points. Pick $a,b,c \in \mathbb{R}$ with $b - a = d(A,B)$ and $c - b = d(B,C)$ and choose three straight geodesic curves ϕ, λ, μ in X with

$$\phi(a) = A\ ,\ \phi(c) = C,\ \lambda(a) = A,\ \lambda(b) = B,\ \mu(b) = B,\ \mu(c) = C$$

Let $\sigma: \mathbb{R} \to X$ denote the curve with restriction ϕ to $]-\infty,a]$, restriction λ to $[a,b]$, restriction μ to $[b,c]$ and restriction ϕ to $[c,+\infty[$. We shall prove that $\sigma: \mathbb{R} \to X$ is a straight geodesic curve. The triangle inequality gives us (compare VI.2.6)

$$d(\sigma(x)),\sigma(y)) \le y-x \qquad ; \quad x,y \in \mathbb{R}, \quad x \le y$$

In order to prove the opposite inequality, choose $r \in \mathbb{R}$ smaller than a and x and choose $w \in \mathbb{R}$ bigger than c and y to get that

$$d(\sigma(r),\sigma(w)) \le d(\sigma(r),\sigma(x))+d(\sigma(x),\sigma(y))+d(\sigma(y)),\sigma(w)) \le x-r+d(\sigma(x),\sigma(y)+w-y$$

Observe that $d(\sigma(r),\sigma(w)) = w-r$ and deduce that $y-x \le d(\sigma(x),\sigma(y))$. □

For two points A and B of the plane we define the segment

$$[A,B] = \{ P \mid d(A,P)+d(P,B) = d(A,B) \}$$

We proceed using the reflection axiom. For convenience we shall use the word sides for the connected components of the complement of a line k.

Proposition A.2 Given a point P in the plane outside the line k. There exists a unique point $P^k \in k$ such that

$$d(P,P^k) < d(P,K) \qquad ; \quad K \in k, \quad K \ne P^k$$

Proof Let κ be an isometry of the plane which fixes k pointwise but interchanges the sides of k. For a $P \notin k$, the points $\kappa(P)$ and P lie on different sides of k. It follows that the line n through P and $\kappa(P)$ will meet k in a point P^k say.

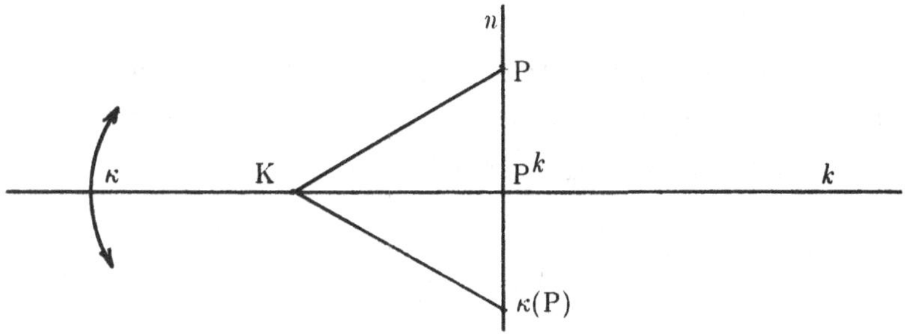

Notice, that the point P^k is fixed by κ. It follows that the line n is stable under κ since the points P and P^k are mapped into n. By examination of the restriction of κ to n we conclude that $\kappa(\kappa(P)) = P$ and that P^k is the midpoint of P and $\kappa(P)$.

Let us now investigate the distance from P to an arbitrary point $K \in k$,

$K \neq P^k$. From the sharp triangle inequality A.1 it follows that

$$d(P,\kappa(P)) < d(P,K) + d(K,\kappa(P)) = 2d(P,K)$$

The left hand side of this inequality can be rewritten as

$$d(P,\kappa(P)) = d(P,P^k) + d(P^k,\kappa(P)) = 2d(P,P^k)$$

and we conclude that $d(P,P^k) < d(P,K)$ as required. □

The point $P^k \in k$ nearest P is called the <u>perpendicular</u> <u>projection</u> of P onto the line k. Let us make the general remark that for any isometry μ of the plane

A.3 $$\mu(P^n) = \mu(P)^{\mu(n)}$$

The isometry κ whose existence is postulated in the reflection axiom is unique and is called <u>reflection</u> in the line k.

Proposition A.4 An isometry σ of the plane which fixes the points of the line k is either reflection κ in the line k or the identity ι.

Proof Let κ denote reflection in k. If σ interchanges the sides of k we have $\sigma = \kappa$ according to the remarks above. If σ stabilises the sides of κ, then $\kappa\sigma$ will interchange the sides of k and therefore $\kappa\sigma = \kappa$, and we find that $\sigma = \iota$. □

We say that a line n is <u>perpendicular</u> to a line k, $n \perp k$, if n is stable under reflection in k but different from k. If $n \perp k$, then for any point $N \in n$ we have that $N^k \in n$ as follows from the proof of proposition A.2.

Proposition A.5 Given lines n and k in the plane. If $n \perp k$ then $k \perp n$.

Proof Assume that $n \perp k$ and pick a point P of k but not on n. Let κ be reflection in k and conclude from A.3 that $\kappa(P^n) = \kappa(P)^{\kappa(n)} = P^n$. This shows that $P^n \in k$ and it follows that $k \perp n$. □

Saccheri's inequality A.6 Let A and B be points outside a given line k. Their perpendicular projections A^k and B^k satisfy $d(A,B) \geq d(A^k,B^k)$.

Proof Let us put $A_0 = A$, $P_0 = A^k$, $A_1 = B$, $P_1 = B^k$ and reflect our quadrangle in the line A_1P_1 to get a new quadrangle $P_1A_1A_2P_2$. Reflect this in the line A_2P_2 to get a new quadrangle $P_2A_2A_3P_3$...

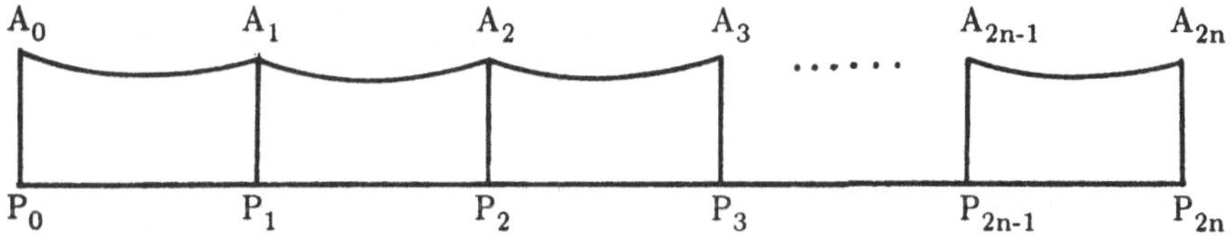

From the triangle inequality we find that

$$d(P_0,P_{2n}) \leq d(P_0,A_0) + d(A_0,A_1) + d(A_1,A_2) + \ldots + d(A_{2n-1},A_{2n}) + d(A_{2n},P_{2n})$$

Using that reflections are distance preserving we find that

$$2n\ d(P_0,P_1) \leq 2n\ d(A_0,A_1) + 2\ d(P_0,A_0)$$

This may be rewritten as

$$d(A^k,B^k) - d(A,B) \leq\ 1/n\ d(A,A^k)$$

Saccheri's inequality follows from this by letting n tend to infinity. □

Proposition A.7 For any point K on the line k there exists a point P off k with $P^k = K$.

Proof Let us fix one of the sides S of k and observe that perpendicular projection $p:S \to k$ is continuous as a consequence of Saccheri's inequality. Using that S is connected, we conclude the image p(S) is a connected subset of k. It remains to prove that p(S) is dense in k, i.e. for $K \in k$ and $\epsilon > 0$ given, there exists $Q \in S$ with $d(K,Q^k) \leq \epsilon$. Observe that for $K \in k$ and $Q \in S$ we get from A.5 that k is perpendicular to the line n through Q, Q^k. It follows that $d(K,Q^k) \leq d(K,Q)$. Finally, observe that for $K \in k$ and $\epsilon > 0$ given there exists a point $Q \in S$ with $d(Q,K) \leq \epsilon$: such a point can be found on any line through K which meets S. □

Let us fix a point S of the plane and define a half-turn σ around S as follows: we put $\sigma(S) = S$ and for $P \neq S$ we draw the line through P and S and determine $\sigma(P)$ such that S is the midpoint between P and $\sigma(P)$. The construction of the map σ requires only the incidence axiom. Taking the reflection axiom into account we shall show that σ is an isometry.

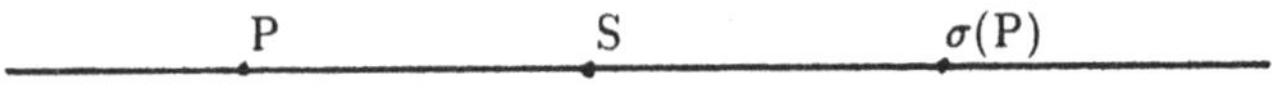

Proposition A.8 A half-turn with respect to a point S of the plane is an isometry of the plane.

Proof Let us use A.7 to pick two perpendicular lines m and n through S and let μ and ν be reflections in m and n respectively. We intend to show that $\mu\nu$ is a half-turn with respect to S. It is easy to check that the isometry $\mu\nu\mu\nu$ is the identity on m and on n. We conclude from A.4 that $\mu\nu\mu\nu = \iota$, the identity. It follows that $\sigma = \mu\nu$ is an involution with fixed point S. We proceed to show that S is the only fixed point for σ. It is easily checked that S is the only fixed point for σ on $m \cup n$. Let us turn to the complement of $m \cup n$. Reflection ν in n leaves the two sides M_1 and M_2 of m invariant: from $n \perp m$ it follows that M_1 and M_2 both meet n, and we conclude that ν has a fixed point on both of them. In the same way we conclude that μ leaves the two sides N_1 and N_2 of n invariant. The complement of $m \cup n$ is decomposed into the union of four disjoint "quadrants"

$$M_1 \cap N_1, \; M_1 \cap N_2, \; M_2 \cap N_1, \; M_2 \cap N_2$$

Observe that σ interchanges the first quadrant with the third and the second quadrant with the fourth. It follows that S is the only fixed point for σ.

From the fact that σ is an involution with S as the only fixed point it follows that σ is a half-turn with respect to S: For a given point $P \neq S$ the two points P and $\sigma(P)$ are interchanged by σ. It follows that the line h through P and $\sigma(P)$ is stable under σ. Second, we deduce that the midpoint of P and $\sigma(P)$ is a fixed point for σ. In conclusion, the point S is the midpoint for P and $\sigma(P)$. □

Theorem A.9 A line h in the plane which meets a second line k must meet both sides of k.

Proof Let us show that a half-turn σ around a point $S \in k$ interchanges the two sides of k: pick a line n through S perpendicular to k and observe that σ and reflection κ in k have the same restriction to n. Let us now assume that h and k meet at the point $S \in h$. Pick a point $H \in h$ belonging to one side of k and conclude that $\sigma(H) \in h$ belongs to the other side of k. □

Corollary A.10 Let A and B be two points of the plane outside a line k. Then A and B lie on different sides of k if and only if k meets [A,B].[1]

Proof If k does not meet [A,B], then the connected set [A,B] is contained in the complement of k and must be contained in one of its connected components. If k meets [A,B] in S, then S divides the line h through A and B into two open half-lines a and b with $A \in a$ and $B \in b$. Since a and b are connected, they are entirely contained in the connected components of the complement of k. It follows from A.9 that they are contained in different components. □

Corollary A.11 Through a point P of the plane there passes a unique line perpendicular to a given line k.

Proof If P is outside k, a line n through P perpendicular to k must pass through P^k, which makes it unique. When $P \in k$, let us analyse a fixed line n through P perpendicular to k. Reflection κ in k must stabilize the two sides of n since k meets both of them, A.5. For a point $S \notin n$ we conclude that S and $\kappa(S)$ lie on the same side of n and we conclude from A.10 that n is disjoint from $[S,\kappa(S)]$. In particular $S^k \notin n$ and therefore $S^k \neq P$. □

---

[1] As a consequence of this we find that the relation " $[A,B] \cap k = \emptyset$ " is an equivalence relation on the complement of k. This is the exact content of Pasch's axiom (M.Pasch 1843 – 1930) which played a key role in Hilbert's axioms.

Let A and B be two distinct points of the plane. The line n through the midpoint of A and B and perpendicular to the line through A and B is called the perpendicular bisector for A and B.

Proposition A.12 Given two distinct points A and B of the plane. A point P of the plane belongs to the perpendicular bisector n of A and B if and only if $d(A,P) = d(B,P)$.

Proof Reflection ν in n interchanges A and B as follows from A.5. Thus

$$d(P,A) = d(\nu(P),\nu(A)) = d(P,B) \qquad ; P \in n$$

In order to prove the converse, we shall show that a point $P \notin n$ on the same side of n as B satisfies $d(A,P) > d(B,P)$.

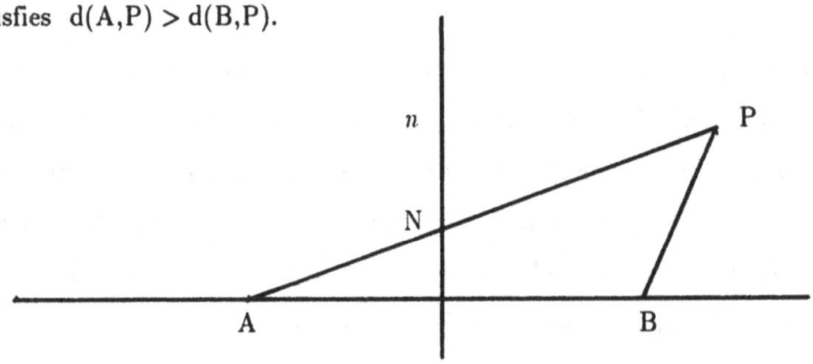

To this end, let N be the point of intersection for [A,P] and n. Since P and B are on the same side of n we have $N \notin [P,B]$ and the sharp triangle inequality gives

$$d(P,B) < d(P,N) + d(N,B) = d(P,N) + d(N,A) = d(P,A)$$

The last equality follows from $N \in [P,A]$. □

This is a good place to stop. Readers with a sense of humour could once again consult the opening chapter of [Coxeter] on Pons Asinorum. Now we have proved "Pasch's axiom", a proof of the classification theorem, presented in the introduction to the book, can be found in [Borsuk, Szmielew] or [Martin].

SOURCES

Let me try to make a list of the sources which have been most useful to me in preparing the book. At the same time I hope to give an introduction to the bibliography and some suggestions to further reading.

Chapter I The two volumes of [Berger], the book of [Coxeter[2]] and the text of [Wilker]. For more information on orthogonal groups: [Deheuvels].

Chapter II The notes of [Rees] and the book of [Nikulin, Shafarevich].

Chapter III Inspired by the book of [Fenchel].

Chapter IV Nielsen and Fenchel, *On discontinuous groups of isometric transformations....* Courant Anniversary Volume. Reprinted in [Nielsen].

Chapter V Most of the material is taken from [Beardon]. The section on cusps is inspired from [Shimura].

Chapter VI The book of [Nikulin, Shafarevich] contains a lot of ideas fundamental to the geometric concepts in this book. For further information see the collected works of [Nielsen] and [Casson, Bleiler].

Chapter VII The proof of Poincaré's theorem leans heavily on its predecessors [de Rham], [Maskit[1]], [Beardon]. For the organization of the proof in case of non orientation preserving transformations, it has been beneficial to conform to the principles of combinatorial topology. Here the book of [Zieschang, Vogt, Coldewey] have been very useful.

Chapter VIII The book [Fenchel] and a paper by [Coxeter[1]]. For the section on fixed points [Lyndon] and [Beardon].

BIBLIOGRAPHY

A.F. Beardon, *The geometry of discrete groups.*
Springer Verlag, Heidelberg 1983.

M.Berger, *Geometry I and II.*
Springer Verlag, Berlin 1987.

K.Borsuk, W.Szmielew, *Foundation of geometry.*
North-Holland Publishing Co. Amsterdam 1960.

N.Bourbaki, *Algèbre III, Algebre multilineaire.* Éléments de Mathématiques,
Hermann, Paris 1958.

N.Bourbaki, *Groupes et Algèbres de Lie* IV,V,VI. Éléments de Mathématiques,
Hermann, Paris 1968.

A.J.Casson, S.A.Bleiler, *Automorphisms of surfaces after Nielsen and Thurston.*
London Math. Soc. Student Texts 9. Cambridge.Univ.Press 1988.

D.J.Collins, H.Zieschang, *Combinatorial group theory and fundamental groups.*
Preprint. I.H.E.S., Bures sur Yvette 1989.

H.S.M. Coxeter, *The inversive plane and hyperbolic space.*
Abh.Math.Sem.Univ.Hamburg 29(1966),217-142.

H.S.M. Coxeter, *Regular polytopes,*
Dover, New York 3rd edition , 1973

H.S.M. Coxeter, W.O.J Moser, *Generators and relations for discrete groups.*
Springer Verlag, Berlin 1957.

R. Deheuvels, *Formes quadratiques et groupes classiques.*
Presses Universitaires de France, Paris 1981.

H.M.Farkas, I.Kra, *Riemann surfaces.*
Springer Verlag, Berlin 1980.

A.Fathi, F.Laudenbach, V.Poenaru, *Travaux de Thursten sur les surfaces.*
Asterisque 66-67 (1979).

W. Fenchel, *Elementary geometry in hyperbolic space.*
Walter de Gruyter, Berlin 1989.

J.E.Gilbert, M.A.M.Murray, *Clifford algebras and Dirac operators in harmoic analysis.* Cambridge University Press. Cambridge 1991.

M.J.Greenberg, *Euclidean and non-Euclidean geometries.* Freeman. San Francisco 1973.

H.B.Griffiths, *Surfaces.* Cambridge University Press, Cambridge 1976.

Klein,Fricke , *Vorlesungen über die Theorie der elliptischen Modulfunctionen. Vol 1,2.* Teubner, Leibzig 1890.

S.Lang, *Introduction to modular forms.* Grundlehren 222. Springer Verlag, Berlin 1976.

J.Lehner, *A short course in automorphic functions.* Holt, Rinehart & Winston. New York 1966.

R.C.Lyndon, *Groups and geometry.* London Math.Soc.Lecture Note Series 101. Cambridge University Press 1985.

R.C.Lyndon, P.E.Schupp, *Combinatorial group theory.* Ergebn.Math.Grenzgeb. 89 Springer Verlag, Berlin 1977.

A.M.Macbeath, *The classification of non-Euclidean plane crystallographic groups.* Canadian Journal Mathematics, 19(1967) 1192-1205.

A.M.Macbeath, A.H.M.Hoare, *Groups of hyperbolic crystallography.* Math.Proc.Cambridge Phil.Soc.79(1976), 235-249.

W.Magnus, *Noneuclidean tesselations and their groups.* Academic Press. New York 1974.

Y.I.Manin, *Three dimensional hyperbolic geometry as ∞−adic Arakelov geometry.* Preprint. I.H.E.S. Bures sur Yvette 1990.

G.E.Martin, *The foundations of geometry and the non-Euclidean plane.* Springer Verlag, Berlin 1975.

B.Maskit, *On Poincaré's theorem for fundamental polygones.* Advances in Math. 7(1971)219-230.

B.Maskit, *Kleinian groups.* Springer Verlag, Berlin 1988.

W.S.Massey, *Algebraic topology: an introduction.* Harcourt, Brace & World Inc. New York 1967.

J.Milnor, *Hyperbolic geometry: The first 150 years.* Bull.Amer.Math.Soc.6 (1982)9-24.

M.Newman, *Integral matrices.* Academic Press 1972.

V.V.Nikulin, I.R.Shafarevich, *Geometries and groups.* Springer Verlag, Berlin 1987.

J.Nielsen, *Collected Mathematical Papers.* 2 Vols. Birkhäuser, Basel 1986.

H.Poincaré, *Théorie des groupes fuchsienne.* Acta.Math.1(1882) 1-62.

G. de Rham, *Sur les polygones générateurs des groupes Fuchsians.* l'Ensignement de Math. 17 (1971) 49-62.

E.G.Rees, *Notes on geometry. Universitext,* Springer Verlag, Berlin 1983.

I.R.Shafarevich, *Algebra I.* Encyclopaedia of Mathematical Sciences vol 11. Springer Verlag, Berlin 1990.

G.Shimura, *Introduction to the arithmetic theory of modular forms.* Iwanami Shoten and Princeton University Press, Princeton 1971.

J.H.Silverman, *The arithmetic of elliptic curves.* Springer Verlag, Berlin 1985.

W.P.Thuston, *Three dimensional manifolds, Kleinian groups and hyperbolic geometry.* Bull.Amer. Math. Soc.6 (1982) 357-381.

J.B.Wilker, *Inversive geometry.* The geometric vein, edited by C.Davies, B.Grünbaum, F.A.Scher. Springer Verlag, Heidelberg 1981.

M.C.Wilkie, *On noneuclidean cristallographic groups.* Math. Z. 91, 87-102, 1966.

Zieschang, E.Vogt, H.D.Coldewey, *Surfaces and planar groups.* Lecture Notes in Math. 835, Springer Verlag, Berlin 1980.

SYMBOLS

INDEX

Printed in the United States
By Bookmasters